Properties, Structures and Applications of Metals

Properties, Structures and Applications of Metals

Editor: Peggy Rusk

New York

Published by NY Research Press
118-35 Queens Blvd., Suite 400,
Forest Hills, NY 11375, USA
www.nyresearchpress.com

Properties, Structures and Applications of Metals
Edited by Peggy Rusk

International Standard Book Number: 978-1-63238-555-0 (Hardback)

Cataloging-in-Publication Data

Properties, structures and applications of metals / edited by Peggy Rusk.
p. cm.
Includes bibliographical references and index.
ISBN 978-1-63238-555-0
1. Metals. 2. Metals--Analysis. 3. Metallurgy. I. Rusk, Peggy.
TA459 .P76 2017
620.16--dc23

Printed in the United States of America.

Contents

Preface

Metals play an important role in the interdisciplinary field of materials science. With advancements in science and technology, it has been possible to produce metals with low chemical reactivity, high strength-to-weight ratios, electromagnetic shielding, etc. This has led to major innovations in aerospace, automotive and construction industries. Some of the diverse topics covered in this book address the varied branches that fall under this category. The various advancements in metals are glanced at and their applications as well as ramifications are looked at in detail. This book provides an extensive analysis of the properties, structures and applications of metals. Scientists and students actively engaged in this field will find this book full of crucial and unexplored concepts.

This book is the end result of constructive efforts and intensive research done by experts in this field. The aim of this book is to enlighten the readers with recent information in this area of research. The information provided in this profound book would serve as a valuable reference to students and researchers in this field.

At the end, I would like to thank all the authors for devoting their precious time and providing their valuable contributions to this book. I would also like to express my gratitude to my fellow colleagues who encouraged me throughout the process.

Editor

1

Evolution of the Annealing Twin Density during δ-Supersolvus Grain Growth in the Nickel-Based Superalloy Inconel™ 718

Yuan Jin [1,*], Marc Bernacki [1], Andrea Agnoli [1], Brian Lin [2], Gregory S. Rohrer [2], Anthony D. Rollett [2] and Nathalie Bozzolo [1,*]

Academic Editor: Johan Moverare

1 MINES ParisTech, PSL—Research University, CEMEF—Centre de mise en forme des matériaux, CNRS UMR 7635, CS 10207 rue Claude Daunesse, Sophia Antipolis Cedex 06904, France; marc.bernacki@mines-paristech.fr (M.B.); andrea.agnoli.job@gmail.com (A.A.)

2 Department of Materials Science and Engineering, Carnegie Mellon University, 5000 Forbes Avenue, Pittsburgh, PA 15213, USA; brianklin@gmail.com (B.L.); gr20@andrew.cmu.edu (G.S.R.); rollett@andrew.cmu.edu (A.D.R.)

* Correspondence: jyuan915@gmail.com (Y.J.); nathalie.bozzolo@mines-paristech.fr (N.B.)

Abstract: Grain growth experiments were performed on Inconel™ 718 to investigate the possible correlation of the annealing twin density with grain size and with annealing temperature. Those experiments were conducted at different temperatures in the δ supersolvus domain and under such conditions that only capillarity forces were involved in the grain boundary migration process. In the investigated range, there is a strong inverse correlation of the twin density with the average grain size. On the other hand, the twin density at a given average grain size is not sensitive to annealing temperature. Consistent with previous results for pure nickel, the twin density evolution in Inconel™ 718 is likely to be mainly controlled by the propagation of the pre-existing twins of the growing grains; *i.e.*, the largest ones of the initial microstructure. Almost no new twin boundaries are created during the grain growth process itself. Therefore, the twin density at a given average grain size is mainly dependent on the twin density in the largest grains of the initial microstructure and independent of the temperature at which grains grow. Based on the observations, a mean field model is proposed to predict annealing twin density as a function of grain size during grain growth.

Keywords: annealing twin; grain growth; EBSD; mean field model

---

1. Introduction

Nickel-based superalloys are used for aeronautical component manufacturing because of their performance at high temperature. Grain boundary engineering (GBE) is a possible route to improve the properties, especially those related to intergranular damage [1]. Because of their low energy [2], annealing twin boundaries, observed in almost all deformed and subsequently annealed face-centered-cubic (FCC) metals with low to medium stacking fault energy, are fundamental for GBE [3–5]. Even though these crystalline defects have been known for a long time [6], the mechanisms by which they appear are still not fully understood. Being able to predict the twin density obtained after a given thermomechanical path, which has been made on an empirical basis for now, would be very valuable for developing GBE routes.

The growth accident model, which asserts that a coherent twin boundary forms at a migrating grain boundary due to a stacking error, is most commonly used to explain annealing twin formation [7–9]. In the growth accident model, the amplitude of the driving force acting on grain boundary migration and the resulting migration velocity are promoting factors for the generation of annealing twins. However, few studies in the literature address the direct effect of the grain boundary migration rate. The aim of the present work is to contribute to filling this gap. Grain growth experiments are performed on a nickel-based superalloy (Inconel™ 718) at different temperatures, so that grain boundaries migrate at different rates. The twin content evolution is discussed quantitatively in relation to temperature, grain size increase and grain boundary migration rate. A mean field model is proposed based on the observations to describe the average twin density evolution as a function of grain size during grain growth.

2. Experimental Details for Quantitative Analysis of Twin Content

Grain growth experiments were performed on semi-cylindrical samples machined from an Inconel™ 718 billet at three different temperatures above the solvus for the delta phase and for different times (Table 1).

Table 1. Annealing conditions (annealing times for each temperature) applied on the initial microstructure shown in Figure 1 to obtain those shown in Figure 2.

Temperature	Applied Annealing Time		
1025 °C	10 min	25 min	
1065 °C	3 min	6 min	9 min
1100 °C	1 min	2 min	3 min

All of the analyzed samples (longitudinal section) were metallographically prepared with a prolonged final mechanical polish using a 0.5-µm colloidal SiO_2 suspension. The electron backscatter diffraction (EBSD) maps were obtained with a ZEISS SUPRA 40 FEG SEM (Jena, Germany) equipped with a Bruker CrystAlign EBSD system (Berlin, Germany). The step size for EBSD map acquisition was set to 0.46 µm for the initial microstructure. However, as the average grain size in the annealed samples became larger, the step size was accordingly increased to 1.44 µm, which appeared as a good compromise for measuring a sufficient number of grains, but still completing each map within a reasonable acquisition time. The mapped area of the initial microstructure contains more than 5000 grains. The EBSD maps recorded after grain growth include 150 to 800 grains (twins being excluded in grain counting). The OIM™ software (EDAX, Mahwah, NJ, USA) was used to analyze the EBSD data.

Grain boundaries were defined as boundaries with a misorientation angle above 5°. Annealing twin boundaries are defined by a misorientation of 60° about the <111> axis with a tolerance of 8.66°, according to Brandon's criterion [10], regardless of their coherent *versus* incoherent character. The Brandon's criterion was applied because an F.C.C. twin can also be depicted as an Σ3 coincidence site lattice. Two different quantities were used to quantify annealing twins: the annealing twin density (N_L) and the number of annealing twin boundaries per grain (N_G), respectively defined as:

$$N_L = \frac{L_{tb}}{S} \times \frac{2}{\pi} \quad (1)$$

$$N_G = \frac{N_2 - N_1}{N_1} \quad (2)$$

where L_{tb} is the twin boundary length detected in a given sample section area S, N_1 is the number of grains ignoring twin boundaries in the grain detection procedure and N_2 is the number of grains by considering twin boundaries as grain boundaries. The formulation to calculate the twin density (N_L) is derived from [11].

The recrystallized grains were identified in the EBSD maps using a criterion that the grain orientation spread (GOS) was less than 1° [12]. Twin boundaries were ignored in the grain detection procedure (except for determining the N_2 value), and grains with an area smaller than three pixels were not considered. No additional clean-up was performed. The number-weighted average grain diameter (D) was used to quantify average grain size (equivalent grain diameter) in the following analyses.

3. Experimental Results and Discussion of the Underlying Mechanisms

The initial microstructure (Figure 1a) is almost fully recrystallized (more than 93%) as indicated by the very low GOS values (Figure 1b). The average size of the recrystallized grains is 13 μm. As depicted by the dashed line in Figure 1d, the twin density in the initial microstructure first increases with grain size, is maximal for grains of about the average grain size and then slowly decreases at larger grain sizes. The twin density in the grains of a given grain size class is calculated with the twin boundary length in these grains and the surface occupied by these grains. This initial microstructure results from the prior recrystallization process. It has been established that the twin density after recrystallization is mainly controlled by the stored energy level [13,14] and that the twin formation event occurs more often when the grains are growing faster [9,15]. The migration rate of the recrystallization front decreases during the recrystallization process because the remaining stored energy level in the deformed matrix is decreasing. Therefore, the growth rate of a recrystallized grain decreases, and fewer twins are formed. Consequently, the twin density decreases with increasing the size of the recrystallized grains. The reason why small grains also have a lower twin density is different. The small grains observed in a 2D section after complete recrystallization are either truly small (in 3D) and have a nucleated late in the recrystallization process or they are sections near the edge of a large grain. In both cases, they correspond to a volume that has been swept late, and thus, relatively slowly, by the recrystallization front because of the low level of remaining stored energy.

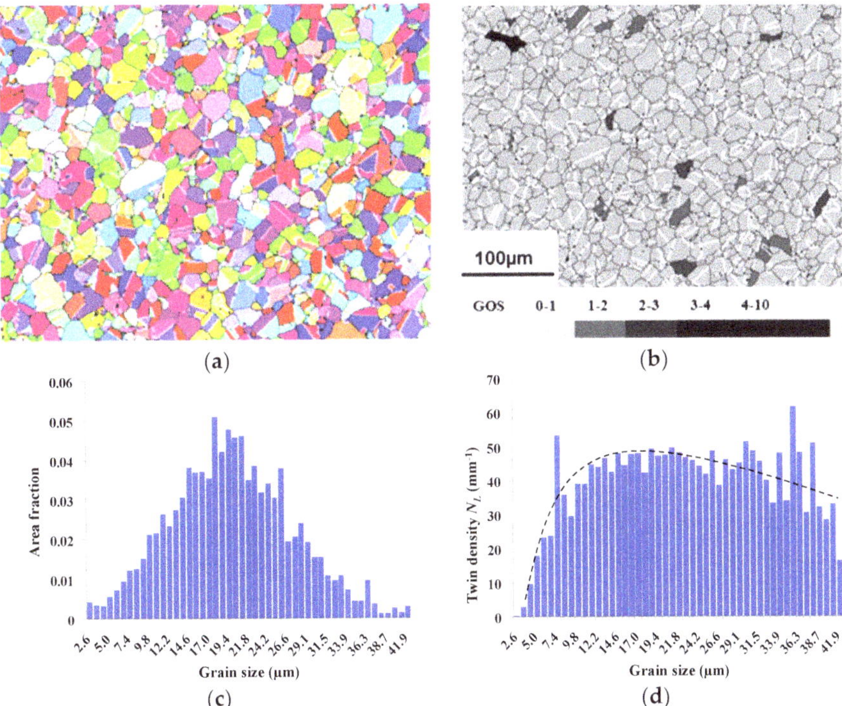

Figure 1. The initial microstructure: (**a**) orientation color-coded EBSD map (vertical direction of the map projected in the standard triangle); (**b**) grain orientation spread (GOS) map; grain boundaries in black and twin boundaries in white; (**c**) grain size distribution histogram (average grain size = 13 μm); (**d**) twin density as a function of grain size; the dashed line is a tendency guideline.

The microstructure evolution at the three different annealing temperatures is shown in Figure 2. Grain growth kinetics are described in Figure 3. As expected, grain growth is faster when increasing the annealing temperature. The annealing times were adapted accordingly to result in similar grain sizes (in each column of Figure 2).

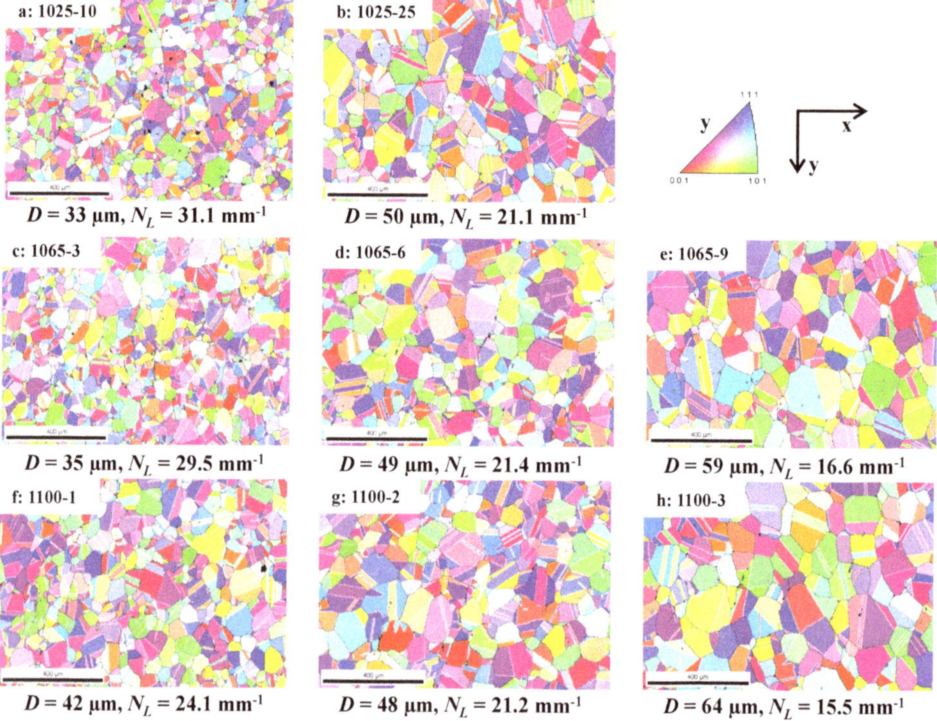

Figure 2. Microstructure evolution after annealing (same color code as in Figure 1a). (**a**) at 1025 °C for 10 min, (**b**) at 1025 °C for 25 min, (**c**) at 1065 °C for 3 min, (**d**) at 1065 °C for 6 min, (**e**) at 1065 °C for 9 min,(**f**) at 1100 °C for 1 min, (**g**) at 1100 °C for 2 min, (**h**) at 1100 °C for 3 min. The corresponding average grain size and annealing twin density are specified under each map. The white lines denote twin boundaries; the black lines are grain boundaries with a disorientation greater than 5°.

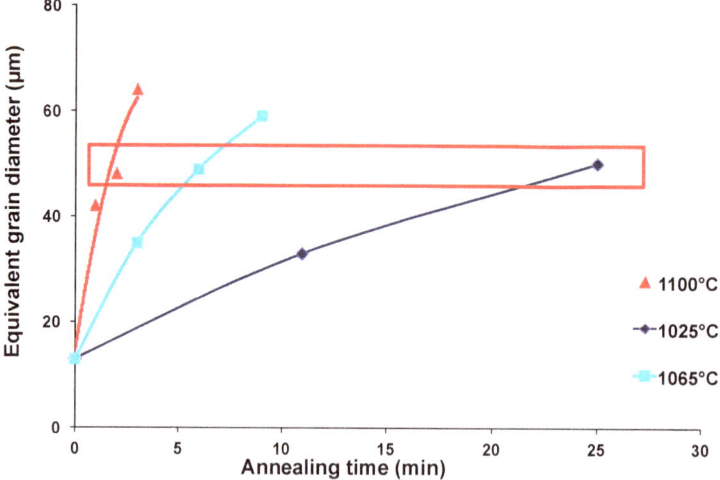

Figure 3. Grain growth kinetics at different annealing temperatures.

The twin density is plotted as a function of the average grain size in the whole microstructure in Figure 4. The total twin boundary length and the surface of the overall microstructure are used to calculate the overall twin density via Equation (1). There is a strong inverse correlation of the twin density with the average grain size; similar trends have already been reported in the literature for different FCC metals and alloys [7,15–18].

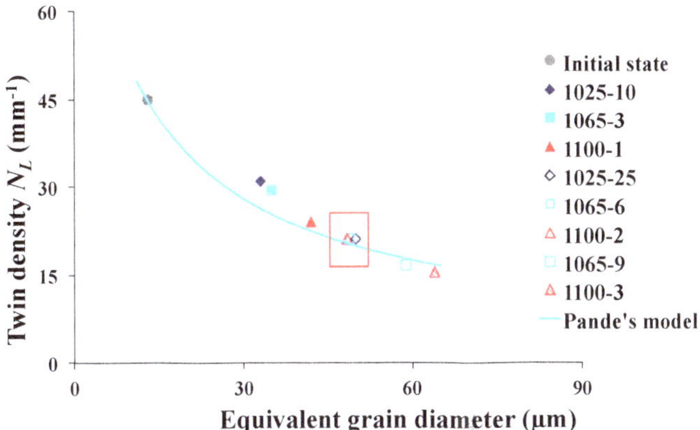

Figure 4. Twin density evolution compared to Pande's model (best fit to Equation (3) with $\gamma_{gb} \approx 1\ \mathrm{J \cdot m^{-2}}$).

According to Pande [18], the annealing twin evolution depends on the grain boundary migration driving force and on the grain boundary migration distance (and therefore, on grain size). In order to test the influence of grain size on annealing twin development, microstructures with an identical average grain size of about 50 µm, but obtained at different annealing times and temperatures, were compared (highlighted by the red rectangle in Figures 3 and 4). Despite the difference in grain growth kinetics at the three temperatures, the twin densities for the same average grain size (50 µm) are identical. Therefore, the twin density evolution appears to be independent of grain growth kinetics within the considered range. This observation is consistent with experimental data reported in the early literature [19].

In addition to the twin density, the number of annealing twin boundaries per grain (N_G) was also used for annealing twin quantification. N_G is calculated using the total grain numbers N_2 with and N_1 without considering twin boundaries as grain boundaries in the overall microstructure. In order to determine the density of twins formed during a given process, *i.e.*, to describe the occurrence of twin formation events, the number of twin boundaries per developing grain provides a better indicator than any of the other quantification parameters. During mean curvature-driven grain growth, the bigger grains grow at the expense of the smaller ones. Therefore, twin formation in the largest grains (about 100 grains) in each analyzed sample was investigated. In the sample with the largest average grain size (1100-3), the 105 largest grains occupied more than 95% of the EBSD mapped area (Figure 5). The number of twin boundaries per grain in the largest grains does not increase during grain growth (Figure 6); if anything, it decreases, although the counts are not large enough to have high confidence in this conclusion. This observation is consistent with the results of *in situ* annealing on pure nickel reported in [15]. Indeed, in this previous work, most annealing twins were formed during recrystallization, and very few new twins were observed during grain growth. In the grain growth regime, small grains containing several twin boundaries are consumed by large grains, which grow, but only very rarely produce new twins. A new mesoscopic model, which relates annealing twin formation to the topology of the moving grain boundaries, was proposed in [13,15,16] to explain this phenomenon. According to this model, coherent twins can nucleate only along grain boundary segments that migrate opposite to their curvature. During grain growth, this may happen only at triple junctions: in other words, only triple junctions can be potential nucleation sites for annealing twins. This idea was confirmed by a recent 3D *in situ* study [20]. In this study, the near-field high-energy X-ray diffraction microscopy (nf-HEDM) technique was used to follow the microstructure evolution during grain growth in a pure nickel sample. In this 3D experiment, only very few twin formation events were observed and all of them were at triple junctions.

The fact that the large grains develop almost without creating new twins, allied with the consumption of the twinned small grains, contributes to the inverse correlation of the twin density with the increasing average grain size in the work on Inconel™ 718 shown in Figure 4. Furthermore, incoherent twin boundaries may migrate in a direction that reduces twin length and total interfacial energy, which is an additional reason why the twin density decreases during grain growth (it is worth mentioning here that this additional mechanism is possibly material and temperature dependent). In addition, the fact that the twin evolution is independent of grain growth kinetics is consistent with the above-described mechanism where twin density evolution is mainly controlled by the extension of pre-existing twins (those formed during prior recrystallization and existing in the large grains at the beginning of the grain growth regime). How much the twins extend depends on how much the grains grow, but not on how fast they grow.

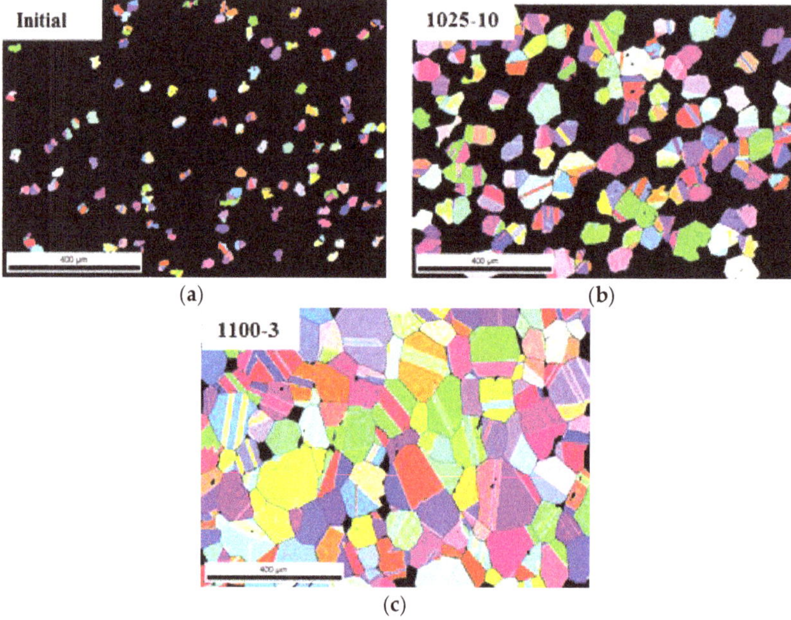

Figure 5. The (about 100) largest grains in (**a**) the initial microstructure, (**b**) the sample annealed at 1025 °C for 10 min and (**c**) the sample annealed at 1100 °C for 3 min. Same color code as in Figure 1.

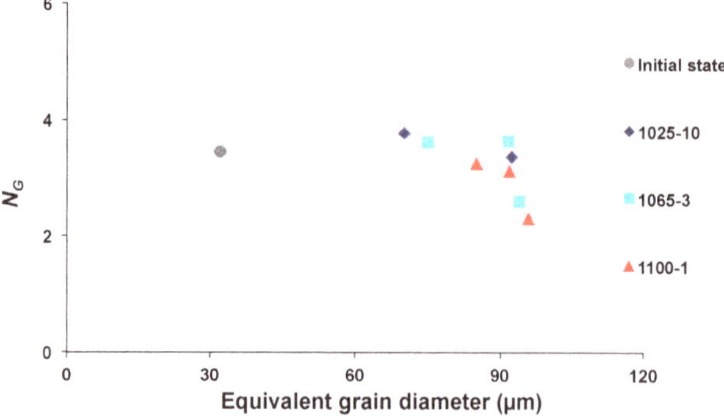

Figure 6. Number of twin boundaries per grain in the 100 largest grains expressed as a function of the average grain size.

4. Literature Models for Predicting Twin Density Evolution

Two main mathematical models to predict twin density exist in the literature, namely Gleiter's model [21] and Pande's model [18], which are both consistent with the growth accident theory. In Gleiter's model, derived from classical nucleation theory, the twin formation probability mainly depends on two factors: the activation enthalpy for migration and the grain boundary migration driving force. However, it is shown in [16] that contrary to the experimental results described in this study and in other works in the literature [19], the twin density predicted by this model is very sensitive to temperature. Meanwhile, Pande's model is shown to be consistent with various experimental data [7,17,18]. Consequently, Pande's model is most often used to compare to the experimental data in the literature. Originally formulated to consider curvature-driven growth, Pande's model has also been adapted to account for the stored energy contribution to the grain boundary migration driving force [7].

Pande's model [18] is derived from the assumption that the increment of annealing twin boundary number per grain (ΔN_G) is proportional to the grain boundary migration driving force (F) and the increase in grain size (ΔD). Therefore, the model assumes that new twins are formed (N_G increasing) while grains grow (D increasing). This is not consistent with our observations within the grain growth regime, presented here for Inconel™ 718 and in the previous paper for pure nickel [15].

The correlation between the annealing twin density and the average grain size can nevertheless be fitted quite well (see Figure 4) by the formula derived from Pande's model (Equation (3)) [18]:

$$P = \frac{1}{D} K \gamma_{gb} \ln \frac{D}{D_0} \tag{3}$$

with $\gamma_{gb} \approx 1\ \mathrm{J \cdot m^{-2}}$, $K \approx 0.3\ \mathrm{m^3 \cdot J^{-1}}$ and $D_0 \approx 2\ \mu\mathrm{m}$, where the parameter values were identified based on experimental data (Figure 4) by an inverse method. A MATLAB function "fminsearch" was used to identify the parameters that minimize the cost function, which is defined as the sum of the squared residuals. A residual is defined here as the difference between a value in the experimental data and the corresponding value predicted by Pande's model. Even though Pande's model can indeed describe the correlation between the annealing twin density and the average grain size, the underlying modeling assumptions are not consistent with the experimental observation of annealing twin evolution. Accordingly, a mean field model will be attempted in the next section to address this gap.

5. Description of the New Mean Field Model

As presented in the Section 3, there are almost no annealing twins formed in the grain growth regime. This is the basis statement for the new model proposed here that aims at describing the average twin density evolution in the overall microstructure as a function of the average grain size during grain growth.

In the present study and for most of the experimental data in the literature, twin boundaries are quantified in 2D. Accordingly, for the sake of brevity, only the 2D version of the mean field model is presented here. The possibility of the 3D extension will be discussed in Section 7.

The microstructure is discretized into n categories of representative grains G_i, $i \in \{1, \ldots, n\}$, based on the same principle applied by [22], but here, representative grains are defined by two variables: the grain diameter D_i and the assumed related twin boundary length L^i_{tb}. Representative grains G_i stand for the average state of a number N_i of assumed identical (in terms of diameter and, so, twin boundary length) circular grains present in all of the microstructure.

In mean curvature-driven grain growth with isotropic mobility and grain boundary energy (which can be assumed since twin boundaries are not considered as grain boundaries, but as internal defects of the growing grains), the two-dimensional average growth rate of the representative grain

G_i, *i.e.*, of the N_i grains belonging to the i-th category, can be approximated using Hillert's classical mean field model [23]:

$$\frac{dD_i}{dt} = 2 \cdot M\gamma_{gb}\left(\frac{1}{D} - \frac{1}{D_i}\right) \quad (4)$$

where M is the grain boundary mobility and D is the average grain size of the microstructure.

In this context, the representative grains that are larger than the average size grow, and those smaller than the average size shrink. This deterministic growth rate thus represents the average behavior of large populations of grains.

As presented previously, essentially no twins formed during grain growth; thus, the twin evolution is mainly controlled by the evolution of the pre-existing twins. More concretely, annealing twin boundaries intersecting grain boundaries are extended or reduced in length as the grain boundary migrates. The basic principle of the model is to describe by how much the twin length in a grain changes when the grain size either increases or decreases. The real topology of twin boundaries can be complex [24], especially that of incoherent twins, which obviously differ from coherent ones [25]. It would thus be quite complicated, if not impossible, to derive a proper analytical description of the change in twin boundary length associated with a change in grain size. Instead, we have tested a rough, but simple assumption, which is as follows. The change in twin boundary length L^i_{tb} is considered as proportional to the change in grain size D_i $\forall i \in \{1, \ldots, n\}$ and in each time increment.

In mathematical terms, the assumptions of the model become:

$$\frac{L^i_{tb}(t+\Delta t) - L^i_{tb}(t)}{L^i_{tb}(t)} = k\frac{D_i(t+\Delta t) - D_i(t)}{D_i(t)}, \forall i \in \{1, \ldots, n\} \quad (5)$$

In the first approximation, k will be assumed to be constant, *i.e.*, identical for all of the categories and identical for growing grains and shrinking grains, which again are strong assumptions.

When a grain category G_k is fully consumed by other categories, *i.e.*, $D_k(t+\Delta t) = 0$, its corresponding twin boundary length, $L^k_{tb}(t+\Delta t)$ is fixed to zero (which is not automatically obtained using Equation (5) for $k \neq 1$). For idealized circular grains, the proportionality factor k seems mainly influenced by two factors, illustrated in Figure 7:

- For a coherent twin boundary spanning to the opposite side of the representative grain, since $\Delta L^i_{tb} \geqslant \Delta D_i$ (equality occurs when L^i_{tb} is a diameter of the considered circular representative grains) and $L^i_{tb} \leqslant D_i$, we have $\Delta L^i_{tb}/L^i_{tb} \geqslant \Delta D_i/D_i$; thus k should in principle be higher than one. However, if there is more than one coherent twin boundary inside the representative grain, the total twin boundary length is not necessarily smaller than the representative grain diameter. In this case, the value of k depends also on the initial ratio between the twin boundary length and the representative grain diameter.
- Incoherent twin boundary segments may migrate inside representative grains to decrease the total twin boundary length. This may lead to the shortening of the considered twin boundary, even though the representative grain boundary migration tends to lengthen it, which is a reason for which k might be smaller than one.

The possible material and temperature dependencies of incoherent twin boundary migration behavior are implicitly considered in the parameter k. In the following, the constant k will be identified based on the experimental results presented in Section 3. The algorithm of the mean field model can be summarized as follows.

During a time step:

- The diameter change of each representative grain is calculated using Equation (4). The volume conservation is naturally verified.

- The change in twin length for each category $N_i \Delta L^i_{\text{tb}}$ is calculated by Equation (5) using the twin boundary length in this category $N_i L^i_{\text{tb}}$, since $N_i \Delta L^i_{\text{tb}} / N_i L^i_{\text{tb}} = \Delta L^i_{\text{tb}} / L^i_{\text{tb}}$ (when a grain category G_i is fully consumed by other categories, *i.e.*, $D_i = 0$, its corresponding twin boundary length, L^i_{tb}, is fixed to zero).
- The twin density in each category (N^i_{L}) is calculated via the twin boundary length in this category ($N_i L^i_{\text{tb}}$) and its area ($N_i S_i$) as follows:

$$N^i_{\text{L}} = \frac{2}{\pi} \cdot \frac{N_i L^i_{\text{tb}}}{\pi N_i \left(\frac{D_i}{2}\right)^2} = \frac{2}{\pi} \cdot \frac{N_i L^i_{\text{tb}}}{N_i S_i} \quad (6)$$

- The average twin density in the overall microstructure is calculated from the summation of twin boundary lengths of each category and the overall area (S) as follows:

$$N_{\text{L}} = \frac{2}{\pi} \cdot \frac{\sum_i N_i L^i_{\text{tb}}}{\sum_i N_i S_i} = \frac{2}{\pi} \cdot \frac{\sum_i N_i L^i_{\text{tb}}}{S} \quad (7)$$

Taken together, to predict the evolution of the average twin density, we need a grain size distribution with the area and the twin boundary length of each category as the input.

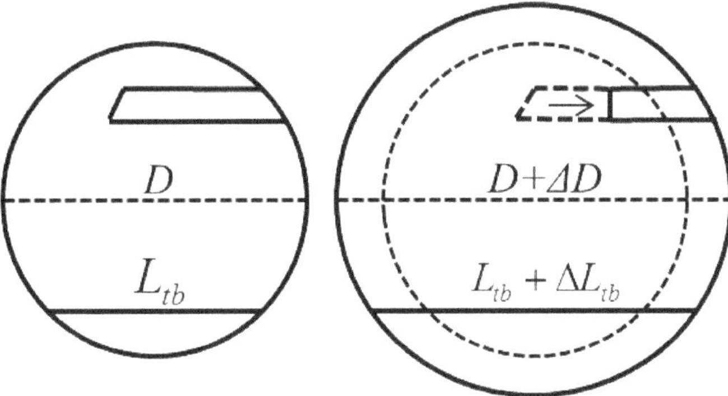

Figure 7. Annealing twin evolution mechanisms during grain growth.

6. Mean Field Modeling of Twin Density Evolution during Grain Growth in Inconel™ 718

The grain size distribution and the twin density in each grain size class of the initial microstructure of Inconel™ 718, shown in Figure 1c,d, was used as input. The EBSD data were discretized into 49 grain size categories (Figure 1c), and the twin density was measured for each category (Figure 1d).

With regard to the physical parameters of the model, the grain boundary mobility M and the grain boundary energy γ_{gb} values were chosen within the range of typical values for metals ($M\gamma_{gb} = 8.22 \times 10^{-13}$ m^2/s). This value therefore does not refer to a specific material. Since this model aims at predicting the twin density evolution as a function of grain size, but not as a function of time, it is not necessary to stick to the real grain growth kinetics of a particular material. Furthermore, for a given material, the twin density evolution is not influenced by the grain growth kinetics (within the present modeling assumptions and consistent with our experimental observations).

The average annealing twin density evolution in the overall microstructure modeled by the mean field model with the time step $dt = 1$ s is compared in Figure 8 to the Inconel™ 718 grain growth experimental data. The value of 0.9 for k was determined as providing the best fit between the model results and the experimental data. With that value of k, the present experimental data can be well described by the new mean field model, which thus sounds promising. Additional experimental data would be relevant to fill the gap between the initial grain size (13 μm) and 30 μm. This work on Inconel™ 718 will be completed in the future; and the model will be compared to other experimental grain growth data for other materials.

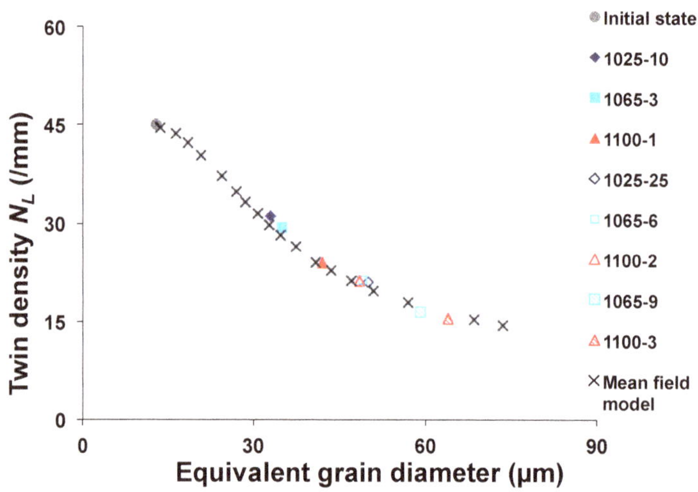

Figure 8. Average twin density (N_L) evolution compared to the new mean field model.

In addition, for the sake of determining the possible impact of the time step on the modeling results, the evolution of the grain size and twin density distributions at different stages of the mean field modeling using dt = 1 s and 0.5 s are shown in Figure 9. The modeling results using these two time steps are almost identical, which shows the convergence of the calculation.

During the modeling process, small grains are consumed by large grains. The disappearance of small representative grains leads to a decrease in the total number of representative grains, as illustrated by Figure 9 (the decrease in grain category number). The area fraction of the i-th category is calculated by the ratio between $N_i S_i$ and S.

In Figure 9, the twin density related to the large representative grains is quite homogenously distributed over the whole range of grain sizes and slightly decreasing as it was in the initial microstructure (Figure 1d). However, the twin density related to the smallest representative grain is much higher. According to Equation (6), for each representative grain, the twin density is inversely correlated to the square of the corresponding grain size. At the same time, the change in twin boundary length L^i_{tb} is proportional to the change in grain size D_i in each time increment, as indicated in Equation (5). For the smallest representative grain, consumed by the other grain families, its grain size shrinks much compared to its initial grain size. Therefore, the high twin density related to the smallest representative grain is numerically caused by the substantial decrease in its size. For example, in Figure 9, at t = 3150 s, the grain size of the smallest representative grain is 14.7 μm, and the related twin density is about 90 mm^{-1}. At t = 375 s, the grain size of the corresponding grain family (marked in Figure 9a) is about 47 μm, and the related twin density is about 26 mm^{-1}. The behavior of the smallest grain size category is physically controlled by the shrinkage of pre-existing grains. This apparently odd behavior of the model for the smallest grains may be due to one of the model assumptions that k has the same value for growing and shrinking grains.

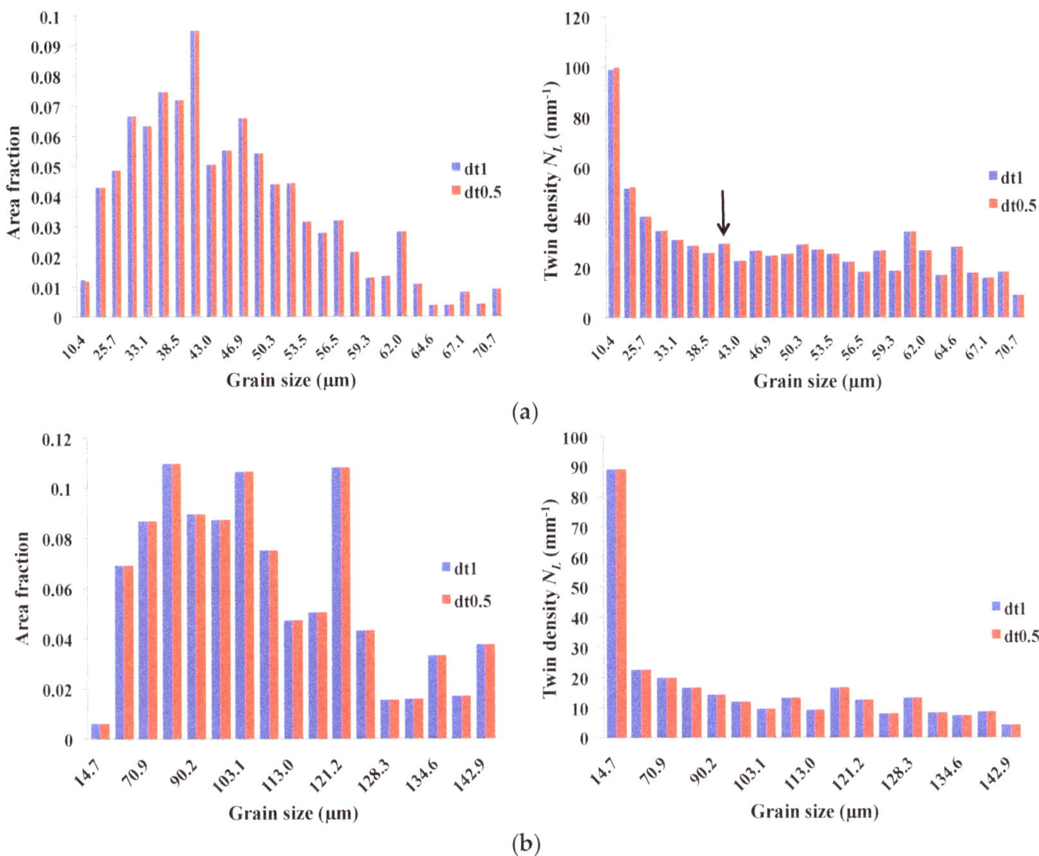

Figure 9. Mean field modelling result at (**a**) $t = 375$ s (average grain size = 32.7 μm) and (**b**) $t = 3150$ s (average grain size = 73.6 μm) obtained using two different time steps (dt), *i.e.*, one and 0.5. Left: Grain size distributions; right: twin density as a function of grain size. The grain family at $t = 375$ s corresponding to the smallest representative grain at $t = 3150$ s is pointed out by the black arrow in (**a**). (**a**) Mean field modeling results at $t = 375$ s obtained using two different time steps (dt), *i.e.*, one and 0.5; (**b**) mean field modelling results at $t = 3150$ s.

7. Discussion of the Relevance of the New Mean Field Model

The model proposed here accounts for the observed mechanism of annealing twin evolution during grain growth: basically, no new twins are formed, but only the pre-existing ones evolve (either by expanding or shortening while the related grain boundary moves or shortening if an incoherent segment migrates). Despite the assumptions made in the model for the dependence of the twin length in a grain on the size of that grain (simple proportionality), the experimental data (grain growth in Inconel™ 718) could be well described by the model. The most remarkable point is that there is only one parameter to be identified, contrary to the other existing models, which have at least two parameters. On the other hand, the new model requires a richer description of the input microstructure (twin length per grain in addition to grain size).

For the sake of comparison, Pande's model has questionable assumptions with regards to the underlying physical mechanisms. Moreover, two parameters must be identified, and this model cannot fit the obtained experimental data as well as the proposed new mean field model (Figure 4 *vs.* Figure 8).

The model proposed here appears then to be promising, even though it needs to be further validated by comparison with more experimental data. Possible improvements would be to check whether a unique value of k can really describe the behavior of all types of grains and to check how far the proportionality relationship of Equation (5) is valid. Both issues are non-trivial, but could be investigated using either 3D microscopy, for example near-field HEDM [20], or by generating synthetic microstructures with a controlled initial twin topology and then evolving them with a full field numerical approach, for example in a finite element context, as already implemented for grain growth in a level-set framework [26–28].

Furthermore, a 3D extension of the present mean field model could be considered, but will require 3D data as an input. The two-dimensional grain size distribution can be reasonably converted to an approached three-dimensional distribution [29]. However, the two-dimensional annealing twin boundary content cannot easily be converted to three-dimensional data. The application of the mean field model on the 3D dataset in [20] will be considered as a perspective of the present work.

8. Conclusions

Grain growth experiments were performed at different temperatures on Inconel™ 718 to study the influence of grain growth rate on annealing twin density during grain growth and led to the following conclusions:

1. Annealing twin density evolution in Inconel™ 718 during grain growth is independent of temperature and, thus, also of the average grain growth rate, in the observed temperature range (1025 to 1100 °C).
2. Annealing twin density exhibits an inverse correlation with the average grain size.
3. The number of annealing twin boundaries per grain does not increase with the average grain size. This observation, which is consistent with our previous results in pure nickel, suggests that almost no new twins are formed during grain growth. On the contrary, it may slightly decrease, notably due to the migration of incoherent twin boundary segments.
4. A mean field model with only one parameter to be determined is proposed to predict twin density evolution as a function of grain size during grain growth, based on the observed behavior of annealing twins during grain growth.

Acknowledgments: This work was funded by the French National Research Agency (ANR project called FORMATING: ANR-11-NS09-001-01) and the Materials World Network of the U.S. National Science Foundation under Grant Number DMR-1107986.

Author Contributions: Y.J. performed the experiments and wrote the paper. A.A. helped the set-up of the experiment. N.B. proposed the idea of the mean field model and supervised the work with M.B. and A.D.R. All results were discussed with G.S.R. and B.L.

Conflicts of Interest: The authors declare no conflict of interest.

References

1. Wantanabe, T. Grain boundary design and control. *Res. Mech.* **1984**, *11*, 47–84.
2. Olmsted, D.L.; Foiles, S.M.; Holm, E.A. Survey of computed grain boundary properties in face-centered cubic metals: I. Grain boundary energy. *Acta Mater.* **2009**, *57*, 3694–3703. [CrossRef]
3. Palumbo, G.; Lehockey, E.M.; Lin, P. Applications for grain boundary engineered materials. *JOM* **1998**, *50*, 40–43. [CrossRef]
4. Randle, V. Twinning-related grain boundary engineering. *Acta Mater.* **2004**, *52*, 4067–4081. [CrossRef]
5. Kumar, M.; Schwartz, A.J.; King, W.E. Microstructural evolution during grain boundary engineering of low to medium stacking fault energy fcc materials. *Acta Mater.* **2002**, *50*, 2599–2612. [CrossRef]
6. Carpenter, H.; Tamura, S. The formation of twinned metallic crystals. *Proc. R. Soc. Lond. A* **1926**, *113*, 161–182. [CrossRef]
7. Cahoon, J.R.; Li, Q.; Richards, N.L. Microstructural and processing factors influencing the formation of annealing twins. *Mater. Sci. Eng. A* **2009**, *526*, 56–61. [CrossRef]
8. Li, Q.; Cahoon, J.R.; Richards, N.L. On the calculation of annealing twin density. *Scr. Mater.* **2006**, *55*, 1155–1158. [CrossRef]

9. Wang, W.; Lartigue-Korinek, S.; Brisset, F.; Helbert, A.L.; Bourgon, J.; Baudin, T. Formation of annealing twin during primary recrystallization of two low stacking fault energy Ni-based alloys. *J. Mater. Sci.* **2015**, *50*, 2167–2177. [CrossRef]
10. Brandon, D.G. The structure of high-angle grain boundaries. *Acta Metall.* **1966**, *14*, 1479–1484. [CrossRef]
11. Underwood, E.E. *Quantitative Stereology*; Addison-Wesley Publishing Company: Boston, MA, USA, 1970.
12. Alvi, M.H.; Cheong, S.W.; Suni, J.P.; Weiland, H.; Rollett, A.D. Cube texture in hot-rolled aluminum alloy 1050 (AA1050)—Nucleation and growth behavior. *Acta Mater.* **2008**, *56*, 3098–3108. [CrossRef]
13. Jin, Y.; Lin, B.; Rollett, A.D.; Rohrer, G.S.; Bernacki, M.; Bozzolo, N. Thermo-mechanical factors influencing annealing twin development in nickel during recrystallization. *J. Mater. Sci.* **2015**, *50*, 5191–5203. [CrossRef]
14. Wang, W.; Brisset, F.; Helbert, A.L.; Solas, D.; Drouelle, I.; Mathon, M.H.; Baudin, T. Influence of stored energy on twin formation during primary recrystallization. *Mater. Sci. Eng. A* **2014**, *589*, 112–118. [CrossRef]
15. Jin, Y.; Lin, B.; Bernacki, M.; Rohrer, G.S.; Rollett, A.D.; Bozzolo, N. Annealing twin development during recrystallization and grain growth in pure nickel. *Mater. Sci. Eng. A* **2014**, *597*, 295–303. [CrossRef]
16. Jin, Y. Annealing Twin Formation Mechanism. Ph.D. Thesis, Mines-ParisTech, Paris, France, December 2014.
17. Detrois, M.; Goetz, R.L.; Helmink, R.C.; Tin, S. Modeling the effect of thermal-mechanical processing parameters on the density and length fraction of twin boundaries in Ni-base superalloy RR1000. *Mater. Sci. Eng. A* **2015**, *647*, 157–162. [CrossRef]
18. Pande, C.S.; Imam, M.A.; Rath, B.B. Study of annealing twins in FCC metals and alloys. *Metall. Trans. A* **1990**, *21*, 2891–2896. [CrossRef]
19. Hu, H.; Smith, C.S. The formation of low-energy interfaces during grain growth in alpha and alpha-beta brasses. *Acta Metall.* **1956**, *4*, 638–646. [CrossRef]
20. Lin, B.; Jin, Y.; Hefferan, C.M.; Li, S.F.; Lind, J.; Suter, R.M.; Bernacki, M.; Bozzolo, N.; Rollett, A.D.; Rohrer, G.S. Observation of annealing twin nucleation at triple lines in nickel during grain growth. *Acta Mater.* **2015**, *99*, 63–68. [CrossRef]
21. Gleiter, H. The formation of annealing twins. *Acta Metall.* **1969**, *17*, 1421–1428. [CrossRef]
22. Bernard, P.; Bag, S.; Huang, K.; Logé, R.E. A two-site mean field model of discontinuous dynamic recrystallization. *Mater. Sci. Eng. A* **2011**, *528*, 7357–7367.
23. Hillert, M. On the theory of normal and abnormal grain growth. *Acta Metall.* **1965**, *13*, 227–238. [CrossRef]
24. Bystrzycki, J.; Przetakiewicz, W.; Kurzydłowski, K.J. Study of annealing twins and island grains in FCC alloy. *Acta Metall. Mater.* **1993**, *41*, 2639–2649. [CrossRef]
25. Song, K.H.; Chun, Y.B.; Hwang, S.K. Direct observation of annealing twin formation in a Pb-base alloy. *Mater. Sci. Eng. A* **2007**, *454–455*, 629–636. [CrossRef]
26. Bernacki, M.; Logé, R.E.; Coupez, T. Level set framework for the finite element modeling of recrystallization and grain growth in polycrystalline materials. *Scr. Mater.* **2011**, *64*, 525–528. [CrossRef]
27. Jin, Y.; Bozzolo, N.; Rollett, A.D.; Bernacki, M. 2D finite element modeling of misorientation dependent anisotropic grain growth in polycrystalline materials: Level set *versus* multi-phase-field method. *Comput. Mater. Sci.* **2015**, *104*, 108–123. [CrossRef]
28. Scholtes, B.; Shakoor, M.; Settefrati, A.; Bouchard, P.O.; Bozzolo, N.; Bernacki, M. New finite element developments for the full field modeling of microstructural evolutions using the level-set method. *Comput. Mater. Sci.* **2015**, *109*, 388–398. [CrossRef]
29. Saltykov, S.A. The determination of the size distribution of particles in an opaque material from a measurement of the size distribution of their sections. In Proceedings of the Second International Congress for Stereology, Chicago, IL, USA, 8–13 April 1967; pp. 163–173.

2

Effects of Cryogenic Forging and Anodization on the Mechanical Properties and Corrosion Resistance of AA6066–T6 Aluminum Alloys

Teng-Shih Shih [1,*], **Hwa-Sheng Yong** [1,2] **and Wen-Nong Hsu** [1,2]

[1] Department of Mechanical Engineering, National Central University, Jhongli District, Taoyuan City 32001, Taiwan; nite-star@hotmail.com (H.-S.Y.); nong88@yam.com (W.-N.H.)

[2] Graduate Student, National Central University, Jhongli District, Taoyuan City 32001, Taiwan

* Correspondence: t330001@cc.ncu.edu.tw

Academic Editor: Nong Gao

Abstract: In this study, AA6066 alloy samples were cryogenically forged after annealing and then subjected to solution and aging treatments. Compared with conventional 6066-T6 alloy samples, the cryoforged samples exhibited a 34% increase in elongation but sacrificed about 8%–12% in ultimate tensile strength (UTS) and yield stress (YS). Such difference was affected by the constituent phases that changed in the samples' matrix. Anodization and sealing did minor effect on tensile strength of the 6066-T6 samples with/without cryoforging but it decreased samples' elongation about 8%–10%. The anodized/sealed anodic aluminum oxide (AAO) film enhanced the corrosion resistance of the cryoforged samples.

Keywords: cryoforging; anodization; tensile properties; corrosion resistance

---

1. Introduction

Al–*x*Mg–*y*Si alloys (6xxx series Al alloys) are commonly used as extruded shapes and forged for making bicycle parts. Their characteristics include ample formability, machinability, weldability, and corrosion resistance, as well as good strength and elongation after heat treatment. These alloys are also readily available on the market.

Aluminum possesses a high stacking fault energy and readily undergoes dynamic recovery during deformation. Plastic deformation at low temperatures, such as cryorolling, is beneficial for refining grains in an aluminum alloy matrix [1,2]. Chen studied equal-channel deformation of an Al–Mg alloy at cryogenic temperatures and found that high-density dislocations distorted grains to a refining grain size [3]. Lee *et al.* also found that cryorolling 5083 alloy could obtain 200 nm fine grains to increase the ultimate tensile strength (UTS) from 315 to 522 MPa [4]. For an Al–Mg–Si alloy, cryorolling with a 90% reduction could cause heavy plastic deformation to produce nanosized extra-fine grains [5–7].

Aluminum alloys that become deformed at cryogenic temperatures could suppress the dynamic recovery that occurs during plastic deformation, and could acquire fine grains featuring high-angle grain boundaries [8,9]. The interaction of high-density dislocations enhanced the precipitation capability for producing a high density of nanosized precipitates [10]. Yin *et al.* found nanograins that were less than 500 nm in size in 7075 alloy samples subjected to compression forging at cryogenic temperatures [11].

Krishina *et al.* [8,12] produced ultrafine-grained Al–4Zn–2Mg alloys by cryorolling and indicated that the driving force for precipitation could be enhanced by differential scanning calorimetry. As a result, the precipitates of the η phase became finer compared with conventional aging treatment. Sarma [13] found that cryorolling significantly changed aging behavior, leading to a reduction in the aging temperature from 190 to 125 °C and in aging time from 24 to 8 h for treating 2219 alloy (Al–Cu alloy).

Jayaganthan [14] used X-ray analyses to study the aging behavior of 6061 alloy subjected to solution treatment then cryorolling. They found that increasing true strain in cryorolling tended to enhance the dissolution of alloys in the alloy matrix and promoted the driving force for precipitation. The UTS was improved from 300 to 365 MPa and elongation was raised from 11% to 13%. Cryorolling apparently reduced the size of intermetallic compounds contained in the matrix of 7075–T73 alloys. As a result, the corrosion resistance of anodized and sealed 7075–T73 alloy could be significantly enhanced [15].

Anodization and sealing improves the corrosion resistance of Al alloys by forming amorphous alumina and hydrate alumina in the anodized film. This process has been widely used in industry. Forging is a common process used for making bicycle and automobile parts. Determining the influence of cryoforging on the corrosion resistance of Al alloys with or without anodization should provide more values for designing and using Al alloys.

Copper was added as an alloying element in a 6xxx series alloy to enhance mechanical strength by precipitation hardening after heat treatment. For example, 6066 Al alloy contains some high-strength Cu (0.8–1.4 Mg, 0.9–1.8 Si, 0.6–1.1 Mn, and 0.7–1.2 Cu) to obtain a UTS of 393 MPa and a yield stress (YS) of 359 MPa after a T6 treatment. This study introduced cryoforging to further improve the toughness of 6066–T6 alloys and their corrosion resistance. The effects of anodization and sealing on the tensile properties and corrosion resistance of AA6066–T6 with and without cryoforging were also evaluated.

2. Experimental Procedures

2.1. Materials

As-extruded 6066 alloy bars, ∅42 × 100 mm in size, were supplied by Tzan Wei Aluminum Co., Ltd. (Tainan, Taiwan). The chemical compositions of the alloys (in wt. %) were 1.38 Si, 0.15 Fe, 1.18 Cu, 1.00 Mn, 1.09 Mg, 0.16 Cr, 0.04 Zn, and 0.02 Ti.

After annealing at 688 K for 120 min, the samples were divided into two groups. The first group was subjected to a solution treatment (803 K for 120 min) and artificial aging (450 K for 480 min); these samples were coded as T6 samples. The second group was subjected first to cryogenic forging, achieving a 40% reduced thickness (from 28 to 16.8 mm), then immersed in liquid nitrogen again, rotated by 90°, and subjected to a second round of cryogenic forging. A 500-ton hydraulic press equipped with one open die set was used to conduct compression forging. After forging, the second group of samples was subjected to the solution to get CFT4 sample and followed by artificial aging treatments: 450 K for 480 min for CFT6a samples and 540 min for CFT6b samples.

2.2. Tensile and Fatigue Tests

The specimens used for testing tensile and rotating bending fatigue strengths were machined from heat-treated samples according to the ASTM B557 [16] (gage diameter: 6 mm) and JIS Z2274 [17] (gage diameter: 8 mm) specifications as shown in Figure 1a,b. The machined test bars were polished by a series of abrasive papers (2000 grit) and an alumina slurry to achieve a surface roughness of less than 0.1 μm (*Ra*).

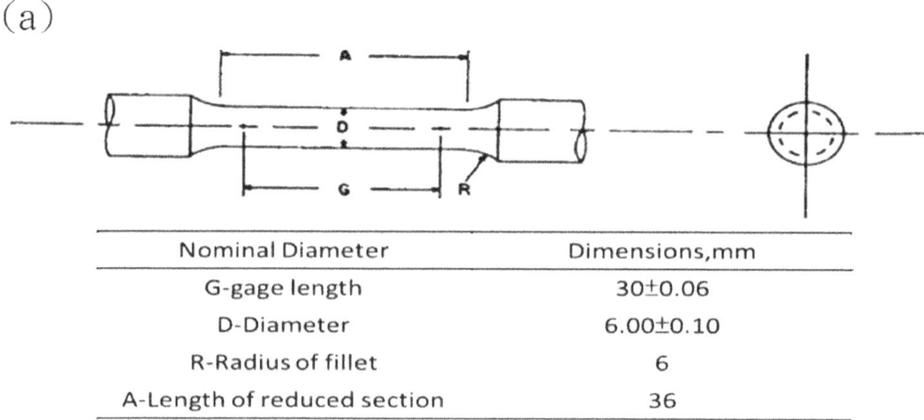

Nominal Diameter	Dimensions,mm
G-gage length	30±0.06
D-Diameter	6.00±0.10
R-Radius of fillet	6
A-Length of reduced section	36

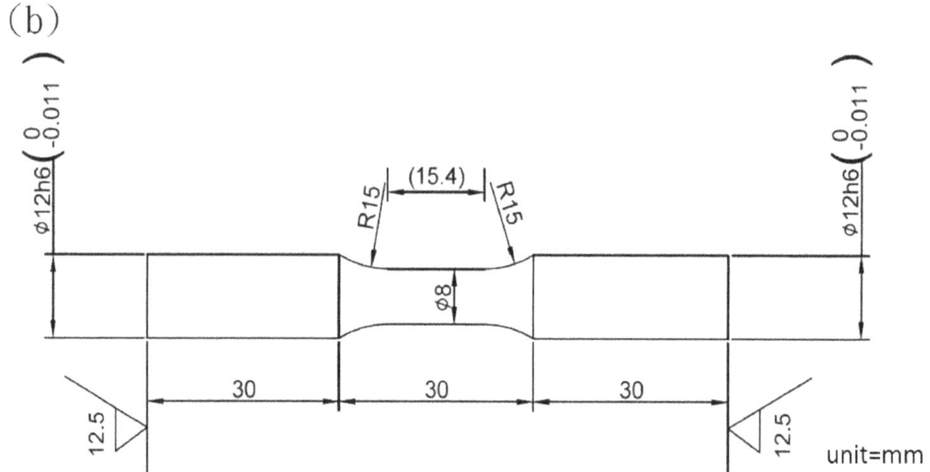

Figure 1. The dimensions of specimens used for (**a**) tensile test (gage diameter: 6 mm); and (**b**) rotating bending test (gage diameter: 8 mm).

2.3. X-ray Tests

X-ray diffraction (XRD) measurements were performed by using a NANO-Viewer Advance (Rigaku, Tokyo, Japan) equipped with a Cu target to identify the precipitates (or second-phase particles) formed in the matrix of different samples, including the T6, CFT4, and CFT6a samples. The cryoforged samples were solution-treated to get a CFT4 sample and followed by aging treatment to acquire a CFT6a sample. A power of 30 KV and current of 10 mA were used in this study. The sample sizes were 10 × 10 × 1 mm.

2.4. Anodization Process

Before anodization, all samples were polished to a surface roughness of approximately $Ra \leqslant 0.1$ μm and then dipped into methanol and ultrasonically vibrated. The specimens were initially degreased by immersion in an alkaline solution (5 mass % NaOH) at 60 °C for 30 s and were then rinsed with water for 1–2 min. For the pickling process, specimens were submerged in an aqueous solution of HNO_3 (30 vol. %) for 90 s at room temperature and then rinsed with water for 1–2 min. The anodization was conducted at 15 $mA \cdot cm^{-2}$ at 15 °C for 900 s in a 15 mass % sulfuric acid solution. The anodized samples were sealed in hot water at 95 °C for 1200 s [18]. After anodization, scanning electron microscopy was conducted and determined that the anodic film was 12–14 μm thick. The anodized samples were sealed in hot water at 368 K for 20 min.

3. Results and Discussion

3.1. Microstructure Observation and Tensile Properties

Table 1 shows the measured mechanical properties of different samples. The CFT6b samples, which aged for 60 min longer than did the CFT6a samples, had increased strength but reduced elongation. The CFT6a samples exhibited a similar strength but superior elongation to the CFT6b samples, and they were adopted in this study for evaluation of their corrosion resistance and fatigue strength. In addition, the CFT6a samples obtained a matrix with a uniform hardness (HV minimum deviation: 1.7).

Table 1. Mechanical properties of T6, CFT6a and CFT6b samples.

Sample	Tensile Strengths			Vickers Hardness (HV)
	UTS (MPa)	YS (MPa)	Elongation (%)	
T6	460 (0.5)	438 (1.4)	10.6 (1.7)	133.6 (2.9)
CFT6a	431 (6.6)	394 (9.2)	14.2 (0.2)	136.6 (1.7)
CFT6b	435 (5.4)	404 (10.1)	13.4 (0.9)	137.1 (2.9)

Note: the deviations are listed in parentheses.

Figure 2a,b show the second-phase (SP) and/or intermetallic compounds (IMC) located at the cross-sections in the transverse direction of the T6 and CFT6a samples. Different sizes of particles were counted and are listed in Table 2. The T6 samples exhibited more coarse SP/IMC particles than the CFT6a samples did as revealed by arrows in Figure 2. During the cryogenic forge, a shear stress was generated to act on the samples' matrix, breaking down the particles. As a result, the particles became finer and dissolution was enhanced during the solution treatment. The total SP/IMC particle count decreased from 864 to 734 counts/mm^2.

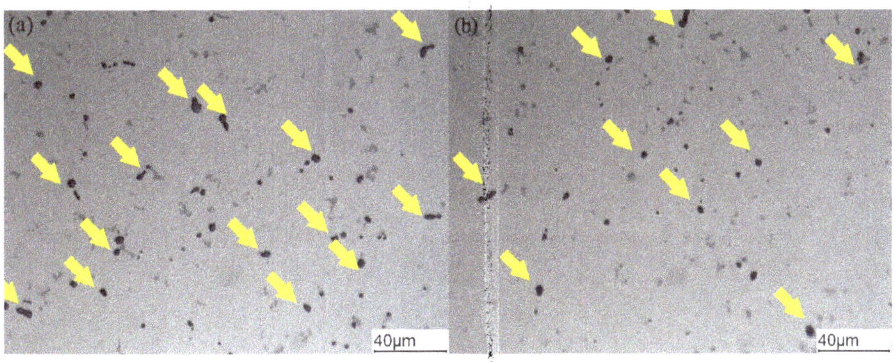

Figure 2. Optical micrographs show second-phase particles located in the matrix of (**a**) T6; (**b**) CFT6a sample; non-etched cross-section in transverse direction.

Table 2. The SP and/or IMC particles measured from T6 and CFT6a samples; maximum particle size and count population were included.

Sample	Second Phase (Coarse Precipitates) Counts/mm^2				Max. Diameter of Particle, μm
	1–10	11–20	>20	Total Count	
T6	838	25	0	864	17
CFT6a	719	15	0	734	17

Electron backscattered diffraction (SUPRA ULTRA 55 field emission scanning electron microscope, ZEISS, Jena, Germany) was performed to measure the misorientation angles of grain boundaries in the longitudinal direction of the tensile test bar samples, as illustrated in Figure 3a,b for the CFT6a and T6 samples, respectively. The CFT6a sample had a high fraction of high-angle grain boundaries (HAGBs). The matrix of the T6 sample had grains featuring mainly low-angle grain boundaries, as shown in Figure 3b. The cyclic loaded sample after being subjected to 250 MPa and fractured at 2.56×10^5 life cycles is shown in Figure 3c, revealing further increasing HAGBs compared with those in Figure 3a. This increase could have been affected by dynamic recrystallization during cyclic loading.

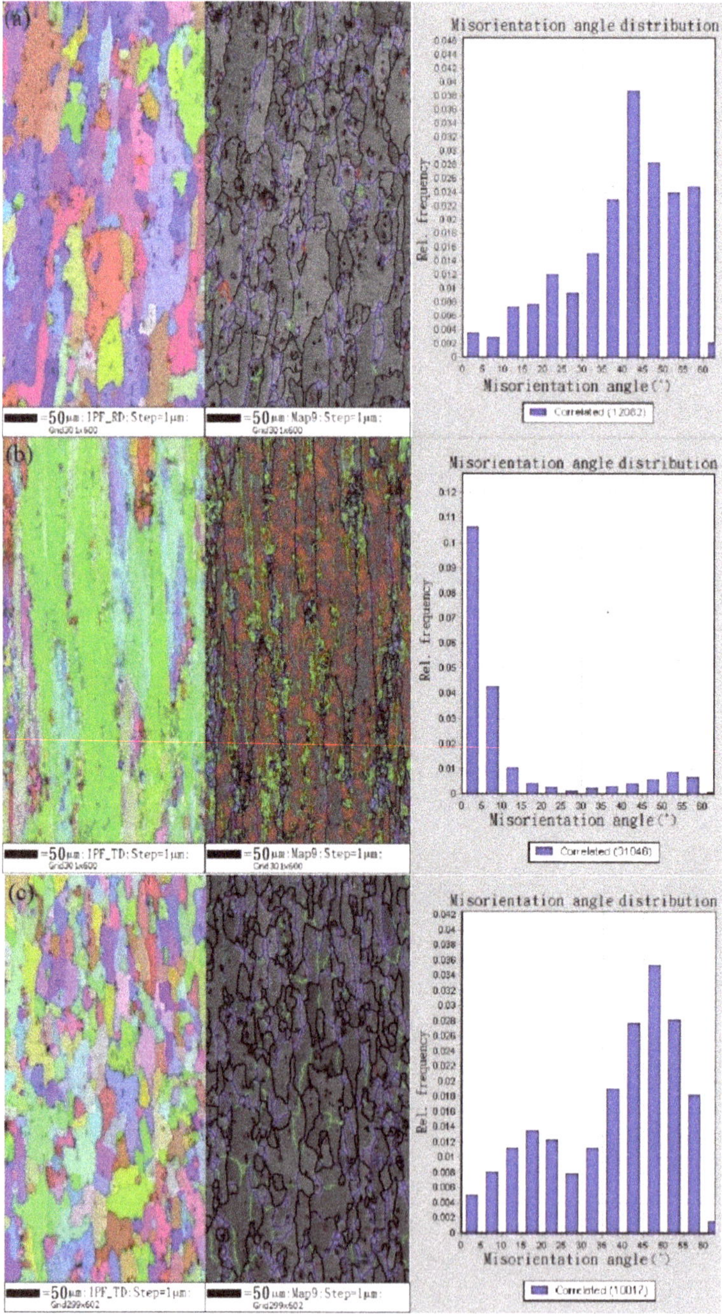

Figure 3. Inverse pole figure maps obtained from EBSD data and measured misorientation angle of grain boundaries from (**a**) CFT6a tensile tested sample; (**b**) T6 tensile tested sample; and (**c**) CFT6a sample after subjected to 250 MPa and fractured at 2.56×10^5 life cycles.

The T6 sample comprised α-Al, *Q*-(AlCuMgSi), Al(Mn,Fe)Si, and some Mg_2Si phases, as shown in Figure 4. After solution treatment, the cryoforged sample was tested and indicated that SP/IMC particles were highly dissolved in its matrix, as confirmed by the CFT4 sample in Figure 4. After artificial aging, the intensity of the *Q* phase was notably decreased, presenting complex phases of Mg_2Si, Q-(AlCuMgSi), and Al(Mn,Fe)Si, as well as some Cu_9Al_4 and CuMgSi phases; see CFT6a sample. Kim *et al.* annealed copper wire and an aluminum pad at 150 to 300 °C and found a Cu_9Al_4 phase with an X-ray spectra peak at 43.9° [19].

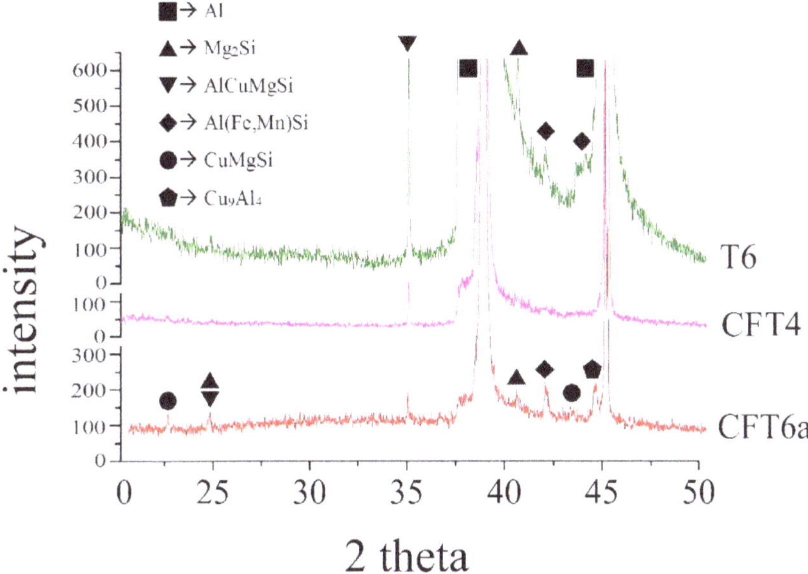

Figure 4. XRD patterns of 6066 alloy samples; including solution and aging T6 sample; cryoforged samples after solution treatment CFT4 and aging CFT6a, respectively.

The diffusivity of alloying elements in the aluminum matrix is in the order of $D_{Cu/Al}$ (4.44×10^{-5} m^2/s), $D_{Mg/Al}$ (1.49×10^{-5} m^2/s), and $D_{Si/Al}$ (1.38×10^{-5} m^2/s) in face-centered cubic Al [20]. Cu atoms tend to segregate around the metastable phase in the Al–Mg–Si alloy, move to grain boundaries during the solution treatment, and diffuse to Mg–Si nanoparticles located at or near grain boundaries to finally form type-C precipitates [21] and the Q phase [22]. As a result, in this study, the T6 sample mainly contained the *Q* phase and Mg_2Si precipitates after aging. The present XRD spectra did not detect the *S*-Al_2CuMg phase from the T6 sample. Neither could Vieira find the *S*-Al_2CuMg phase in his study, in which two age-hardening heat treatments of Al–10Si–4.5Cu–2Mg were completed [23].

The CFT6a sample increased the HAGBs in its matrix to provide more potent sites for accommodating Cu atoms after the solution treatment. During quench or aging in room temperature, Cu diffused to tie up with Al forming Cu_9Al_4 phase. During artificial aging, a Mg–Si cluster formed *in situ*; this either led to the formation of CuMgSi precipitates through movement of the Cu atoms, or the Cu–Mg clusters formed first and subsequently incubated the CuMgSi precipitate in the matrix. Figure 5a,b shows the transmission electron microscopy photos of the T6 and CFT6a samples, respectively. For a given solution and aging conditions, the CFT6a sample achieved finer precipitates (less than 100 nm) than the T6 sample did. The main constituted phases likely included Al(Mn,Fe)Si in block and plate shapes, and some fine particles (less than 100 nm) likely being Cu_9Al_4 and CuMgSi precipitates or Cu atoms surrounding fine Mg_2Si phases, as shown in Figure 5b. The α-Al(Mn,Fe)Si dispersoids could have a block-shaped or plate-shaped morphology in the size of 50–200 nm, as reported by Li *et al.* [24].

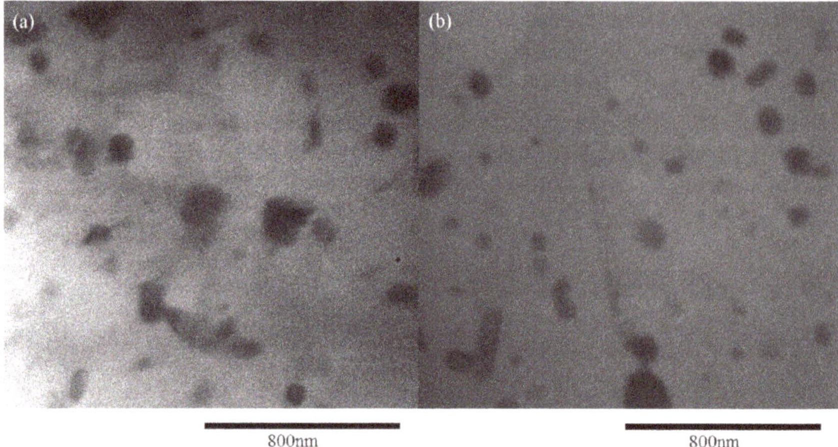

Figure 5. TEM photo show the plate-shape and lump-shape precipitate in the matrix of (**a**) T6 sample, and (**b**) CFT6a sample.

An electron probe X-ray microanalyzer (JXA-8200, JEOL USA, Inc., Peabody, MA, USA) was used to obtain images of Mg, Si, Mn, and Cu mappings from the matrix of the CFT6a sample. The white aggregates in Figure 6a are the Al(Mn,Fe)Si and *Q* phase, and the black lumps are Mg_2Si particles. The arrows indicate the locations of *Q*-phase precipitates. The CuMgSi and Cu_9Al_4 precipitates are nanoscale and were difficult to detect by mapping. In Figure 6b, the white particles shown in the matrix of the T6 samples are Al(Mn,Fe)Si and *Q*-AlCuMgSi phases. An Al(MnFe)Si particle attached with Cu atoms could be distinguished and are indicated by an arrow. The black spots are the Mg_2Si phase.

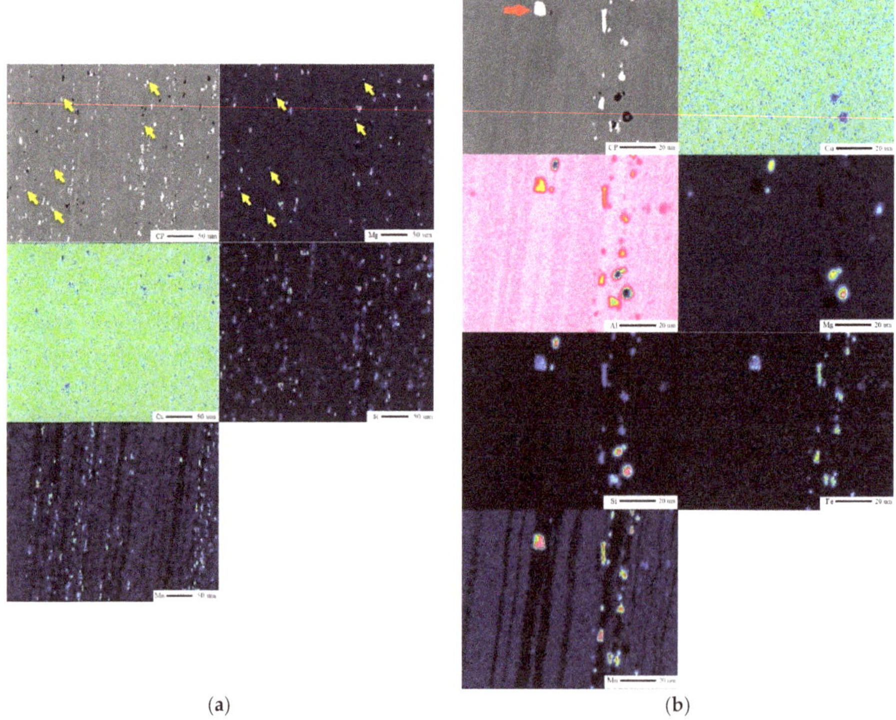

Figure 6. SEM photo and alloying elements mappings obtained from (**a**) T6 sample and (**b**) CFT6a sample.

The *Q* phase, Al(Mn,Fe)Si, and Mg_2Si (larger than 80 nm) contained in the aluminum alloy sample are non-shearable particles in the α-Al matrix [25]. Therefore, the T6 samples gained high strength. By contrast, the CFT6a samples contained Al(Mn,Fe)Si, Mg_2Si, fine CuMgSi, and Cu_9Al_4 precipitates. The fine Mg_2Si or CuMgSi and Cu_9Al_4 precipitates are shearable. Figure 7 illustrates the dislocations (marked as 1-1 and 2-2) intersected with two fine precipitates. In addition, the SP/IMC particle counts (Figure 2) were lower in the matrix of the CFT6a samples. As a result, the CFT6a sample had decreased numbers of barrier sites for tangling dislocations to reduce strength but enhance elongation.

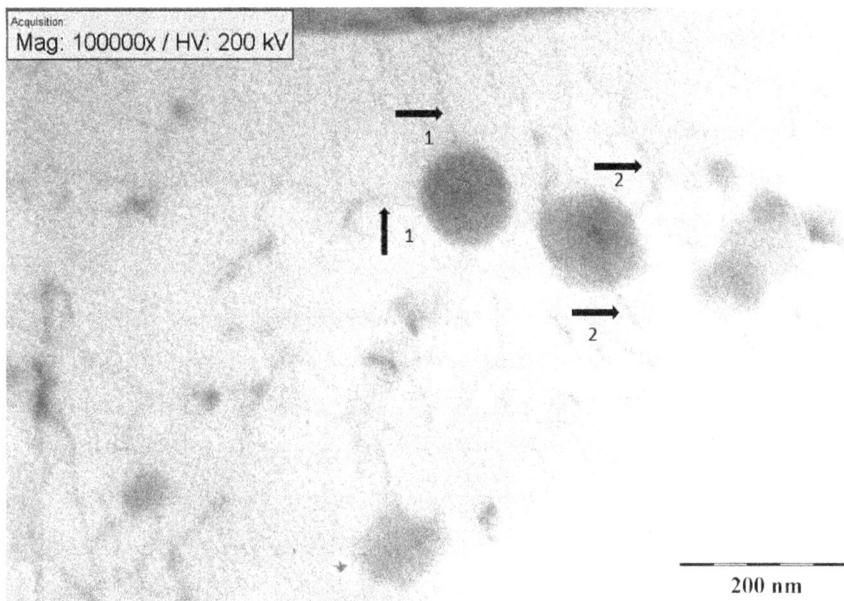

Figure 7. TEM photo shows the intersection of dislocation 1-1 and 2-2 with two fine precipitates in the matrix of CFT6a sample, respectively.

Table 3 compares the surface roughness of T6 and CFT6a samples before and after anodization/sealing. Compared with the samples without anodization, these anodized and sealed samples showed an approximately 8%–10% reduction in elongation. Such a decrease is likely caused by the dissolution of SP/IMC at the film/matrix interface leading to degrade surface roughness and elongation. Before anodization, the surface roughness of the T6 and CFT6a sample was approximately 0.07 (0.04) μm; after anodization and sealing, the surface roughness became 0.16 (0.03) and 0.13 (0.06) μm.

Table 3. Measured surface roughness of different samples before and after anodization.

Sample	Surface Roughness (μm)
T6; CFT6a	0.07 (0.04)
T6-A	0.16 (0.03)
CFT6a-A	0.13 (0.06)

Note: the deviations are listed in parentheses.

3.2. Fatigue and Corrosion Tests

Applying the cryoforge before the solution treatment reduced the SP/IMC particle counts and transformed the *Q* phase into nanoscale CuMgSi and Cu_9Al_4 precipitates, which reduced UTS and YS but increased elongation. Figure 8 shows that both the T6 and CFT6a samples obtained fatigue strength of 180 MPa at 1 × 10^7 life cycles. As shown in Figure 3a,c, we found that cyclic loading functioned to shift the peak of relative frequency from the misorientation angle of 40°–45° to 45°–50°, and increase some grain boundaries at angles of 10°–20°. During cyclic loading, shear stress drove part of the dislocations to conduct polygonization and/or reorganization.

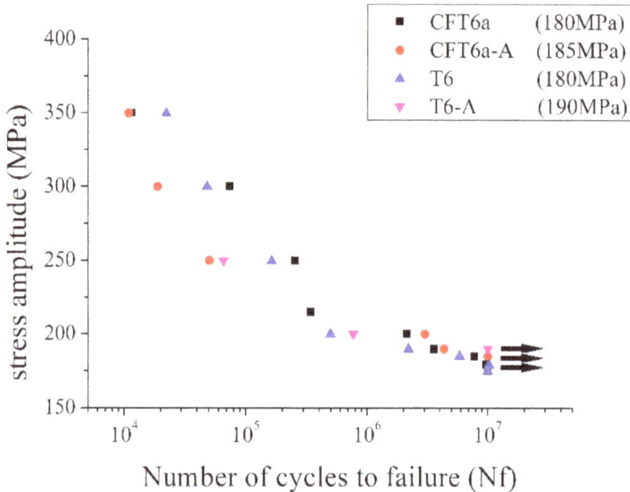

Figure 8. S–N curves for different samples; including T6 and CFT6a with/without anodization/sealed treatment; "A" represented anodization/sealed sample.

Figure 9a,b show the fractured surface of the T6 and CFT6a samples, both of which were subjected to 185 MPa and which performed 5.8 × 10^6 and 7.7 × 10^6 life cycles, respectively. The CFT6a sample exhibited narrower striation spacing than did the T6 samples and achieved a longer fatigue life. The T6 sample obtained fracture steps, as revealed in Figure 9a, which can be attributed to the Q phase located at or near grain boundaries that served to strengthen the grain boundaries [22,26]. Increasing fine precipitates in the matrix of the CFT6a sample likely more effectively to consume the crack propagation energy and thus slightly enhanced the life cycles.

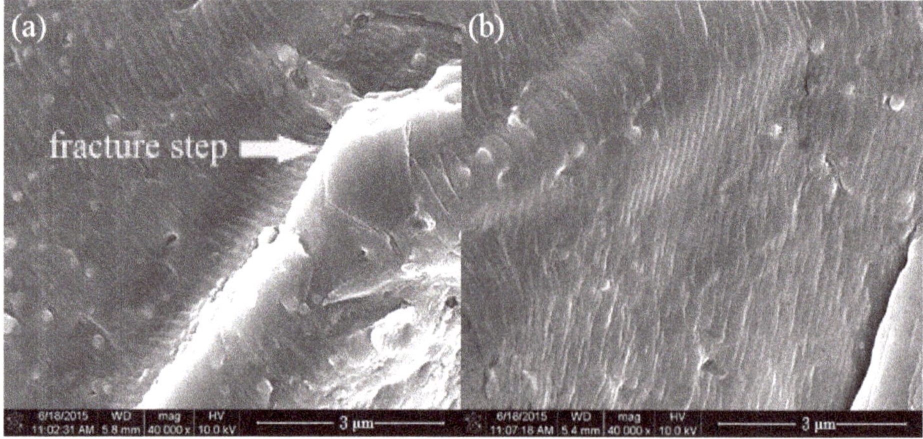

Figure 9. Fracture surface prepared from (**a**) T6; (**b**) CFT6a samples; both subjected to 185 MPa and fractured at 5.8 × 10^6 and 7.7 × 10^6 life cycles, respectively.

The SP/IMC particles were small to be less than 17 µm. Anodization did not yield the deteriorated effect of reducing fatigue strength. The two anodized and sealed samples (T6-A and CFT6a-A) showed similar fatigue strength (185–190 MPa) at 1×10^7 life cycles as did those without anodization and sealing (180 MPa).

The polarization curves for all the samples with and without sealed anodic aluminum oxide (AAO) films are shown in Figure 10. The corrosion behavior of the samples was affected by the constituent phases of each. The T6 samples contained AlCuMgSi, Al(FeMn)Si, and Mg_2Si phases, whereas the CFT6a samples acquired mainly Al(FeMn)Si and Mg_2Si phases as well as some CuMgSi and Cu_9Al_4 precipitates. Mg_2Si is anodic relative to the aluminum matrix, but Al(FeMn)Si and AlCuMgSi are cathodic. The Cu–Al precipitates are more vulnerable to attack during the immersion test, compared with the *Q* phase [23]. The CFT6a samples exhibited slightly inferior E_{corr} (−1.0 V) and I_{corr} (4.7×10^{-6} A/cm^2) than the T6 sample did (−0.93 V and 2.6×10^{-6} A/cm^2, respectively).

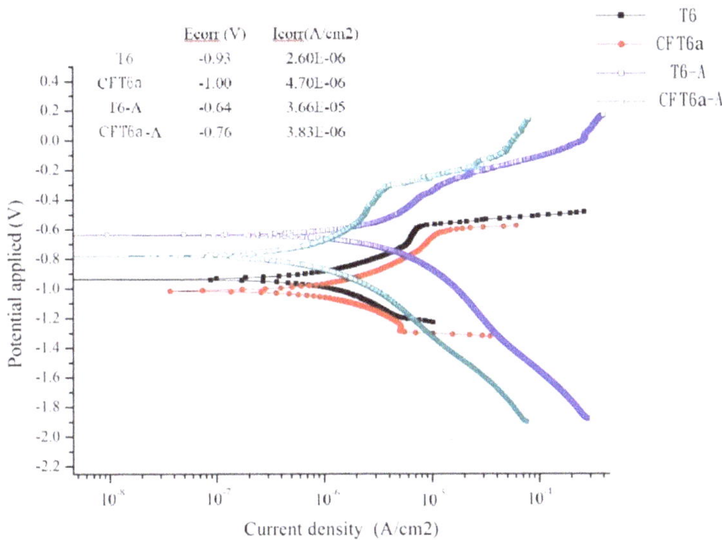

Figure 10. Polarization curves obtained from immersion tests of T6 and CFT6a samples with/without anodization/sealing treatment.

After anodization and sealing, the anodized films on the aluminum alloy samples contained mainly amorphous alumina and a few hydrated alumina [27]. These AAO films could significantly enhance the corrosion resistance of the aluminum alloys [28]. Therefore, the anodized films deposited on the T6 and CFT6 samples improved E_{corr} from (−0.9 to −1.0 V) to (−0.64 to −0.76 V).

Affected by a greater number of coarse SP/IMC (Table 2) and/or a higher particle population, the anodized and sealed film on the T6 sample was entrapped with particles along with air pockets, as shown in Figure 11a. By contrast, the film on the CFT6a sample is relatively sound with few trapped particles; see Figure 11b. The trapped particles could function as corrosion channels to accelerate chloride attacks. Therefore, the anodized and sealed T6 sample obtained a higher current density (I_{corr} of 3.6×10^{-5} A/cm^2) than did the anodized and sealed CFT6a sample (I_{corr} of 3.8×10^{-6} A/cm^2).

The dissolution of SP/IMC at the film–matrix interface drove aluminum and magnesium ions to move toward electrolytes leaving the Si particles remaining in the film, as shown in Figure 11c [27]. Consequently, the anodized and sealed T6 sample obtained inferior corrosion current density to the bare T6 sample (3.6×10^{-5} A/cm^2 *vs.* 2.6×10^{-6} A/cm^2). Decreasing SP/IMC particle size and counts also reduced the size and counts of the Si particles that remained in the anodized film to undergo a minor change in corrosion current density (4.7×10^{-6} *vs.* 3.86×10^{-6} A/cm^2) in the bare CFT6a and anodized/sealed CFT6a samples.

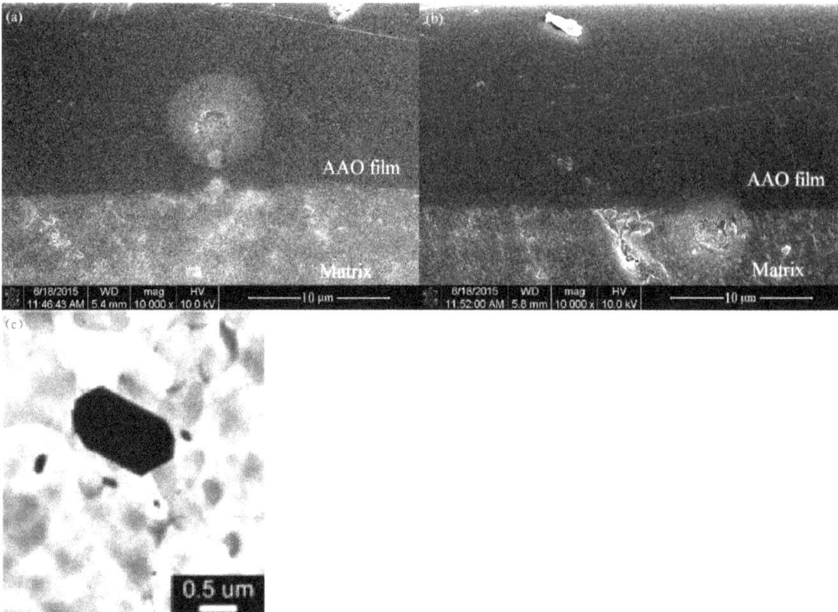

Figure 11. SEM photos show the cross-section of deposited anodized/sealed films on (**a**) T6 sample; (**b**) CFT6a sample; and (**c**) anodized/sealed 6063-T6 samples [27].

4. Conclusions

Adding cryoforging to the process changed the microstructure and mechanical properties of the AA6066–T6 alloy samples; specifically, it decreased SP/IMC particle counts and reduced the *Q* phase but increased fine CuMgSi and Cu_9Al_4 precipitates in the matrix of the CFT6a sample. As a result, the tensile property of elongation was increased by approximately 34% comparing with T6 sample. The anodized and sealed AAO film on the CFT6a sample improved its corrosion resistance but decreased its elongation by approximately 10% (from 14.2% to 12.7%).

Acknowledgments: We gratefully acknowledge the financial support from the Ministry of Science and Technology of the Republic of China (MOST 103-2221-E-008-026-MY2). Many thanks also to National Central University for providing the SEM and TEM tests and to National Sun-Yat Sen University for the EBSD analysis.

Author Contributions: Hwa-Sheng Yong and Wen-Nong Hsu ran experiments of this study and did OM, SEM, TEM observation and EBSD analyses. Main contribution also included experimental data collection. All authors provided equal contribution.

Conflicts of Interest: The authors declare no conflict of interest.

References

1. Wang, Y.M.; Ma, E.; Chen, M.W. Enhanced tensile ductility and toughness in nanostructured Cu. *Appl. Phys. Lett.* **2002**, *80*, 2395–2397. [CrossRef]
2. Rangaraju, N.; Raghuram, T.; Krishna, B.V.; Rao, K.P.; Venugopal, P. Effect of cryo-rolling and annealing on microstructure and properties of commercially pure aluminium. *Mater. Sci. Eng. A Struct.* **2005**, *398*, 246–251. [CrossRef]
3. Chen, Y.J.; Roven, H.J.; Gireesh, S.S.; Skaret, P.C.; Hjelen, J. Quantitative study of grain refinement in Al–Mg alloy processed by equal channel angular pressing at cryogenic temperature. *Mater. Lett.* **2011**, *65*, 3472–3475. [CrossRef]
4. Lee, Y.B.; Shin, D.H.; Park, K.T.; Nam, W.J. Effect of annealing temperature on microstructures and mechanical properties of a 5083 Al alloy deformed at cryogenic temperature. *Scr. Mater.* **2004**, *51*, 355–359. [CrossRef]
5. Panigrahi, S.K.; Jayagathan, R. A study on the mechanical properties of cryorolled Al–Mg–Si alloy. *Mater. Sci. Eng. A Struct.* **2008**, *480*, 299–305. [CrossRef]

6. Panigrahi, S.K.; Jayagathan, R.; Chawla, V. Effect of cryorolling on microstructure of Al–Mg–Si alloy. *Mater. Lett.* **2008**, *62*, 2626–2639. [CrossRef]
7. Panigrahi, S.K.; Jayagathan, R. Development of ultrafine-grained Al 6063 alloy by cryorolling with the optimized initial heat treatment conditions. *Mater. Des.* **2011**, *32*, 2172–2180. [CrossRef]
8. Krishna, K.G.; Sivaprasad, K.; Venkateswarlu, K.; Kumar, K.C.H. Microstructural evolution and aging behavior of cryorolled Al–4Zn–2Mg alloy. *Mater. Sci. Eng. A Struct.* **2012**, *535*, 129–135. [CrossRef]
9. Das, P.; Jayaganthan, R.; Singh, I.V. Tensile and impact-toughness behavior of cryorolled Al 7075 alloy. *Mater. Des.* **2011**, *32*, 1298–1305. [CrossRef]
10. Panigrahi, S.K.; Jayaganthan, R. Effect of ageing on microstructure and mechanical properties of bulk, cryorolled, and room temperature rolled Al 7075 alloy. *J. Alloy. Compd.* **2011**, *509*, 9609–9016. [CrossRef]
11. Yin, J.; Lu, J.; Ma, H.; Zhang, P. Nanostructural formation of fine grained aluminum alloy by severe plastic deformation at cryogenic temperature. *J. Mater. Sci.* **2004**, *39*, 2851–2854. [CrossRef]
12. Krishna, K.G.; Sivaprasad, K.; Narayanana, T.S.N.S.; Kumar, K.C. Localized corrosion of an ultrafine grained Al–4Zn–2Mg alloy produced by cryorolling. *Corros. Sci.* **2012**, *60*, 82–89. [CrossRef]
13. Shanmugasundaram, T.; Murty, B.S.; Sarma, V.S. Development of ultrafine grained high strength Al–Cu alloy by cryorolling. *Scr. Mater.* **2006**, *54*, 2013–2017. [CrossRef]
14. Rao, P.N.; Jayaganthan, R. Effects of warm rolling and ageing after cryogenic rolling on mechanical properties and microstructure of Al 6061 alloy. *Mater. Des.* **2012**, *39*, 226–233.
15. Huang, Y.S.; Shih, T.S.; Chou, J.H. Electrochemical behavior of anodized AA7075-T73 alloys affected by matrix structures. *Appl. Surf. Sci.* **2013**, *283*, 249–257. [CrossRef]
16. *ASTM B557-15, Standard Test Methods for Tension Testing Wrought and Cast Aluminum- and Magnesium-Alloy Products*; ASTM International: West Conshohocken, PA, USA, 2015. [CrossRef]
17. *JIS Z2274, Method of Rotating Bending Fatigue Testing of Metals*; Japanese Standards Association: Tokyo, Japan, 1978.
18. Shih, T.S.; Lee, T.H.; Jhou, Y.J. The effects of anodization treatment on the microstructure and fatigue behavior of 7075-T73 aluminum alloy. *Mater. Trans.* **2014**, *55*, 1280–1285. [CrossRef]
19. Kim, H.J.; Lee, J.Y.; Paik, K.W.; Koh, K.W.; Won, J.; Choe, S.; Lee, J.; Moon, J.T.; Park, Y.J. Effects of Cu/Al intermetallic compound (IMC) on copper wire and aluminum pad bondability. *IEEE Trans. Pack. Technol.* **2003**, *26*, 367–374.
20. Du, Y.; Chang, Y.A.; Huang, B.; Gong, W.; Jin, Z.; Xu, H.; Yuan, Z.; Liu, Y.; He, Y.; Xie, F.Y. Diffusion coefficients of some solutes in fcc and liquid Al: Critical evaluation and correlation. *Mater. Sci. Eng. A Struct.* **2003**, *363*, 140–151. [CrossRef]
21. Matsuda, K.; Teguri, D.; Sato, T.; Uetani, Y.; Ikeno, S. Cu Segregation around Metastable Phase in Al–Mg–Si Alloy with Cu. *Mater. Trans.* **2007**, *48*, 967–974. [CrossRef]
22. Svenningsen, G.; Larsen, M.H.; Walmsley, J.C.; Nordlien, J.H.; Nisancioglu, K. Effect of artificial aging on intergranular corrosion of extruded AlMgSi alloy with small Cu content. *Corros. Sci.* **2006**, *48*, 1528–1543. [CrossRef]
23. Vieira, A.C.; Pinto, A.M.; Rocha, L.A.; Mischler, S. Effect of Al_2Cu precipitates size and mass transport on the polarisation behaviour of age-hardened Al–Si–Cu–Mg alloys in 0.05 M NaCl. *Electrochim. Acta* **2011**, *56*, 3821–3828. [CrossRef]
24. Li, Y.J.; Muggerud, A.M.F.; Olsen, A.; Furu, T. Precipitation of partially coherent α-Al(Mn,Fe)Si dispersoids and their strengthening effect in AA 3003 alloy. *Acta Mater.* **2012**, *60*, 1004–1014. [CrossRef]
25. Cabibbo, M. Microstructure strengthening mechanisms in different equal channel angular pressed aluminum alloys. *Mater. Sci. Eng. A Struct.* **2013**, *560*, 413–432. [CrossRef]
26. Zhaia, T.; Jianga, X.P.; Lia, J.X.; Garratt, M.; Bray, G.H. The grain boundary geometry for optimum resistance to growth of short fatigue cracks in high strength Al-alloys. *Int. J. Fatigue* **2005**, *27*, 1202–1209. [CrossRef]
27. Huang, Y.S.; Shih, T.S.; Wu, C.E. Electrochemical behavior of anodized AA6063-T6 alloys affected by matrix structures. *Appl. Surf. Sci.* **2013**, *264*, 410–418. [CrossRef]
28. Shih, T.S.; Chiu, Y.W. Corrosion resistance and high-cycle fatigue strength of anodized/sealed AA7050 and AA7075 alloys. *Appl. Surf. Sci.* **2015**, *351*, 997–1003. [CrossRef]

3

Ti/Al Multi-Layered Sheets: Accumulative Roll Bonding (Part A)

Jan Romberg [1,2], Jens Freudenberger [1,2,3,*], Hansjörg Bauder [4], Georg Plattner [4], Hans Krug [4], Frank Holländer [5], Juliane Scharnweber [6], Andy Eschke [6], Uta Kühn [1], Hansjörg Klauß[1], Carl-Georg Oertel [6], Werner Skrotzki [6], Jürgen Eckert [7,8] and Ludwig Schultz [1,2]

1 Leibniz Institut für Festkörper- und Werkstoffforschung Dresden (IFW Dresden), Helmholtzstr. 20, Dresden D-01069, Germany; janromberg@yahoo.de (J.R.); u.kuehn@ifw-dresden.de (U.K.); h.j.klauss@ifw-dresden.de (H.K.); l.schultz@ifw-dresden.de (L.S.)
2 Institut für Werkstoffwissenschaft, Technische Universität Dresden, Dresden D-01062, Germany
3 Institut für Werkstoffwissenschaft, Technische Universität Bergakademie Freiberg, Gustav-Zeuner-Str. 5, Freiberg D-09599, Germany
4 Carl Wezel KG, Industriestr. 95, Mühlacker D-75417, Germany; carl_wezel_kg@t-online.de (H.B.); carl_wezel_kg@t-online.de (G.P.); hk.krug@gmx.net (H.K.)
5 Lehrstuhl Metallkunde und Werkstofftechnik, Brandenburgische Technische Universität Cottbus, Konrad-Wachsmann-Allee 17, Cottbus D-03046, Germany; hollaend@tu-cottbus.de
6 Institut für Strukturphysik, Technische Universität Dresden, Dresden D-01062, Germany; juliane.scharnweber@tu-dresden.de (J.S.); andy.eschke@gmx.de (A.E.); oertel@physik.tu-dresden.de (C.-G.O.); werner.skrotzki@physik.tu-dresden.de (W.S.)
7 Erich Schmid Institut für Materialwissenschaft, Österreichische Akademie der Wissenschaft, Jahnstraße 12, Leoben A-8700, Austria; juergen.eckert@unileoben.ac.at
8 Department Materialphysik, Montanuniversität Leoben, Jahnstraße 12, Leoben A-8700, Austria
* Correspondence: j.freudenberger@ifw-dresden.de

Academic Editor: Hugo F. Lopez

Abstract: Co-deformation of Al and Ti by accumulative roll bonding (ARB) with intermediate heat treatments is utilized to prepare multi-layered Ti/Al sheets. These sheets show a high specific strength due to the activation of various hardening mechanisms imposed during deformation, such as: hardening by grain refinement, work hardening and phase boundary hardening. The latter is even enhanced by the confinement of the layers during deformation. The evolution of the microstructure with a special focus on grain refinement and structural integrity is traced, and the correlation to the mechanical properties is shown.

Keywords: accumulative roll bonding; Ti/Al multi-layered composites; grain refinement; microstructure; mechanical properties

1. Introduction

Light weight materials are in the focus of the development of novel structural materials, as they bear the potential to reduce the energy consumption in mobile applications while retaining the functionality. There are numerous approaches to obtain materials with a high specific strength. On the one hand, novel steels are developed to complement the most dominant class of structural materials and to extend their field of usability to light weight applications [1–3]. On the other hand, the "classic" light weight materials for structural applications with densities below 4.5 g/cm^3, such as Ti, Al, Mg and their alloys, are optimized with respect to strength. For this purpose, various means are utilized, including alloy design and the development of novel processing techniques [4–6]. The latter also involves non-common deformation techniques, such as the application of severe plastic deformation (SPD) [7–11] or co-deformation of at least two different materials [12–16].

Co-deformation of two metallic materials, such as Al and Ti, bears the potential to gain improved material properties regarding strength and formability. Both material properties are strongly linked to the microstructure. Their evolution including the formation of texture in these materials represents the key aspect in developing materials with superior properties. In recent years, it has been shown that cold working to large plastic strains can impose ultra-fine grained microstructures [17–19] or even microstructures with features being in the nanometer range [20–22]. The small grain sizes, fine dislocation networks and/or small precipitates cause a significant contribution to the mechanical strength according to Hall–Petch-type, work and precipitation hardening, respectively [8,23–25]. When applied in combination, these hardening mechanisms may even have a synergetic strength effect. When co-deformation is applied repetitively, as is the case for accumulative roll bonding (ARB) [26–28] and for accumulative swaging and bundling (ASB) [16,29–33], the density of phase boundaries increases with applied bonding or bundling cycles, respectively. The increasing number of phase boundaries is beneficial for two reasons. First, the phase boundary itself represents a barrier for dislocation movement [34]. Secondly, these phase boundaries limit possible grain growth. After deformation to very large plastic strains, the multi-phased materials have a mean phase boundary distance in the the sub-micrometer-range. These materials also show considerable strength at elevated temperatures when grain growth occurs in at least one phase, as grain growth is limited to the distance of the phase boundaries. Summarizing, high densities of grain and phase boundaries significantly contribute to the strength of metallic composites.

Although it has already been shown that Ti/Al composite sheets can be obtained by ARB [35–40], this process remains difficult. A crucial issue is to retain individual continuous layers within the composite during accumulative deformation. Once the Ti layers are strain hardened beyond a certain limit, their formability is negligible when compared to the Al sheets. Consequently, necking of the Ti layers is observed, and the stretched Ti pieces remain rather stable in size within the continuously-deforming Al matrix [36,41–43].

In contrast to this, the individual Ti and Al layers remain stable when an intermediate heat treatment (IHT) is applied between the rolling cycles. After eight ARB + IHT cycles, a fine multi-layered composite sheet is obtained. With respect to grain refinement, IHT is counter-productive, as it may cause recovery, recrystallization and grain growth. Consequently, a highly strengthened light weight material cannot be achieved with this procedure. In addition, it is necessary to emphasize that the individual layers become wavy with increasing deformation strain, *i.e.*, the number of ARB cycles, which has been considered as a consequence of the evolution of the texture within the layers [44,45].

To avoid necking, a high work hardening rate and a low strength level are necessary. During the first ARB cycle, these conditions are met, but already for the second cycle, the work hardening ability of Ti is already saturated and, thus, may be responsible for the necking of the Ti layers. An annealing treatment at 723 K for 90 min under vacuum conditions between each ARB cycle lowers the strength, and the strain hardening ability is restored. Optimum strain hardening conditions would possibly be restored above the transus temperature of Ti. However, this is not possible to apply to the composite, as this temperature is above the melting temperature of Al.

At elevated temperatures, successful roll bonding of Al can be achieved using a lower thickness reduction. It can be expected that this would result in reduced necking of the Ti layers. However, the reduction of strength with rising temperature is less pronounced in Ti than in Al. Alternatively, the arrangement of rolls and the geometry of the rolling gap can be altered. In contrast to a four high rolling mill, which is symmetric with respect to the sheet plane, trio rolling with different sizes of the upper and lower rolls is asymmetric. Even in the case when the excenter velocities of the upper and lower rollers are identical, the difference in diameter causes shearing in the deformation zone and, thus, eases bonding and reduces necking. This second possibility of controlling the homogeneity of the layers' appearance is addressed in Part B of the present study (Ti/Al Multi-Layered Sheets: Differential Speed Rolling). This article addresses the influence of ARB on the microstructure, as well as on the appearance and faults of the individual layers; four different solutions to receive finely-layered Ti/Al composite sheets by ARB are discussed: the effect of intermediate heat treatments, temperature, strain rate and that of the confinement in the deformation zone.

2. Experimental Section

ARB was performed on a stack of five layers composed of pure Ti (99.995%) and the aluminum alloy AA5049 (Al, 2% Mg and 0.8% Mn; in the following referred to as Al) sheets with a thickness of 1 mm each. This stack (Al-Ti-Al-Ti-Al) with an initial height of 5 mm was reduced by 50% in thickness within one single rolling pass. Subsequently, the sheets were cut in length into two pieces, wire-brushed after cleaning with ethanol, stacked and further roll bonded. The ARB cycles were repeated up to eight times. To investigate the effect of rolling speed on the formation of the microstructure, two series of sheets have been produced with different rolling speeds. Some sheets have been intermediately heat treated at 723 K for 90 min under vacuum conditions ($p \leq 10^{-5}$ mbar). The deformation rate was determined from the rolling speed assuming constant rolling speed and neglecting shear deformation.

The influence of temperature on ARB was studied for the second and fourth ARB cycle. The initial bonding has always been performed at room temperature (RT). For ARB at elevated temperature, the rolls were heated up to 573 K. The temperature at which the samples have been heated before rolling varied from RT to 573 K.

The microstructure has been observed by scanning electron microscopy (SEM, Zeiss Microscopy, Jena, Germany) and electron backscatter diffraction (EBSD, Oxford Instruments, Abingdon, UK) after metallographic preparation, described in detail elsewhere [43]. EBSD measurements were done in the upper third of the single material sheets and the 160-layer composite sheets. In the 5-layer sheets, EBSD maps were measured within the top layer (aluminum) and the second-top layer (titanium). The step size for EBSD mapping was 200 nm and 40 nm for Al and Ti, respectively. Areas separated by grain boundaries with misorientations of more than 15° are considered as grains. The grain size is given as the median number-weighted and the median area-weighted grain size. The first metric captures predominantly small grains, the second measures particularly large grains, especially in wide spread grain size distributions. Consequently, two values are obtained on the same data, representing lower and upper bounds for the grain size. The latter method implies that 50% of the cross-sectional area is covered by grains with a size larger than the median value. The advantage of this procedure is that the sensitivity to detect changes in the grain size distribution is enhanced, and the effect of rolling on the microstructure can be traced more easily. EBSD also provides information about the orientation distribution in a generally small area of the sheets, *i.e.*, the local texture with a poor statistic is reflected.

Tensile tests were carried out at RT at a constant strain rate of 0.01 s^{-1} utilizing an electro-mechanical Instron 8562 testing machine. The geometry of the test specimens is shown in Figure 1. The thickness of the tensile test specimens varied, as the samples were cut from the as-rolled sheets. The initial sheets had a thickness of 1 mm, and the sheets that have been subjected to ARB had a thickness of about 2.5 mm.

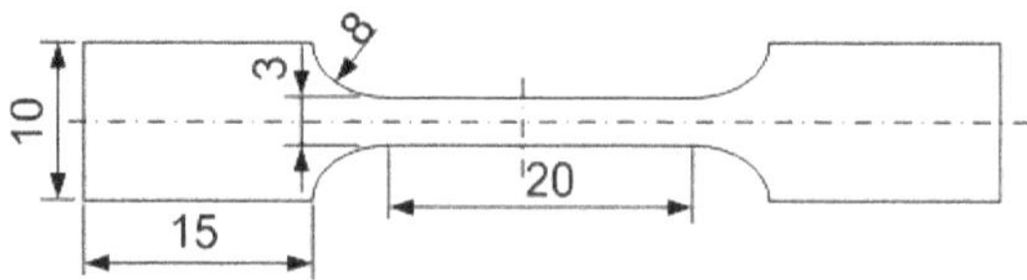

Figure 1. Geometry of the samples used for tensile tests. Dimensions are provided in mm.

3. Results and Discussion

3.1. *Intermediate Heat Treatment*

Roll bonding of five sheets with a thickness reduction of 50% does not significantly affect the grain size. Therefore, the effect of further processing by ARB is addressed in the following. Figure 2 shows the layer and phase distribution of Ti/Al composites after different ARB cycles. It can be seen that the Ti layers (bright grey) are not continuous anymore after four or more ARB cycles at room temperature and no IHT applied. Under the mentioned conditions, Al layers remain continuous. Therefore, they are capable of the observed plasticity. Further argument for this assumption is taken from the layer thickness: Al layers are thinner than the Ti layers, although their initial thickness was the same. Furthermore, the Al flows in between the Ti flakes. Both features support the conclusion that the Ti layers are locally sheared in bands, but not continuously thinned. Indeed, the Ti flakes do not show a considerable change in their size with further deformation.

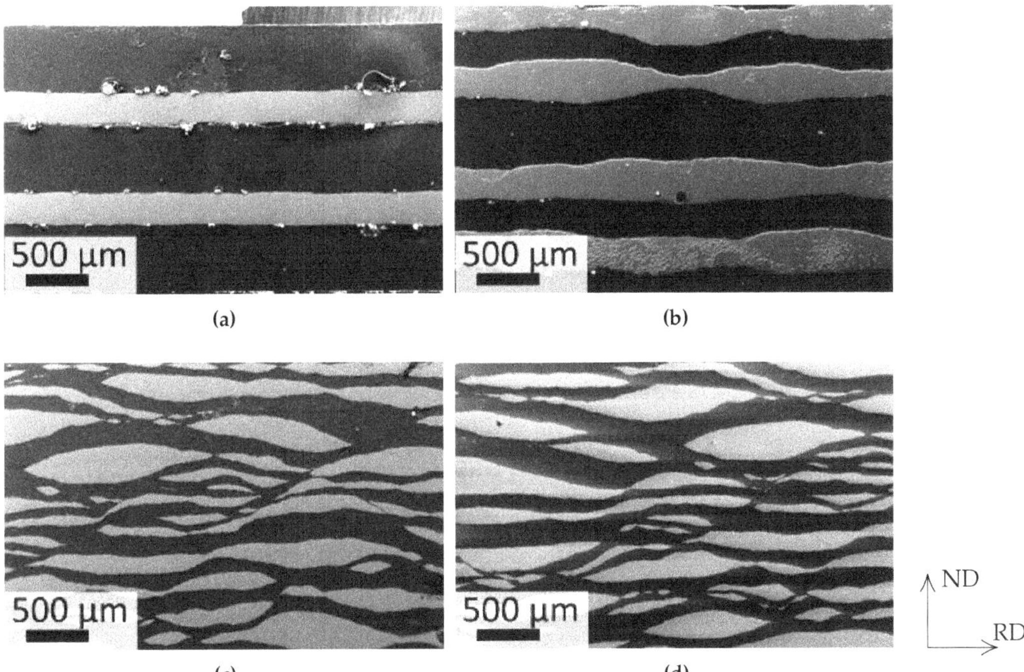

Figure 2. SEM micrographs showing the longitudinal cross-section of Ti/Al composite sheets after accumulatively roll bonding a five-layered composite, *i.e.*, after the first (**a**), second (**b**), fourth (**c**) and sixth (**d**) accumulative roll bonding (ARB) cycle (no intermediate heat treatment (IHT) was applied).

Nevertheless, the mechanical properties of the five-layered composite, as well as the further ARB processed composite are investigated. Figure 3 shows the evolution of the mechanical properties with respect to the number of ARB cycles.

Roll bonding of three Al and two Ti sheets to the five-layered composite (*i.e.*, the first ARB cycle) causes a remarkable decrease of the plastic strain when comparing the composite to pure Ti. This decrease is still significant, but less pronounced when comparing the composite to pure Al. This processing step on the other hand leads to an increase of yield and ultimate tensile strength by a factor of about two. Further processing by ARB does not change the mechanical properties significantly, which is mainly attributed to the discontinuous layers. The work hardening ability of Ti reaches its limit; therefore, the contribution to strain of Ti is shrinking. Furthermore, Al shows dynamic recovery during further processing. Thus, this processing route is not promising for obtaining ultra-fine grained materials with a large number of interfaces, as at a higher number of ARB cycles, the Ti flakes produced are circumvented by Al flow, as can be seen from the similar SEM micrographs obtained on sheets being deformed with four and six ARB cycles (Figure 2c,d). In order to retain the formability of the Ti-layers within the composite, IHTs have been introduced after each roll bonding step. These IHTs were performed at 723 K for 90 min under vacuum conditions ($p \leq 10^{-5}$ mbar) and are applied to retain the formability of the composite. Furthermore, the individual Ti layers remain continuous during the ARB cycles. The corresponding layer structures are shown in Figure 4.

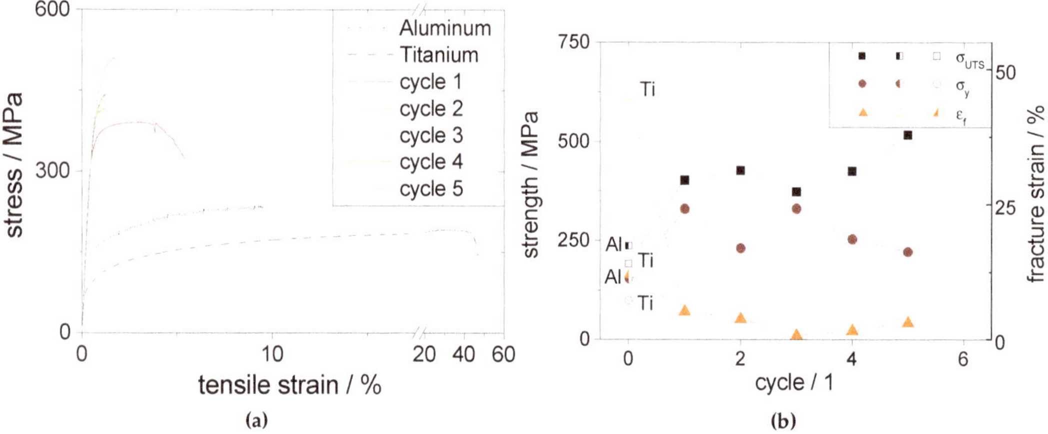

Figure 3. Engineering stress-strain curves of the five-layered Ti/Al composite accumulatively roll bonded with up to five cycles without intermediate heat treatments (**a**) and ultimate tensile strength (σ_{UTS}), yield strength (σ_y) and plastic strain (ε) of Al (half open symbols), Ti (open symbols) and the five-layered composite (filled symbols) (**b**).

Unfortunately, the IHT causes recrystallization of the Al, as well as recovery of the Ti. These processes are reflected by the grain size. Figure 5 shows the evolution of the grain size within the Ti and Al layers during processing by ARB. In order to make this evaluation comparable to others, the weighting methods described in the Experimental Section have been applied.

Without IHT, the mean grain size of both phases is lowered during ARB. This reduction is pronounced within the first to third ARB cycle. Subsequent ARB cycles lead to a steady state of the grain size in the sub-micrometer range.

The grain sizes are also determined after each ARB cycle before and after additional IHTs (pre-IHT and post-IHT). The grain size is larger in the heat-treated samples due to grain growth. Additional ARB does not reduce the grain size to the condition observed without IHTs.

Anyhow, the change in grain size due to the deformation by ARB with and without IHTs is small in the present case. Therefore, it is quite expectable that the mechanical strength of the composites remains at a comparable level. The tensile properties of the Ti/Al composites as prepared by increasing ARB cycles with intermediate heat treatments are shown in Figure 6.

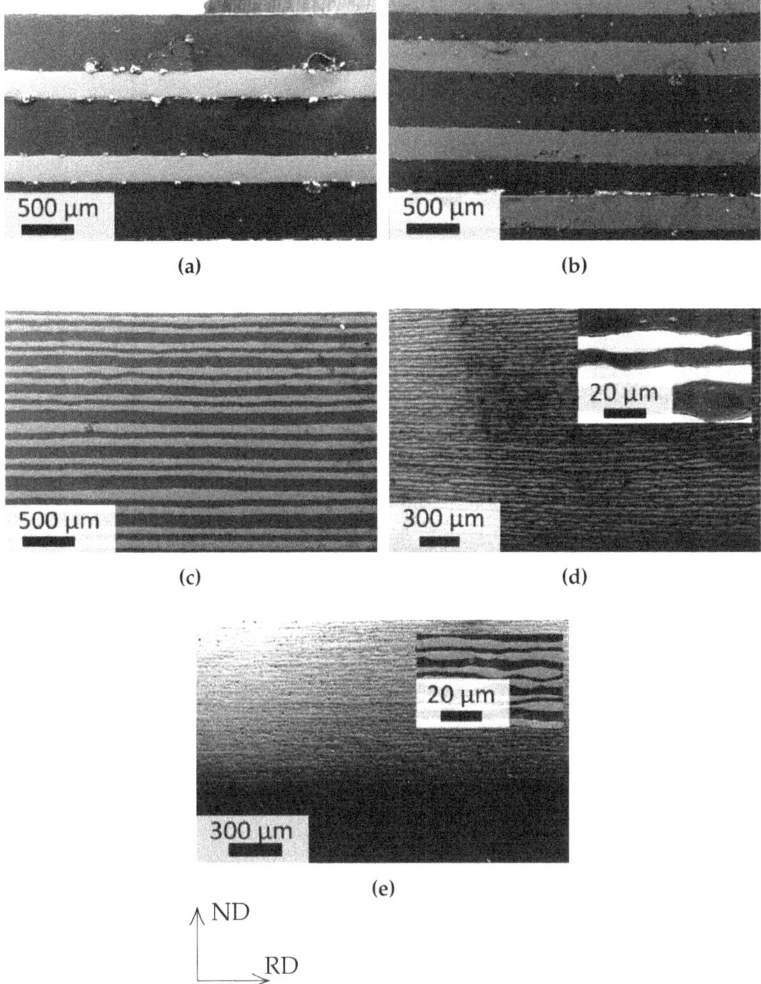

Figure 4. SEM micrographs showing the longitudinal cross-section of the layer structure of a Ti/Al composite sheets after accumulatively roll bonding a five-layered composite after the first (**a**), second (**b**); fourth (**c**), sixth (**d**) and eighth (**e**) ARB cycle with IHTs being applied.

The engineering stress-strain curves reveal ultimate tensile strength values ranging from 375 MPa to 406 MPa. In contrast to the common strength values of the composites, a different behavior is observed for the plastic strain. After the first ARB cycle, the stress-strain curve shows the maximum strength and an uniform elongation of about 3%. Further straining causes remarkable jumps in the stress-strain curve, which are related to the failure of the individual layers of the composite. Obviously, the bonding between these layers is not sufficient at this stage of deformation. Already after the next ARB cycle, just one single step remains visible, indicating that bonding has improved.

The reason for the disappearing of the steps in the stress-strain curve is related to the processing of the composites. In the first ARB step, the Ti and Al layers have to be roll bonded. This is the most difficult processing step of the Ti/Al composite materials. Later ARB cycles only bond outer Al layers, which is much easier. The inner Ti/Al interfaces become more perfect during each rolling cycle, and thus, their bonding strength further increases.

Considering a Ti/Al five-layered composite after the first ARB cycle: when the first layer fails, the fracture toughness of the composite has been reached, causing macroscopic fracture. With further ARB cycles, the number of layers increases. The failure of an individual layer does not cause the shear strength to be larger than the fracture stress, and consequently, the observed plastic strain is larger. This situation, however, changes again with further deformation by ARB, as a further reinforcement caused by the improvement of the bonding strength of the interface due to further rolling is observed. As a consequence, a reduction of the plastic strain is gained. Hence, optimum plasticity is observed for the composite with stabilized interfaces and a certain, but not too small number of layers. A further enlargement of the number of individual layers would not be beneficial with respect to the plasticity at this processing step, unless the material shows a grain size in the nanometer range [7,19,46].

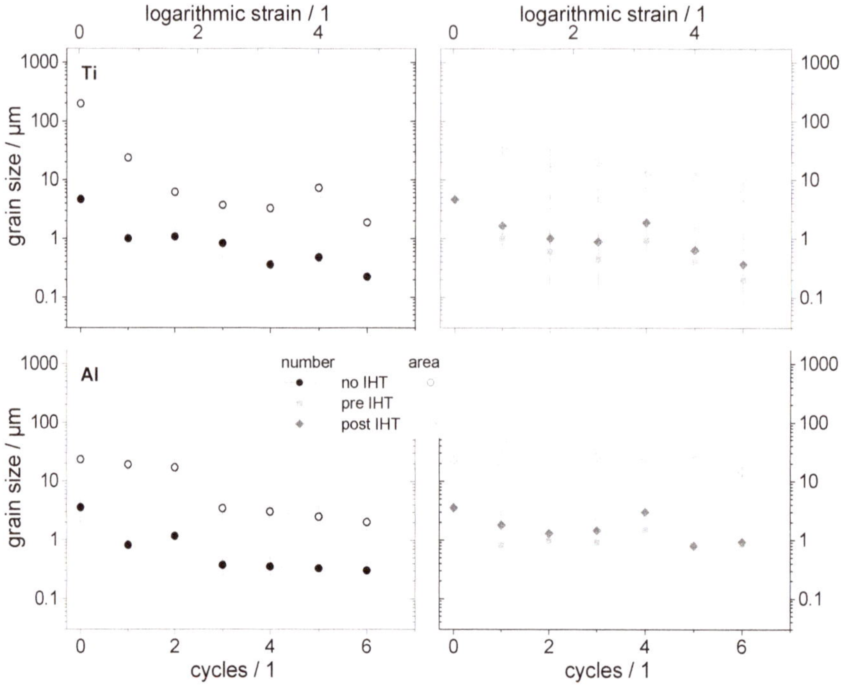

Figure 5. Grain size within the Ti (upper row) and Al layers (lower row) after ARB at room temperature (left), as well as with IHTs applied (right). In the latter case, the grain sizes are provided after the ARB cycles before (pre-IHT) and after IHT (post-IHT). The grain sizes were evaluated by both number and area weighting.

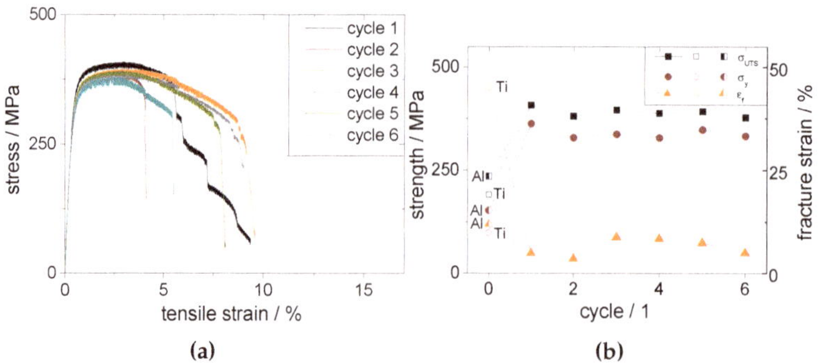

Figure 6. Engineering stress-strain curves of the five-layered Ti/Al composite accumulatively roll bonded with up to six cycles, including intermediate heat treatments (IHT) (**a**) and ultimate tensile strength (σ_{UTS}), yield strength (σ_y) and plastic strain (ε) of Al (half open symbols), Ti (open symbols) and the five-layered composite (filled symbols) (**b**).

3.2. Deformation Speed

ARB has also been performed at different strain rates, established by varying the rolling velocity. Here, results for strain rates of 5.3 s^{-1} and 23.4 s^{-1} are discussed. Figure 7 shows the corresponding layer structure obtained after four ARB cycles. It is observed that slow rolling results in enhanced necking of the Ti layers. The grains within Al and Ti are elongated in the rolling direction.

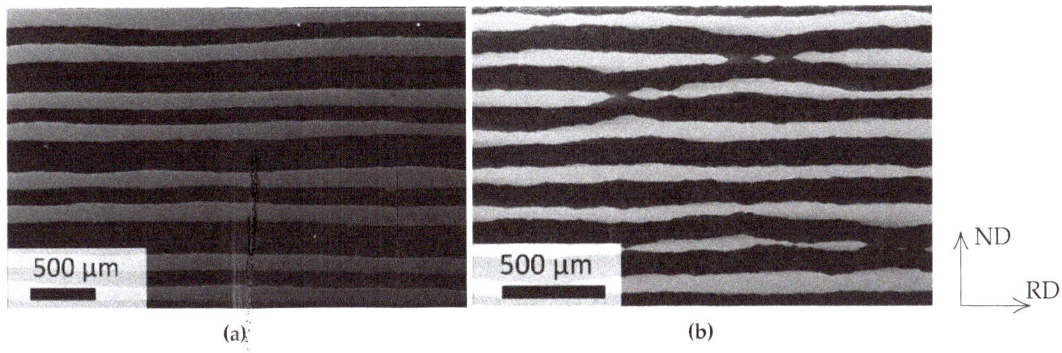

Figure 7. Micrographs showing the longitudinal sections of five-layered composites after four ARB cycles; rolling with a strain rate of 23.4 s^{-1} (**a**) and 5.3 s^{-1} (**b**).

As discussed by Emmens [47], work hardening delays or avoids necking. The effect is more pronounced if the strain rate sensitivity is high, because the necked regions deform with a higher strain rate compared to the rest of the material. Therefore, the necked material is hardened and consequently becomes less deformed. According to this observation, the strain rate sensitivity for Ti is higher for the strain rate of 23.4 s^{-1} than for 5.3 s^{-1}. Consequently, less necking occurs for deformation at large strain rates. Combining IHT and a high strain rate enables more stable deformation conditions, yielding a laminar structure after eight ARB cycles. The microstructure of Al contains a small number of elongated grains, while the majority of the grains is equiaxed. The grain size of most grains is limited by the layer thickness. The Ti grains are smaller than the layer thickness. There is also a mixture of elongated and equiaxed grains, as shown in Figure 8.

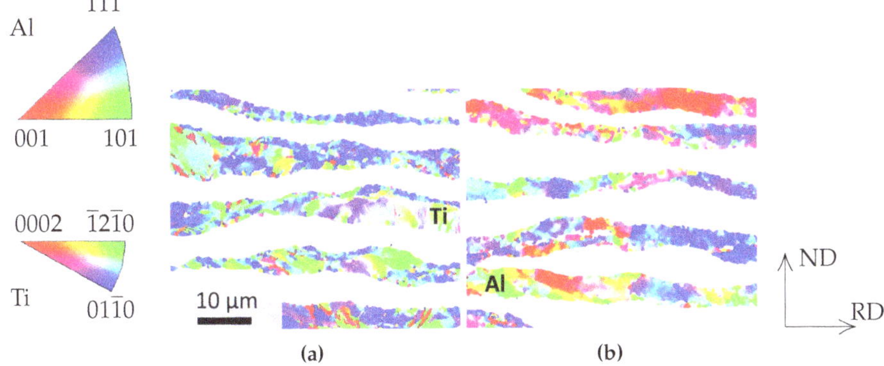

Figure 8. Electron backscatter diffraction (EBSD) maps of a five-layered composite sheet after eight ARB + IHT cycles. The maps are separated for Ti (**a**) and Al (**b**). The color code describes the crystal orientation in the rolling direction according to the inverse pole figure maps (shown leftmost for the cubic (Al, upper part) and hexagonal case (Ti, lower part)).

3.3. ARB at Elevated Temperatures

ARB has also been performed at elevated temperatures. For this purpose, two five-layered composites that were roll bonded at RT were further rolled at higher temperatures. It turned out that this condition requires a modification of the applied strain, as thermal softening is superimposed by strain hardening. The applicability of the process is evaluated after the second ARB cycle. The corresponding results are summarized in Table 1. Micrographs of the corresponding layer structures are shown in Figure 9 for the case of successful processing.

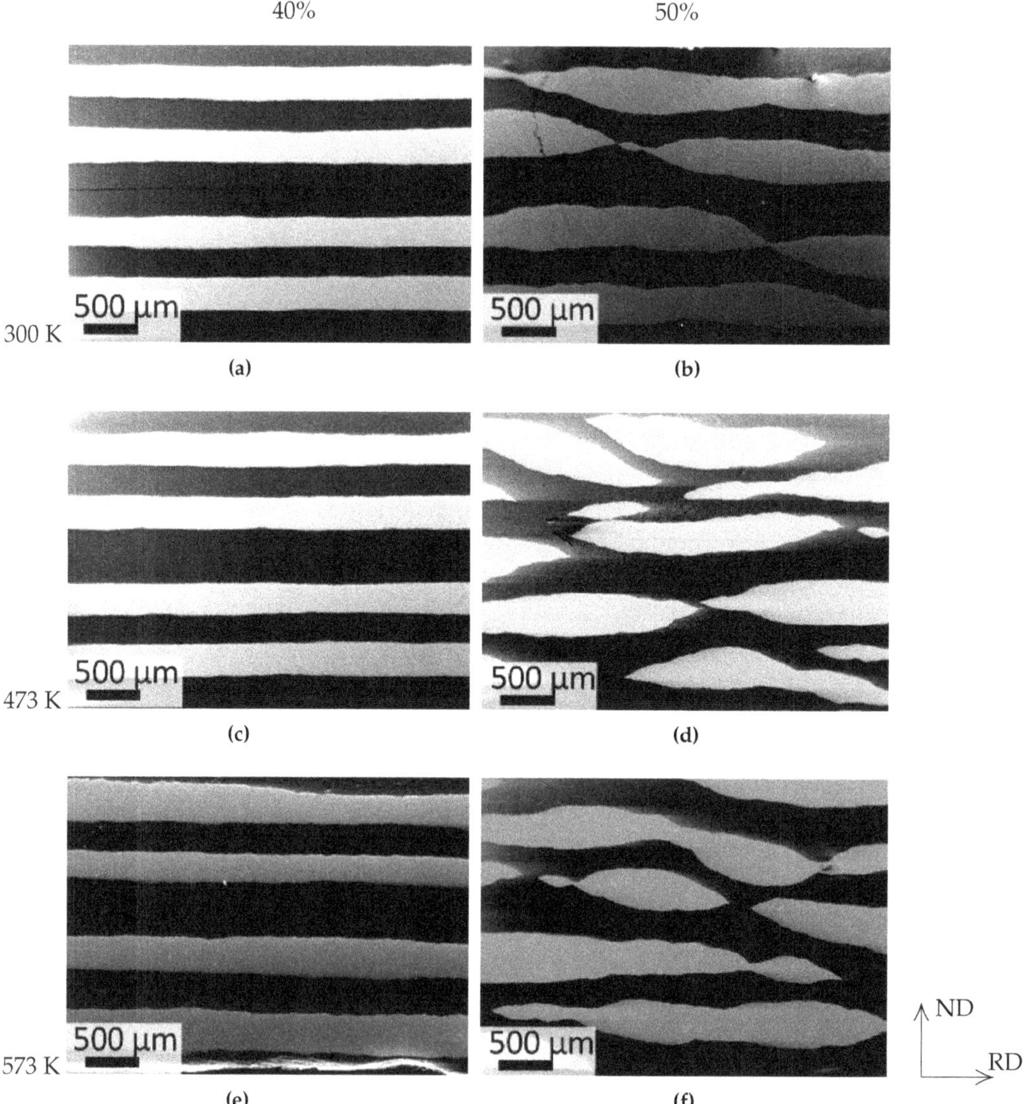

Figure 9. SEM micrographs showing the layer structure of a five-layered composite roll bonded with a single ARB cycle at elevated temperatures. The sheets are rolled with a 40% (**a**,**c**,**e**) or a 50% (**b**,**d**,**f**) thickness reduction during the second cycle.

Table 1. Overview of elevated temperature ARB.

Temperature	Thickness Reduction		
	30%	40%	50%
RT	no bonding	weak bonding laminar layers	strong bonding necking of Ti
473 K	no bonding	sufficient bonding laminar layers	strong bonding necking of Ti
573 K	no bonding	sufficient bonding laminar layers	sufficient bonding necking of Ti

A thickness reduction of 30% is not sufficient for roll bonding independent of the working temperature. In contrast, a thickness reduction of 40% is sufficient for bonding, although the bonding strength may be very weak, as in the case of RT deformation. A thickness reduction of 50% yields strong bonding, but also necking of the Ti layers, as shown in Figure 9. Since roll bonding with 40% thickness reduction has been found as most successful, it is used for the following ARB cycles, as well. Further ARB cycles cause deformation localization, as well as the formation of necks, which is found to be independent of temperature. The necks show a random spatial distribution (not shown here).

An increase in rolling temperature facilitates roll bonding of Ti/Al composites at a lower thickness reduction than 50%, which is generally used in ARB processes. At elevated temperatures, Al becomes significantly softer, and therefore, the difference in yield stress between the Al and Ti increases. The deformation becomes unstable and, consequently, results in an increasing number of necks of the Ti layers. By reducing the applied strain, necking can be reduced, but not excluded. Although the second ARB cycle at elevated temperatures has been successful, further processing results in necking of the Ti layers.

3.4. *Asymmetric ARB*

Asymmetric ARB has been accomplished by means of a four high rolling mill and also with the help of a trio mill. The latter has an asymmetric geometry with respect to the sheet plane, as the working rolls have different diameters, but are operated at the same velocity of the surface. Therefore, the deformation zone is additionally subjected to shear deformation. In order to evaluate the effect of the asymmetric rolling geometry, a sheet obtained from three ARB cycles has been rolled in different rolling mills. Asymmetric rolling has been performed at a trio rolling mill. The diameters of the working rolls were 235 and 290 mm, respectively, and the surface velocity has been set to 3 m/min. The comparative symmetric rolling has been realized utilizing a four high rolling, whose working rolls had a diameter of 110 mm, operated with a surface velocity of 3.5 m/min.

Figure 10 shows the arrangement of the different metallic phases in the cross-section (layer structure) after the fourth ARB cycle with different conditions being applied. It has been found that trio rolling results in a more pronounced necking of the Ti layers.

Thus, at a 50% thickness reduction per cycle, trio rolling is detrimental to achieve continuous lamellar sheets. However, trio rolling enables roll bonding with a 45% thickness reduction, which is not possible by means of a symmetrical four high rolling mill. In order to quantitatively assess the effect of the thickness reduction under asymmetric rolling conditions, a composite was accumulatively roll bonded utilizing a trio mill with a thickness reduction of 50% at each of three cycles with an IHT at 823 K for 90 min being applied after each cycle. In the fourth rolling cycle, the thickness reduction was varied stepwise. Thus, all different deformation states used for further analysis are obtained from one single sheet.

At a reduction of less than 45%, the single sheets do not bond. Between 45% and 50%, the number of necks significantly increases with thickness reduction, as shown in Figure 11. Further thickness reduction does not change the number of necks anymore, but increases the fraction of completely separated layers. However, in comparison to processing with a four high rolling mill, the amount of separated necks shrinks from 0.4 mm^{-2} to 0.1 mm^{-2}. Thus, a low thickness reduction is beneficial for the homogeneity of the deformation process.

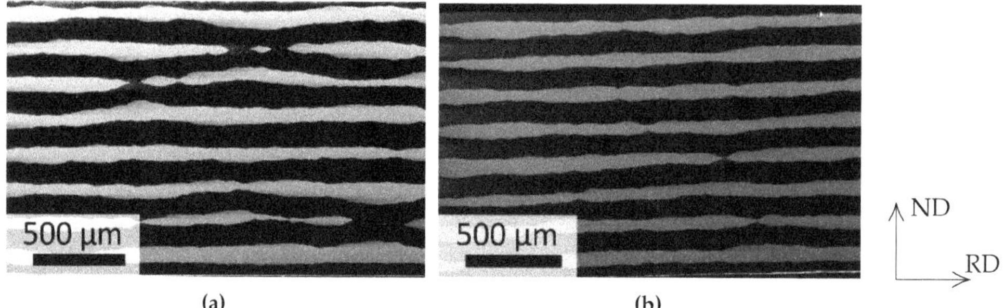

Figure 10. Cross-section of the layer structure after four ARB cycles in a trio (**a**) and four high rolling mill (**b**).

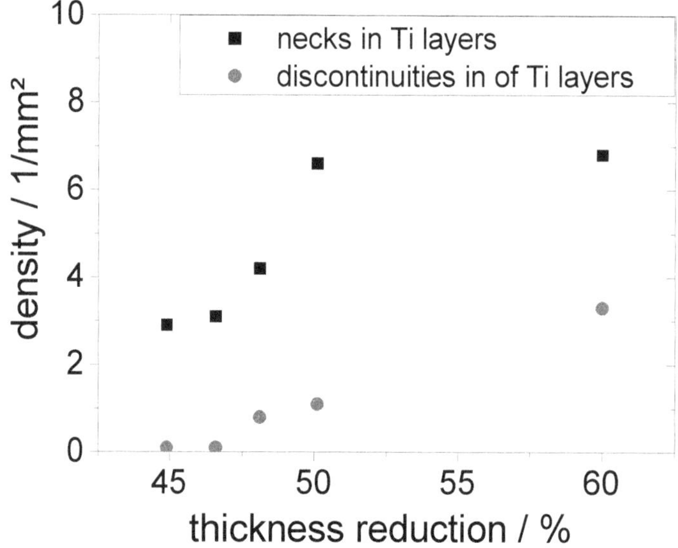

Figure 11. Density of necks and separated necks of a Ti/Al composite sheet after four ARB passes in a trio rolling mill in dependence of the thickness reduction during the last rolling cycle.

4. Summary

Sheets composed of a high number of alternating Ti and Al layers have been successfully prepared by accumulative roll bonding. In order to keep individual continuous layers, necking occurring predominantly in the Ti layers has to be prevented. This study demonstrated that this is feasible if the applied thickness reduction per rolling step is lowered or if the work hardening rate is increased.

A lower thickness reduction can be achieved by rolling at elevated temperatures and, alternatively, by incorporating a shear component within the deformation zone by utilizing a trio mill. A higher work hardening rate can be achieved by an increased rolling velocity. However, this only works for a high strain rate sensitivity of the Ti layers, which lowers with increasing strength. As a consequence, in order to maintain the low strength and high strain rate sensitivity in the Ti layers, the composites have to be thermally treated after each single ARB cycle. Hence, to obtain materials with a high specific strength strain, hardening of the composite cannot be achieved while the process demands for continuous layers.

Acknowledgments: This work has been supported by the European Union and the Free State of Saxonia in the framework of the European Centre for Emerging Materials and Processes (ECEMP), Contract No. 13795/2379.

Author Contributions: Jan Romberg and Jens Freudenberger designed the experiments, collected and interpreted the data, wrote and edited the paper and contributed to all activities. Juliane Scharnweber and Andy Eschke contributed to scanning electron microscopy, including sample preparation, imaging and diffraction; together with Carl-Georg Oertel, they analyzed the results and fostered their interpretation. Hansjörg Bauder, Georg Plattner and Hans Krug performed the accumulative roll bonding experiments; they contributed to the discussion and interpretation of the results and adapted the rolling mill to the present requirements. Frank Holländer performed the hot rolling experiments. Hansjörg Klauß performed the tensile tests. Uta Kühn, Werner Skrotzki, Jürgen Eckert and Ludwig Schultz directed the research and contributed to the discussion and interpretation of the results.

Conflicts of Interest: The authors declare no conflict of interest.

References

1. Kaufmann, H.; Sonsino, C.; Demofonti, G.; Riscifuli, S. High-strength steels in welded state for light-weight constructions under high and variable stress peaks. *Steel Res. Int.* **2008**, *79*, 382–389.
2. Sonsino, C. Light-weight design chances using high-strength steels. *Mater. Und Werkst.* **2007**, *38*, 9–22.
3. Kwon, O.; Lee, K.; Kim, G.; Chin, K. New Trends in Advanced High Strength Steel Developments for Automotive Application. *Mater. Sci. Forum* **2010**, *638–642*, 136–141.
4. Lee, S.; Saito, Y.; Tsuji, N.; Utsunomiya, H.; Sakai, T. Role of shear strain in ultragrain refinement by accumulative roll-bonding ARB process. *Scr. Mater.* **2002**, *46*, 281–285.
5. Shi, M.; Takayama, Y.; Ma, C.; Watanabe, H.; Inoue, H. Microstructure and texture evolution in titanium subjected to friction roll surface processing and subsequent annealing. *Trans. Nonferrous Met. Soc. China* **2012**, *22*, 2616–2627.
6. Belov, N. Sparingly alloyed high-strength aluminum alloys: Principles of optimization of phase composition. *Met. Sci. Heat Treat.* **2012**, *53*, 420–427.
7. Sabirov, I.; Murashkin, M.; Valiev, R. Nanostructured aluminum alloys produced by severe plastic deformation: New horizons in development. *Mater. Sci. Eng. A* **2013**, *560*, 1–24.
8. Valiev, R.; Enikeev, N.; Murashkin, M.; Utyashev, F. Using intensive plastic deformations for manufacturing bulk nanostructure metallic materials. *Mech. Solids* **2012**, *47*, 463–474.
9. Latysh, V.; Semenova, I.; Salimgareeva, G.; Kandarov, I.; Zhu, Y.; Lowe, T.; Valiev, R. *Nanomaterials by Severe Plastic Deformation*; Trans Tech Publications Ltd.: Dürnten, Switzerland, 2006.
10. Sabirov, I.; Perez-Prado, M.; Molina-Aldareguia, J.; Semenova, I.; Salimgareeva, G.; Valiev, R. Anisotropy of mechanical properties in high-strength ultra-fine-grained pure Ti processed via a complex severe plastic deformation route. *Scr. Mater.* **2011**, *64*, 69–72.
11. Bonarski, B.; Mikulowski, B.; Schafler, E.; Holzleithner, C.; Zehetbauer, M. Crystallographic textures of single and polycrystalline pure Mg and Cu subjected to HPT deformation. *Arch. Metall. Mater.* **2008**, *53*, 117–123.
12. Lu, C.; Tieu, K.; Wexler, D. Significant enhancement of bond strength in the accumulative roll bonding process using nano-sized SiO_2 particles. *J. Mater. Process. Technol.* **2009**, *209*, 4830–4834.
13. Peng, J.; Liu, Z.; Xia, P.; Lin, M.; Zeng, S. On the interface and mechanical property of Ti/Al-6%Cu-0.5%Mg-0.4%Ag bimetal composite produced by cold-roll bonding and subsequent annealing treatment. *Mater. Lett.* **2012**, *74*, 89–92.
14. Wang, J.; Hoagland, R.; Misra, A. Mechanics of nanoscale metallic multilayers: From atomic-scale to micro-scale. *Scr. Mater.* **2009**, *60*, 1067–1072.
15. Lapovok, R.; Ng, H.; Tomus, D.; Estrin, Y. Bimetallic Copper-Aluminium Tube by Severe Plastic Deformation. *Scr. Mater.* **2012**, *66*, 1081–1084.
16. Marr, T.; Freudenberger, J.; Kauffmann, A.; Romberg, J.; Okulov, I.; Petters, R.; Scharnweber, J.; Eschke, A.; Oertel, C.G.; Kühn, U.; *et al.* Processing of Intermetallic Titanium Aluminide Wires. *Metals* **2013**, *3*, 188–201.

17. Beausir, B.; Scharnweber, J.; Jaschinski, J.; Brokmeier, H.G.; Oertel, C.G.; Skrotzki, W. Plastic anisotropy of ultrafine grained aluminum alloys produced by accumulative roll bonding. *Mater. Sci. Eng. A* **2010**, *527*, 3271–3278.
18. Sauvage, X.; Wilde, G.; Divinski, S.; Horita, Z.; Valiev, R. Grain boundaries in ultrafine grained materials processed by severe plastic deformation and related phenomena. *Mater. Sci. Eng. A* **2012**, *540*, 1–12.
19. Valiev, R.; Islamgaliev, R.; Alexandrov, I. Bulk nanostructured materials from severe plastic deformation. *Prog. Mater. Sci.* **2000**, *45*, 103–189.
20. Zeipper, L.; Zehetbauer, M.; Holzleithner, C. Defect based micromechanical modelling and simulation of nanoSPD CP-Ti in post-deformation. *Mater. Sci. Eng. A* **2005**, *410–411*, 217–221.
21. Pippan, R.; Wetscher, F.; Hafok, M.; Vorhauer, A.; Sabirov, I. The Limits of Refinement by Severe Plastic Deformation. *Adv. Eng. Mater.* **2006**, *8*, 1046–1056.
22. Zehetbauer, M.; Grossinger, R.; Krenn, H.; Krystian, M.; Pippan, R.; Rogl, P.; Waitz, T.; Wurschum, R. Bulk Nanostructured Functional Materials By Severe Plastic Deformation. *Adv. Eng. Mater.* **2010**, *12*, 692–700.
23. Gottstein, G. *Physikalische Grundlagen der Materialkunde*; Springer: Berlin, Germany, 2001
24. Kang, D.H.; Kim, T. Mechanical behavior and microstructural evolution of commercially pure titanium in enhanced multi-pass equal channel angular pressing and cold extrusion. *Mater. Des.* **2010**, *31*, S54–S60.
25. Langdon, T. *Processing of Aluminum Alloys by Severe Plastic Deformation*; Trans Tech Publications Ltd.: Dürnten, Switzerland, 2006.
26. Saito, Y.; Utsunomiya, H.; Tsuji, N.; Sakai, T. Novel ultra-high straining process for bulk material-development of the accumulative roll-bonding (ARB) process. *Acta Mater.* **1999**, *47*, 579–583.
27. Terada, D.; Inoue, S.; Tsuji, N. Microstructure and mechanical properties of commercial purity titanium severely deformed by ARB process. *J. Mater. Sci.* **2007**, *42*, 1673–1681.
28. Saito, Y.; Tsuji, N.; Utsunomiya, H.; Sakai, T.; Hong, R. Ultra-fine grained bulk aluminum produced by accumulative roll-bonding (ARB) process. *Scr. Mater.* **1998**, *39*, 1221–1227.
29. Marr, T.; Freudenberger, J.; Kauffmann, A.; Scharnweber, J.; Oertel, C.G.; Skrotzki, W.; Siegel, U.; Kühn, U.; Eckert, J.; Martin, U.; *et al.* Damascene Light-Weight Metals. *Adv. Eng. Mater.* **2010**, *12*, 1191–1197.
30. Levi, F. Permanent Magnets Obtained by Drawing Compacts of Parallel Iron Wires. *J. Appl. Phys.* **1960**, *31*, 1469–1471.
31. Latypov, M.; Lee, D.; Jeong, H.; Lee, J.; Kim, H. Design of Hierarchical Cellular Metals Using Accumulative Bundle Extrusion. *Metall. Mater. Trans. A* **2013**, *44A*, 4031–4036.
32. Marr, T.; Freudenberger, J.; Seifert, D.; Klauß, H.; Romberg, J.; Okulov, I.; Scharnweber, J.; Eschke, A.; Oertel, C.G.; Skrotzki, W.; *et al.* Ti-Al Composite Wires with High Specific Strength. *Metals* **2011**, *1*, 79–97.
33. Nzoma, E.; Guillet, A.; Pareige, P. Nanostructured Multifilamentary Carbon-Copper Composites: Fabrication, Microstructural Characterization, and Properties. *J. Nan.* **2012**, *2012*, 360818.
34. Dundurs, J. Elastic Interaction of Dislocations with Inhomogeneities. In *Mathematical Theory of Dislocations*; Mura, T., Ed.; ASME: New York, NY, USA, 1969; pp. 70–115.
35. Yang, D.; Hodgson, P.; Wen, C. The kinetics of two-stage formation of $TiAl_3$ in multilayered Ti/Al foils prepared by accumulative roll bonding. *Intermetallics* **2009**, *17*, 727–732.
36. Maier, V.; Höppel, H.; Göken, M. Nanomechanical behavior of Al-Ti layered composites produced by accumulative roll bonding. *J. Phys. Conf. Ser.* **2010**, *240*, 012108.
37. Yang, D.; Cizek, P.; Hodgson, P.; Wen, C. Ultrafine equiaxed-grain Ti/Al composite produced by accumulative roll bonding. *Scr. Mater.* **2010**, *62*, 321–324.
38. Ng, H.; Przybilla, T.; Schmidt, C.; Lapovok, R.; Orlov, D.; Höppel, H.W.; Göken, M. Asymmetric accumulative roll bonding of aluminum-titanium composite sheets. *Mater. Sci. Eng. A* **2013**, *576*, 306–315.
39. Gu, S.; Fang, H.; Zhou, Z.; Du, J. The evolution of microstructure and mechanical properties of Ti/Al composite synthesized by accumulative roll-bonding. *Acta Phys. Sin.* **2012**, *61*, 186104.
40. Hausöl, T.; Maier, V.; Schmidt, C.; Winkler, M.; Höppel, H.; Göken, M. Tailoring Materials Properties by Accumulative Roll Bonding. *Adv. Eng. Mater.* **2010**, *12*, 740–746.
41. Chaudhari, G.; Acoff, V.L. Titanium aluminide sheets made using roll bonding and reaction annealing. *Intermetallics* **2010**, *18*, 472–478.
42. Gurao, N.P.; Sethuraman, S.; Suwas, S. Evolution of Texture and Microstructure in Commercially Pure Titanium with Change in Strain Path During Rolling. *Metall. Mater. Trans. A* **2013**, *44A*, 1497–1507.

43. Romberg, J.; Freudenberger, J.; Scharnweber, J.; Gaitzsch, U.; Marr, T.; Eschke, A.; Kühn, U.; Oertel, C.G.; Okulov, I.; Petters, R.; *et al.* Metallographic Preparation of Aluminium-Titanium Composites. *Pract. Metall.* **2013**, *50*, 739–753.
44. Bouvier, S.; Benmhenni, N.; Tirry, W.; Gregory, F.; Nixon, M.; Cazacu, O.; Rabet, L. Hardening in relation with microstructure evolution of high purity titanium deformed under monotonic and cyclic simple shear loadings at room temperature. *Mater. Sci. Eng. A* **2012**, *535*, 12–21.
45. Chun, Y.; Yu, S.; Semiatin, S.; Hwang, S. Effect of deformation twinning on microstructure and texture evolution during cold rolling of CP-titanium. *Mater. Sci. Eng. A* **2005**, *398*, 209–219.
46. Valiev, R.; Ivanisenko, Y.; Rauch, E. Structure and deformation behavior of armco iron subjected to severe plastic deformation. *Acta Mater.* **1996**, *44*, 4705–4712.
47. Emmens, W. *Formability: A Review of Parameters and Processes that Control, Limit or Enhance the Formability of Sheet Metal*; Springer: Berlin, Germany, 2011; ISBN: 9783642219047.

4

Ti/Al Multi-Layered Sheets: Differential Speed Rolling (Part B)

Jan Romberg [1,2], Jens Freudenberger [1,2,3,*], Hiroyuki Watanabe [4], Juliane Scharnweber [5], Andy Eschke [5], Uta Kühn [1], Hansjörg Klauß [1], Carl-Georg Oertel [5], Werner Skrotzki [5], Jürgen Eckert [6,7] and Ludwig Schultz [1,2]

1 Leibniz Institute for Solid State and Materials Research Dresden (IFW Dresden), Helmholtzstr. 20, 01069 Dresden, Germany; janromberg@yahoo.de (J.R.); u.kuehn@ifw-dresden.de (U.K.); h.j.klauss@ifw-dresden.de (H.K.); l.schultz@ifw-dresden.de (L.S.)

2 Institut für Werkstoffwissenschaft, Technische Universität Dresden, 01062 Dresden, Germany

3 Institut für Werkstoffwissenschaft, Technische Universität Bergakademie Freiberg, Gustav-Zeuner-Str. 5, 09599 Freiberg, Germany

4 Osaka Municipal Technical Research Institute, 1-6-50 Morinomiya, Joto-ku, Osaka 5368553, Japan; hwata@omtri.or.jp

5 Institut für Strukturphysik, Technische Universität Dresden, 01062 Dresden, Germany; juliane.scharnweber@tu-dresden.de (J.S.); andy.eschke@gmx.de (A.E.); oertel@physik.tu-dresden.de (C.-G.O.); werner.skrotzki@physik.tu-dresden.de (W.S.)

6 Erich Schmid Institut für Materialwissenschaft, Österreichische Akademie der Wissenschaft, Jahnstraße 12, A-8700 Leoben, Austria; juergen.eckert@unileoben.ac.at

7 Department Materialphysik, Montanuniversität Leoben, Jahnstraße 12, A-8700 Leoben, Austria

* Correspondence: j.freudenberger@ifw-dresden.de

Academic Editor: Hugo F. Lopez

Abstract: Differential speed rolling has been applied to multi-layered Ti/Al composite sheets, obtained from accumulative roll bonding with intermediate heat treatments being applied. In comparison to conventional rolling, differential speed rolling is more efficient in strengthening the composite due to the more pronounced grain refinement. Severe plastic deformation by means of rolling becomes feasible if the evolution of common rolling textures in the Ti layers is retarded. In this condition, a maximum strength level of the composites is achieved, *i.e.*, an ultimate tensile strength of 464 MPa, while the strain to failure amounts to 6.8%. The deformation has been observed for multi-layered composites. In combination with the analysis of the microstructure, this has been correlated to the mechanical properties.

Keywords: accumulative roll bonding; differential speed rolling; grain refinement; microstructure; mechanical properties; Ti/Al multi-layered composites

---

1. Introduction

A key aspect in the production of composite materials lies within the possibility of combining different material properties that are attributed to the individual components of the composite. The specific objective in the development of multi-layered Ti/Al composites is the aim to combine (i) a large tensile strength; (ii) a low mass density, yielding a high specific strength and (iii) a sufficient ductility, allowing further forming processes.

One prerequisite for the production of metal composites is a high forming capability. In the present work, accumulative roll bonding (ARB) is utilized to generate composites hardened by grain- and phase-boundaries. When ARB is applied to two different materials, the phase boundary area per volume doubles for each ARB cycle as long as the layers remain continuous. Hence, the stability of the layers is crucial when aiming at maximum phase boundary area. Part A of this study [1] revealed that the stability of the layers can be maintained by introducing an intermediate heat treatment (IHT) after each individual ARB cycle. However, IHT leads to recovery and, therefore, reduces the effect of strain hardening introduced by rolling.

The present study presents a possibility to overcome this issue. ARB with IHT is utilized to generate multi-layered sheets with individual continuous layers. As mentioned before, IHT prevents an accumulation of strain hardening. The composite can be strengthened further when the finally aimed number of layers has been reached. Further cold working by conventional rolling is possible, but the hardening rate is comparably low as the Ti layers fail by necking and do not further deform. Thus, a high degree of deformation would be required, which results in limitations with respect to the sheet thickness. In order to obtain a higher strain hardening rate, the deformation should have a reasonable shear component. A small amount of shear deformation is also observed for conventionally rolled materials. This shear arises from e.g., friction between sheet and rolls. During conventional rolling, the friction is kept at a low level. However, this can be considerably altered by differential speed rolling (DSR). DSR utilizes different velocities of the upper and lower rolls or, alternatively, differently sized rolls (asymmetric rolling) yielding different surface velocities of the rolls. By this means, DSR induces a significant amount of shear deformation within the rolled material yielding a pronounced grain refinement [2–5] and a change in texture [2,6,7].

In order to ease the discussion, conventional rolling with similar velocity and size of the working rolls, here referred to as equal speed rolling (ESR), represents the benchmark and reference system to DSR [8]. For ESR, shear deformation is restricted to the near surface region [8,9]. In the case of DSR, the volume that is affected by shear is strongly material dependent as shown in numerous studies on different materials, e.g., steel, copper, aluminium and niobium [3,10–15]. Special attention is paid to the investigation of DSR of hexagonal metals such as magnesium [4,16–18] and titanium [6,19–21]. Hoi Pang Ng *et al.* combined roll bonding and asymmetric rolling on Ti/Al composite sheets with up to four rolling passes and observed a remarkable higher grain refinement and increasing inter diffusion compared to symmetric ARB [22]. However, the thin Ti layers show considerable necking. Consequently, the layered structure is destroyed and the matrix is capable of deformation. Thus, strengthening of the composite is mainly accomplished by strengthening the Al matrix [22]. In summary, the use of shear deformation in Ti/Al multi-layered sheets is still unrevealed, when (i) the multi-layered architecture has to be preserved during processing and (ii) a high deformation strain has to be achieved.

The aim of this study is to investigate the effect of DSR on multi-layered Ti/Al sheets and to emphasize the influence of the multi-layered structure compared to single material sheets in the response to DSR.

2. Experimental Section

For the present study, pure Ti (99.995%) sheets as well as sheets of the aluminium alloy AA5049 ($AlMg_2Mn_{0.8}$, in the following referred to as Al) with an initial size of 30 mm × 125 mm × 1 mm (width × length × thickness) have been used. For benchmarking purposes, single material sheets were also deformed. All sheets were initially in a recrystallized state. In addition, investigations were performed on a five-layered stack of Al-Ti-Al-Ti-Al that has been roll bonded using a thickness reduction of 50% as well as on a Ti/Al composite being accumulatively roll bonded with six ARB cycles including IHTs at 723 K for 90 min after each ARB cycle under vacuum conditions. This composite is composed of 160 continuous layers. For any details, please refer to part A of this study [1].

DSR experiments were performed at room temperature (RT) utilizing a rolling mill with identical diameters of the working rolls of 250 mm operated at different velocities, *i.e.*, rolling speed ratios, and with different numbers of rolling cycles. The thickness reduction of the sheets per rolling pass was approximately 20%, and deformation was performed with up to four rolling passes, while the ratio of the roll velocities used was 1.1, 1.5, 2.0 and 3.0.

As already mentioned, DSR leads to enhanced shearing. In order to visualize this contribution to deformation, pins of annealed pure copper were inserted into several composite sheets before rolling. This procedure follows the experiments of Sakai [9] and Lee [8], who measured the shearing of metal sheets during rolling and ARB. The response of the material was determined on longitudinal cuts, which also include the pin by scanning electron microscopy (SEM, Zeiss Microscopy, Jena, Germany) and electron backscatter diffraction (EBSD, Oxford Instruments, Abingdon, UK). The metallographic preparation for these studies is described in detail in a previous study [23]. EBSD measurements were done in the upper third of the single material sheets and the 160 layer composite sheets. In the five layer sheets, the EBSD maps were taken within the top layer (Al) and the second-top layer (Ti).

Hardness measurements were executed with two nano hardness tester devices (MTS, Eden Prairie, MN, USA) using identical measurement parameters. The Berkovich indenter penetrates the material at a rate of 0.05 s^{-1}, while a vibration with a frequency of 45 Hz is superimposed. This facilitates a continuous determination of the hardness at any penetration depth. The hardness of each measurement is averaged for depths ranging from 500 nm to 900 nm because of the large fluctuations at lower depths. For each sample and phase, the values of 15 indents are averaged. Tensile tests were carried out at RT; for any detail, please refer to part A of this study.

3. Results and Discussion

3.1. *Differential Speed Rolling of Al and Ti*

The effect of DSR on the microstructure of Al is shown in Figure 1. All investigations were performed on longitudinal cuts of the specimens at the center of the sheets and the second or third layer from the top.

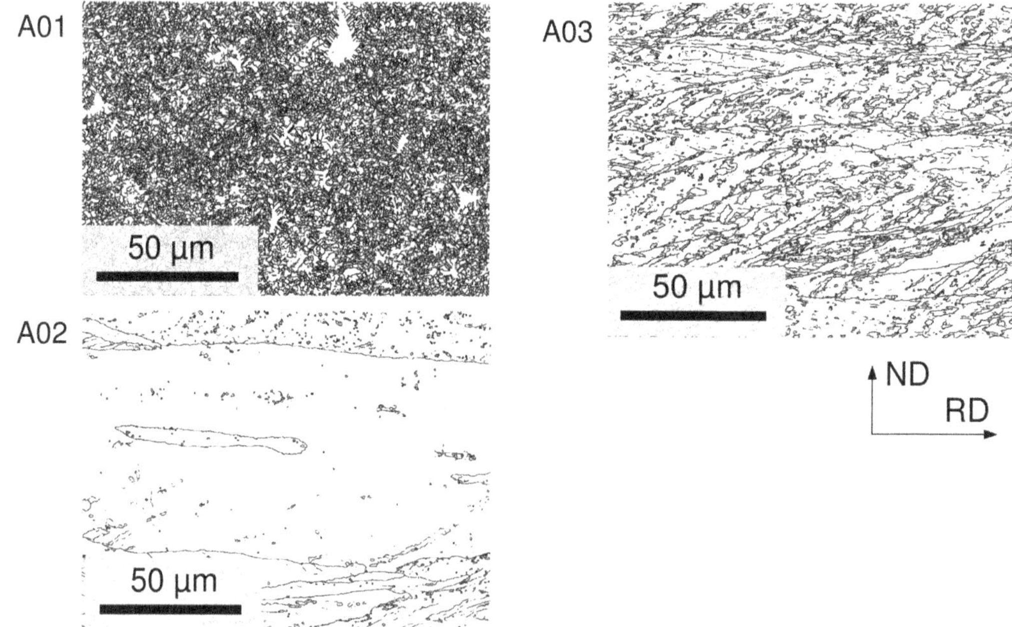

Figure 1. Grain boundary maps of Al, prepared by DSR. A01: one rolling pass, speed ratio: 1:1; A02: 1 rolling pass, speed ratio: 3.0; A03: four rolling passes, speed ratio: 1:1.

The grain boundary maps shown in Figure 1 were calculated from EBSD micrographs and mark all boundaries with a misorientation angle of 15° and above. The Al sheets were deformed by DSR with different speed ratios and number of rolling passes. An increasing shear deformation, which is related to higher speed ratio results in a significant larger mean grain size after one DSR pass. A similar tendency is observed for a higher shear strain due to an increasing number of DSR passes, as shown for the increase of passes from one to four. A quantitative analysis of the grain sizes is provided in Figure 2.

According to Equation (1) the speed ratio can be transferred to the rolling shear strain γ, assuming plane strain deformation. Therein, v_u and v_l are the velocities of the upper and lower roll (measured in m/s), respectively, and d is the inside width between the rolls (measured in m).

$$\gamma = \arctan\left(\frac{\frac{v_u}{v_l} - 1}{d}\right) \tag{1}$$

Figure 2a,b show the quantitative analysis of the grain size of the Al sheets being processed by DSR with different speed ratios as well as increasing strain (*i.e.*, number of DSR passes) at a constant speed ratio of 1.1.

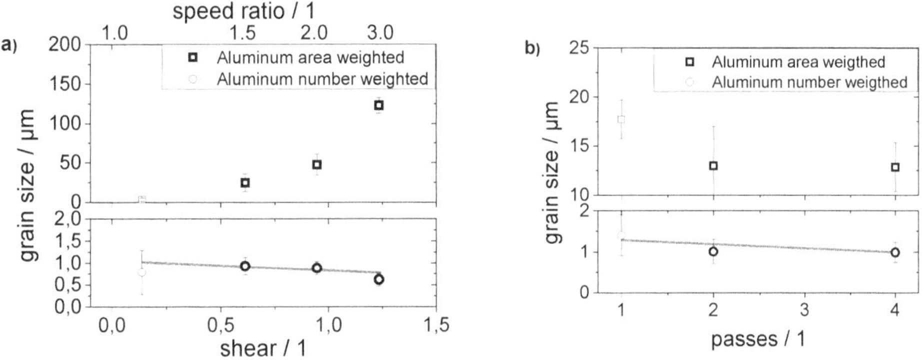

Figure 2. (**a**) Grain size of Al with different weighting methods being applied: grain area with increasing speed ratio, *i.e.*, amount of shear and (**b**) evolution of the grain size with increasing number of rolling passes (DSR) with a constant speed ratio of 1.1. The weighting methods are described in part A of this study.

The analysis shows a significant increase of the grain size with increasing shear deformation when the mean grain size is weighted upon the area. According to Figure 1, a small number of large grains is responsible for this behavior, which is causing an overestimation of the large grains. In contrast, the arithmetic average shows a slight decrease of the grain size with increasing amount of shear deformation. Furthermore, the grain size also only slightly decreases with increasing number of rolling passes at a constant speed ratio of 1.1, no matter the weighting method.

A similar analysis has been made for DSR Ti sheets (The sheets were deformed in the same was as the Al sheets). Figure 3 shows grain boundary maps of Ti.

An increasing shear deformation decreases the mean grain size. This is more evident when comparing the different speed ratios. In order to assess the effect of DSR on the mechanical properties, Al and Ti sheets in the cold rolled condition (as-received), in which the amount of shear deformation can be neglected, have been tensile tested in comparison to those being additionally processed by DSR with a low shear deformation (*i.e.*, speed ratio of 1.1), as shown in Figure 4.

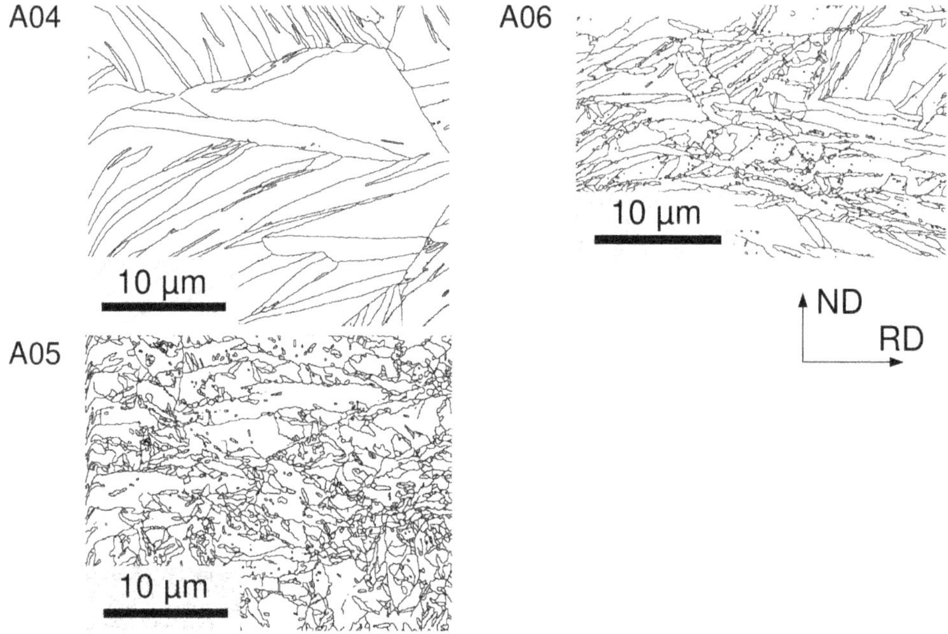

Figure 3. Grain boundary maps (*i.e.*, boundaries with a misorientation angle of 15 degree and above) of DSR Ti. A04: one rolling pass, speed ratio: 1.1; A05: 1 rolling pass, speed ratio: 3.0; A06: four rolling passes, speed ratio: 1.1.

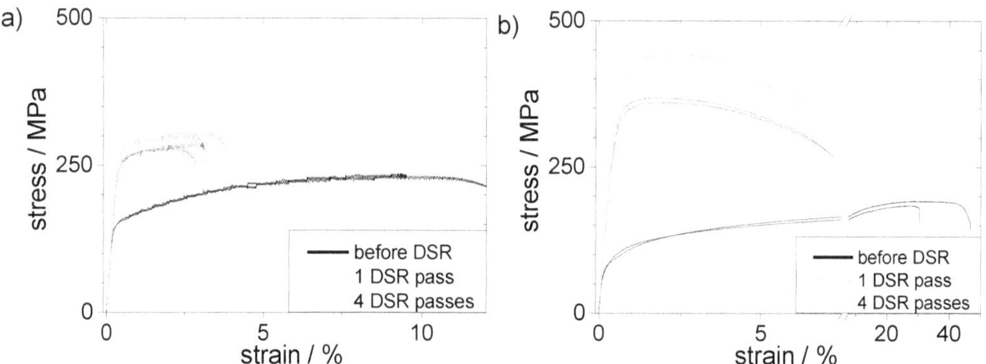

Figure 4. Engineering stress-strain curves for (**a**) Al and (**b**) Ti sheets in the cold rolled condition and after one and four DSR passes at a speed ratio of 1:1. Results for two samples in each condition are shown in order to illustrate the occurring variations.

As can be seen from Figure 4, DSR significantly affects the flow behavior. The increase of strength is likely to be caused by a higher dislocation and twin (Ti) density within the materials, which is reflected in significantly decreasing EBSD pattern quality. A significant contribution to the strength according to the Hall–Petch relationship [24,25] can be excluded due to similar grain sizes in all conditions (see Figure 2). DSR also causes a decreasing strain to failure. However, the reduction of ductility is very pronounced due to the shear deformation itself. The ultimate tensile strength (UTS) of Ti is only 190 MPa in the initial state. A single pass of DSR increases the UTS to about 370 MPa and multi-pass DSR further to 460 MPa. The UTS of Al increase is less pronounced. One single DSR pass causes an increase of the UTS from 240 MPa to 280 MPa and three more passes to 308 MPa.

3.2. Differential Speed Rolling of Ti/Al-Multi-Layers

Multi-layered sheets have been obtained from a stack of three Al and two Ti sheets being alternatingly placed, *i.e.*, Al-Ti-Al-Ti-Al, and accumulatively roll bonded as described in part A of this study [1]. This sheet could be further processed by ARB as the Al is always at the outer side of the composite which eases roll bonding. ARB was performed with up to six ARB cycles. The five and 160-layered composites were further processed by DSR.

Figure 5 shows grain boundary maps for Al and Ti within a five-layered composite. EBSD images were taken within the Ti (left) and Al (right) layers at different deformation states. For the Ti layers within the five-layered composite, a quite similar behavior is observed for the deformed Ti, *i.e.*, a reduction of the grain size with increasing number of DSR passes (compare to Figure 3). In contrast to this, the grain size observed within the aluminum layers also decreases with increasing shear deformation, which has not been observed for the equivalently deformed Al (compare Figure 1). However, the grain size within the Al layers is still larger with no shear deformation being applied when compared to the five-layered composite which has been deformed by DSR up to an equivalent strain. This difference between the conventionally rolled and differentially speed rolled five-layered composite is less pronounced for the Ti layers.

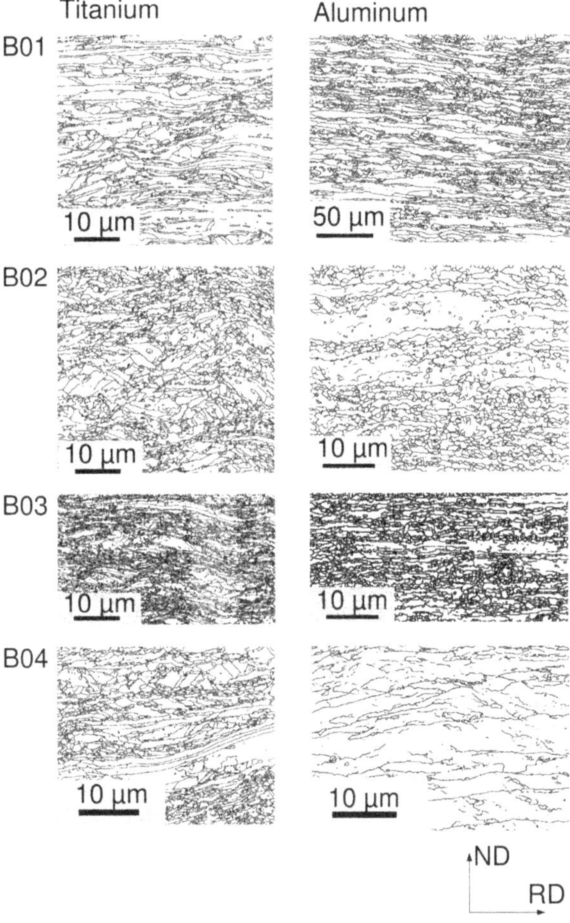

Figure 5. Grain boundary maps (*i.e.*, boundaries with a misorientation angle of 15 degrees and above) for Al and Ti being co-deformed by ARB within a five-layer stack and subsequent DSR. B01: one rolling pass, speed ratio: 1:1; B02: one rolling pass, speed ratio: 1:5; B03: four rolling passes, speed ratio: 1:1; B04: conventional rolling with three rolling passes.

Figure 6a,b show the quantitative analysis of the grain sizes. In contrast to the findings of the pure metals, the present analysis shows a decreasing grain size with increasing shear deformation. This situation is independent on the weighting method. However, the total reduction of the grain size is rather small.

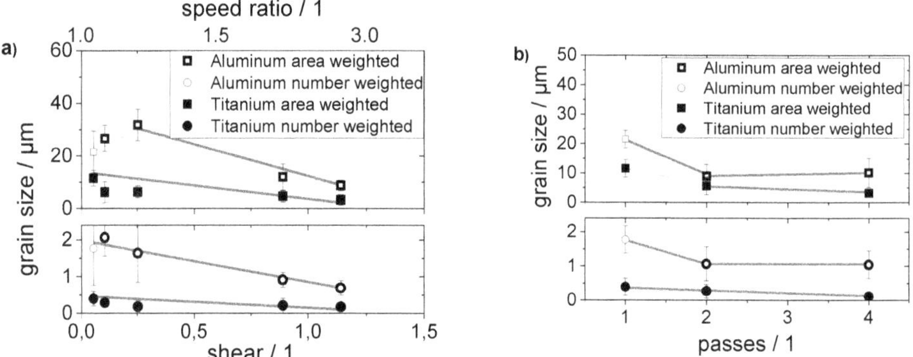

Figure 6. (**a**) Grain size within the Ti and Al layers with different weighting methods being applied: grain area with increasing speed ratio, *i.e.*, amount of shear and (**b**) evolution of the grain size with increasing number of rolling passes (DSR) with a constant speed ratio of 1:1.

The five-layered composite has been generated by roll bonding with a total thickness reduction of 50%, which already imposes a high degree of deformation and causes a significant reduction in ductility when compared to the pure metals. The engineering stress-strain curves of the five-layered composite are shown in Figure 7 and should be compared to those of the pure metals (Figure 4). The large plastic strain at fracture which is observed for pure metals cannot be obtained in the multi-layered sheet. On the other hand, ARB strain hardens the composite and, in consequence, a yield strength of 320 MPa and an ultimate tensile strength of 403 MPa are observed for the five-layered composite. These strength values can be further increased by DSR, whereby a larger accumulated shear deformation by multiple DSR passes further increases the tensile strength.

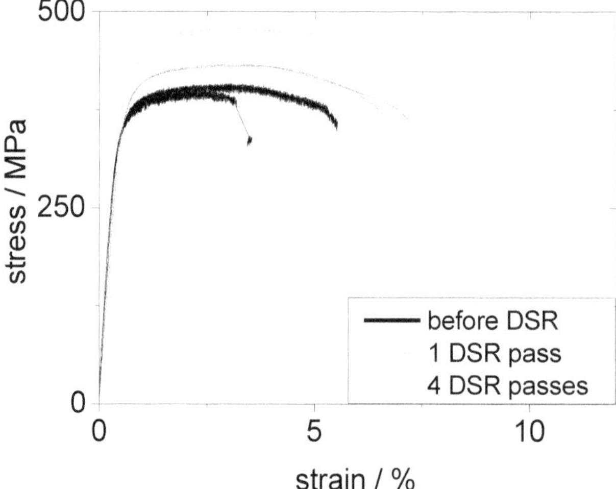

Figure 7. Engineering tensile stress-strain curves of five-layered Ti/Al composites after ARB and further processing by DSR with a speed ratio of 1:1: with one and four DSR passes, respectively. Results for two samples in each condition are shown in order to illustrate the occurring variations.

A reduction of fracture strain which is generally observed with strength increase is not observed after processing by DSR. The ductility is even enhanced by DSR.

In the following, the question of how the accumulatively roll bonded Ti/Al multi-layered sheets react on DSR is addressed, *i.e.*, on the superimposed shear. In order to visualize the shear being incorporated into the composites by DSR, a Cu-pin is utilized. This pin has been put into a drilled hole in a 160-layered Ti/Al composite sheet. The Cu is in a recrystallized state and, therefore, it deforms at lower stress than the composite and even its individual layers. Therefore, the shear strain applied can be easily obtained from the interface between the Cu pin and the composite. For this purpose, longitudinal cross-sectional cuts were prepared by standard metallographic procedures aiming at imaging the largest cross section of the Cu-pin. Figure 8 shows such cross-sectional cuts of the Cu-pin (bright grey) embedded within a 160-layered composite being prepared by ARB and (i) further conventionally rolled as well as (ii) further processed by DSR with a speed ratio of 1.1. Both composites were deformed by three rolling passes, incorporating a total strain of 0.62.

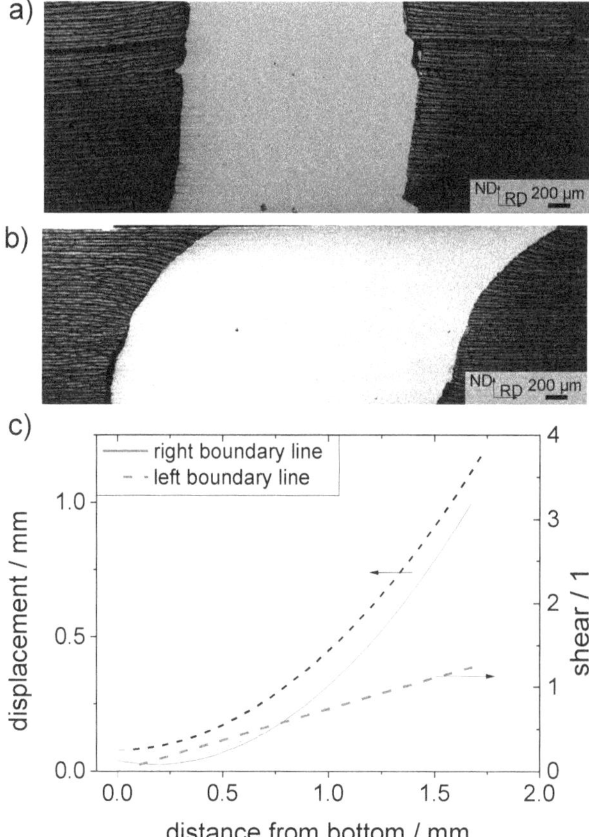

Figure 8. Shear deformation imaged by a Cu-pin inserted into a 160-layered Ti/Al composite. This Cu-pin shows no shearing in the case of conventional rolling (**a**); whereas DSR yields considerable shear (**b**); The left (dotted lines) and right (straight lines) boundaries between the Cu-pin and the Ti/Al-composite have been used to evaluate the relative position of the material as shown in the right chart (**c**). The grey lines represent the calculated amount of shear deformation.

Figure 8a,b reveal the shear deformation incorporated during DSR. The shear within the pin can be addressed to DSR, as no shearing of the Cu-pin is observed for the conventionally rolled 160-layered composite. The amount of shear has been evaluated from the curvature of the interfaces between the Cu-pin and the composite. Figure 8c shows the displacement of the Cu-pin relative to the vertical projection of the lowest position of the interface as a function of the distance from the bottom of the sheet. This evaluation has been performed for the left as well as the right interface in order to increase the accuracy. The effective amount of shear deformation has been calculated upon the displacement. The corresponding shear is shown in Figure 8c.

Assuming that the Cu-pin correctly reflects the amount of shear deformation and taking into account that the interfaces within the 160-layered composite do not fail during DSR, it can be concluded that the composite is also subjected to the same amount of shear deformation.

Figures 9 and 10 show SEM micrographs of the microstructure of the composites processed by conventional rolling and DSR, respectively. Each of these images contain inserts showing grain boundary maps as obtained from EBSD measurements. In the case of the conventionally rolled 160-layered composite, the grain boundary maps are independent of the position at which they were taken.

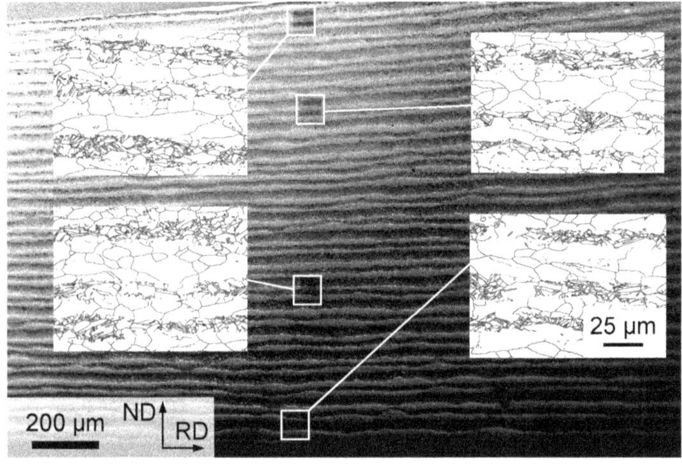

Figure 9. SEM micrograph showing the microstructure of a 160-layered Ti/Al composite obtained from six ARB cycles and subsequent conventional rolling (three passes). The inserts show grain boundary maps that have been evaluated from EBSD measurements at the depicted positions. Lines indicate high angle grain boundaries including twin boundaries.

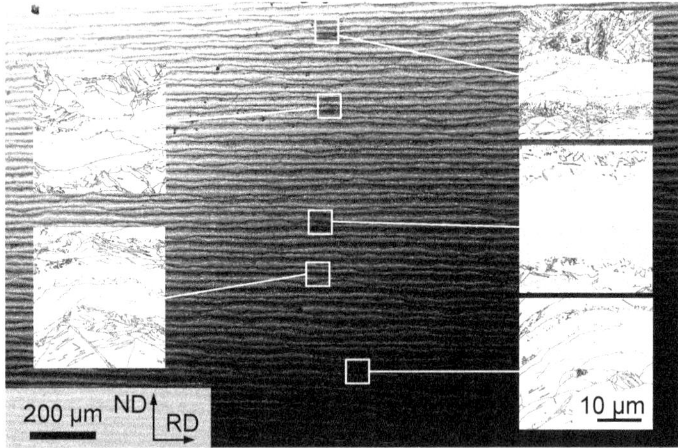

Figure 10. SEM micrograph showing the microstructure of a 160-layered Ti/Al composite obtained from six ARB cycles and subsequent differential speed rolling (three passes). The inserts show grain boundary maps at positions depicted. Lines indicate high angle grain boundaries including twin boundaries.

The situation is considerably different for the case of the 160-layered composite being processed by DSR. As this composite is subjected to an additional shear deformation, which increases with the distance from the lower roll (see Figure 8), a variation in the microstructure across the thickness of the sheets is observed. The macroscopic view of the sample shows the same features as in the case of the conventionally rolled 160-layered composite, *i.e.*, continuous layers, some of them necked, but none failed, and no detachment of the layers occurred. The grain boundary maps show that the grain size shrinks with increasing shearing strain (in Figure 10 from bottom to top). It is noteworthy to mention that no intermetallic phase was detected by means of SEM within any cross section under investigation.

In order to quantitatively assess these images, the grain size has been evaluated upon the grain boundary maps. The corresponding mean grain sizes for the Ti and Al layers are shown for the conventionally rolled 160-layered composites as well as for the composites processed by DSR in Figure 11. The data has been evaluated upon the regions as depicted in Figures 9 and 10. Figure 11b shows the hardness of the phases in both conditions.

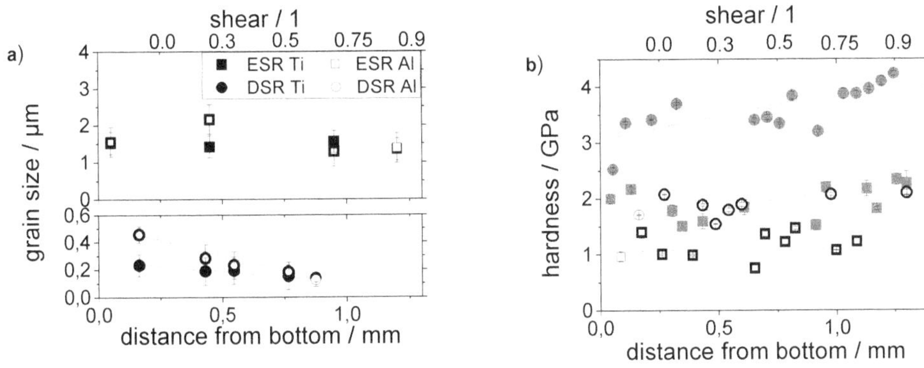

Figure 11. Grain size (**a**) and hardness (**b**) of a 160-layered Ti/Al composite subjected to three passes of conventional rolling (ESR) or differential speed rolling (DSR). The values are taken along the thickness of the sheet and, thus, reflect the amount of shear introduced into the samples.

As already mentioned, when qualitatively discussing the grain boundary maps, the grain size shows no variation along the sheet thickness for the conventionally rolled composite. The Ti and Al grain sizes are about the same. As the composite is homogeneously strengthened by work-, phase- and grain-boundary hardening, it is not surprising that the hardness values are constant across the sample thickness.

This situation is different for the samples processed by DSR. The additionally introduced shear deformation yields a significant grain refinement. As the shear strain varies across the thickness of the sheet, the contribution of the Hall-Petch type hardening also differs. Indeed, the hardness increases with shear deformation as the grains undergo refinement. This is observed for the Ti as well as Al layers.

Up to here, the discussion regarding the effect of DSR on multi-layered Ti/Al composites was restricted to the speed ratio of 1.1. In the following, the effect of increasing speed ratio will be illustrated. Figure 12 shows the effect of different rolling speed ratios on the microstructure of Ti and Al, respectively.

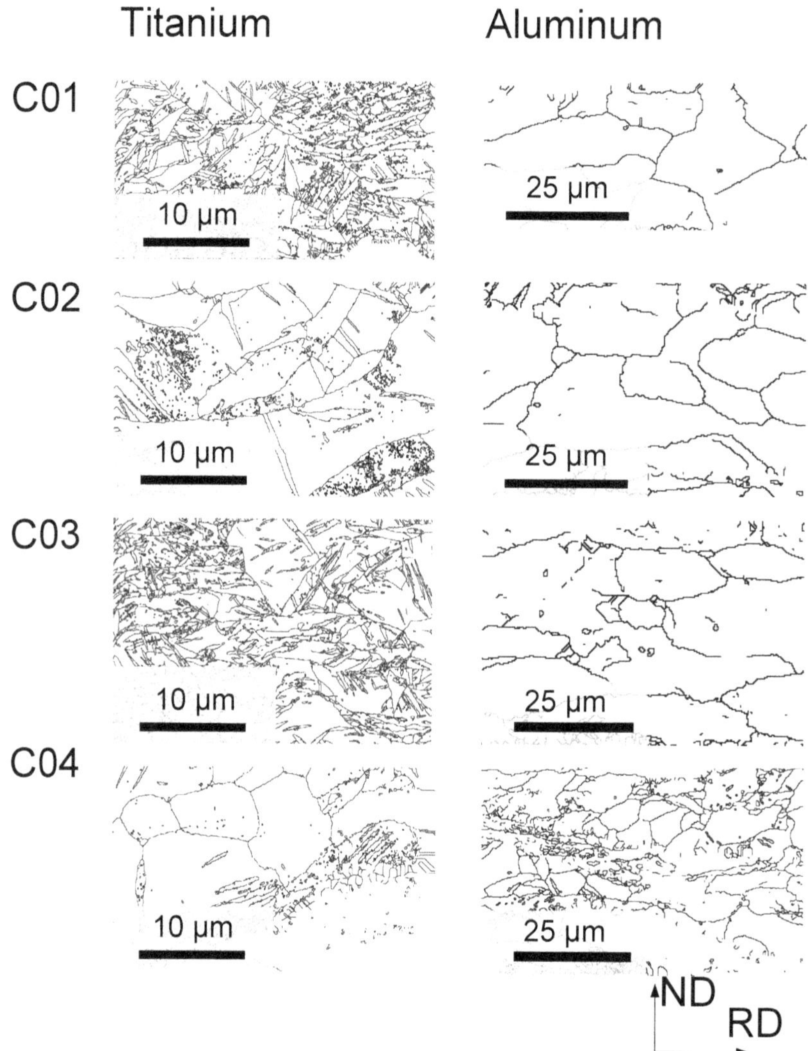

Figure 12. Grain boundary maps (*i.e.*, boundaries with a misorientation angle of 15° and above) of the Ti (left) and Al (right) layers within a 160-layered Ti/Al composite sheet. The sheets are being deformed by rolling with one (C01) and three (C02-04) rolling passes. The speed ratio amounts to 1 (C01 and C02), 1:1 (C03) and 1:5 (C04), respectively.

As previously mentioned, the Al and Ti layers have a widely spread grain size distribution. Figure 13 shows the grain size within the Al and Ti layers evaluated upon the grain boundary maps shown in Figure 12. The effect of conventional rolling with up to three rolling passes on the grain size is not significant.

The smaller Al grains show a nearly constant grain size with increasing rolling shear, whereas the larger grains show a significant decrease in grain size. Both mean values come closer with increasing shear deformation. This is not observed for the grains within the Ti layers. Both values remain nearly constant. In addition, there is a significant change in the grain statistics. With increasing shear deformation, the number of small grains within the Al layers increases.

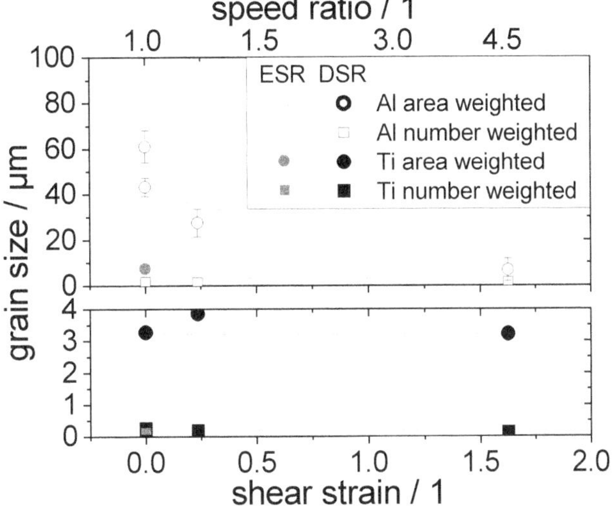

Figure 13. Grain size within the Al and Ti layers of a 160-layered Ti/Al composite sheet prepared by six ARB cycles and with three additional DSR passes with a speed ratio of up to 1:5. Gray colored data points correspond to samples being conventionally rolled up to the same shear strain.

The mechanical properties of the composite are affected by the microstructure. Figure 14 shows the engineering stress-strain curves of the 160-layered Ti/Al composite as well as the curves of composites being additionally deformed by DSR. One DSR pass decreases both UTS and yield strength but increases the strain to failure. However, the application of three further DSR passes increases the UTS and yield strength but also further increases the ductility.

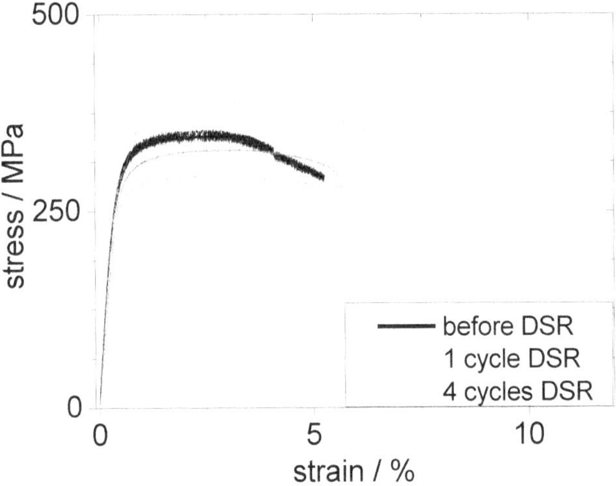

Figure 14. Engineering tensile stress-strain curves of 160-layered Ti/Al composite sheets as well as those with subsequent DSR being applied with a speed ratio of 1.1. Results are shown for one and four DSR passes. Results for two samples in each condition are shown in order to illustrate the occurring variations.

Figure 15 compares the development of the mechanical properties of all sheets examined. The influence of DSR is most remarkable in case of Ti sheets. The UTS increases from 200 MPa to 350 MPa after the first DSR cycle and to about 450 MPa after the fourth DSR cycle. At the same time, the elongation to failure decreases from 46% to 7%. The influence of DSR on Al follows the same tendency: UTS increases from 230 MPa to 280 MPa and later to 310 MPa while the elongation decreases from 12% to 3.8%. In the composite sheets, both the strength and the elongation to failure increase when the composite is deformed by DSR. However, this is only visible after four DSR passes. Whereas the UTS and strain to failure are nearly independent of the number of DSR passes for the five-layered composite, the 160-layered composite shows an increase for a higher number of DSR passes, only. Both values even show a decrease after the first DSR pass when compared to the initial state. This hints at the fact that friction between the layers plays a significant role with respect to the deformation behavior and especially with respect to the observed elongation to failure. The tensile test samples from the 160-layered Ti/Al composite shows noticeable sliding of the layers at the Ti-Al interfaces.

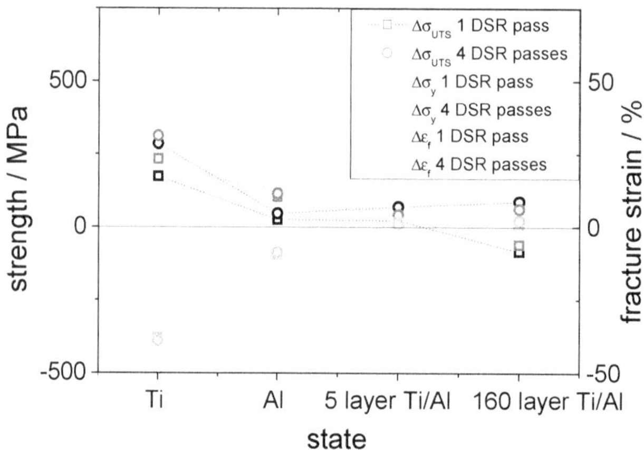

Figure 15. Effect of DSR on the mechanical properties. The chart shows the change of the ultimate tensile strength, yield strength and strain to failure for pure Al and Ti as well as for five-layered and 160-layered Ti/Al composite sheets before and after DSR.

4. Conclusions

Differential speed rolling of Ti/Al composite sheets leads to an increased ductility due to an improved macroscopic shear strain. The different layers of a five-layered composite sheet prepared by a single ARB cycle slide along the phase boundary during DSR. In contrast, a 160-layered composite prepared by six ARB cycles reveals shear deformation of the individual layers. This shear deformation shows a gradient in the direction of sample thickness and increases from the layer next to the slower roll to that next to the faster roll. Furthermore, the shear deformation causes hardening and grain refinement and, consequently, gradients in grain size and hardness. Table 1 qualitatively summarizes the evolution of the grain size during DSR of single material sheets as well as five- and 160-layered composites. An increase in the speed ratio affects the grain size differently compared to an increase in the number of DSR passes. Except for separately rolled Al, the grain size decreases with increasing speed difference. During further DSR passes, the grain size remains constant or even increases. In case of the five-layered composite, the grains further refine when DSR passes are repeated.

The application of DSR facilitates both an increase in UTS and ductility while the laminated structure is retained. The tensile strength increases with the number of DSR passes up to 464 MPa, and the elongation to failure rises to 6.8% for the five-layered composite. Strengthening is caused by grain refinement, strain hardening and an increase of bonding strength between the layers.

Table 1. Qualitative evolution of the grain size during DSR responding to increasing speed ratio and increasing DSR pass number. (A) single material sheets; (B) five-layered composite sheets; (C) 160-layered composite sheets.

Series	Material	Effect of Speed Ratio Small Grains	Effect of Speed Ratio Large Grains	Effect of Pass Number Small Grains	Effect of Pass Number Large Grains
A	Al	→	↗	→	→
A	Ti	↘	↘	→	→
B	Al	↘	↘	↘	↘
B	Ti	↘	↘	↘	↘
C	Al	↘	↘	→	↘
C	Ti	↘	↘	→	→

Acknowledgments: This work has been supported by the European Union and the Free State of Saxonia in the framework of the European Centre for Emerging Materials and Processes (ECEMP), contract No. 13795/2379. The authors gratefully acknowledge support with the hardness measurements from Verena Maier and Mathias Göken (University Nürnberg-Erlangen) as well as Kristina Eichhorn and Konrad Schneider (Leibniz Institute for Polymer Research Dresden).

Author Contributions: Jan Romberg and Jens Freudenberger designed the experiments, collected and interpreted the data, wrote and edited the paper, and contributed to all activities. Juliane Scharnweber and Andy Eschke contributed to electron microscopy, including sample preparation, image and diffraction collection; together with Carl-Georg Oertel, they analysed the results and fostered their interpretation. Hiroyuki Watanabe performed the differential speed rolling experiments and contributed to the interpretation of the results. Hansjörg Klauß performed the tensile tests. Uta Kühn, Werner Skrotzki, Jürgen Eckert and Ludwig Schultz directed the research and contributed to the discussion and interpretation of the results.

Conflicts of Interest: The authors declare no conflict of interest.

References

1. Romberg, J.; Freudenberger, J.; Bauder, H.; Plattner, G.; Krug, H.; Scharnweber, J.; Eschke, A.; Kühn, U.; Klauß, H.; Oertel, C.G.; *et al.* Ti/Al Multilayered Sheets: Accumulative Roll Bonding (part A). *Metals* **2016**, *6*, 30.
2. Kim, W.; Hwang, B.; Lee, M.; Park, Y. Effect of speed-ratio on microstructure, and mechanical properties of Mg-3Al-1Zn alloy, in differential speed rolling. *J. Alloys Compd.* **2011**, *509*, 8510–8517.
3. Ko, Y.G. Microstructure evolution and mechanical properties of severely deformed Al alloy processed by differential speed rolling. *J. Alloys Compd.* **2012**, *536*, S122–S125.
4. Chang, L.; Cho, J.; Kang, S. Microstructure and mechanical properties of AM31 magnesium alloys processed by differential speed rolling. *J. Mater. Process. Technol.* **2011**, *211*, 1527–1533.
5. Watanabe, H.; Mukai, T.; Ishikawa, K. Effect of temperature of differential speed rolling on room temperature mechanical properties and texture in an AZ31 magnesium alloy. *J. Mater. Process. Technol.* **2007**, *182*, 644–647.
6. Huang, X.; Suzuki, K.; Chino, Y. Improvement of stretch formability of pure titanium sheet by differential speed rolling. *Scr. Mater.* **2010**, *63*, 473–476.
7. Kim, W.; Lee, H.; Yoo, S.; Park, Y. Texture and mechanical properties of ultrafine-grained Mg-3Al-1Zn alloy sheets prepared by high-ratio differential speed rolling. *Mater. Sci. Eng. A* **2011**, *528*, 874–879.
8. Lee, S.; Saito, Y.; Tsuji, N.; Utsunomiya, H.; Sakai, T. Role of shear strain in ultragrain refinement by accumulative roll-bonding ARB process. *Scr. Mater.* **2002**, *46*, 281–285.
9. Sakai, T.; Saito, Y.; Hirrano, K.; Kato, K. Deformation and Recrystallization Behavior of Low Carbon Steel in High Speed Hot Rolling. *Trans. Iron Steel Inst. Jpn.* **1988**, *28*, 1028–1035.
10. Kim, W.; Wang, J.; Choi, S.; Choi, H.; Sohn, H. Synthesis of ultra high strength Al-Mg-Si alloy sheets by differential speed rolling. *Mater. Sci. Eng. A* **2009**, *520*, 23–28.
11. Ko, Y.G.; Lee, J.S. Microstructure evolution and mechanical properties of ultrafine grained IF steel via multipass differential speed rolling. *Mater. Sci. Technol.* **2013**, *29*, 553–558.

12. Suharto, J.; Ko, Y.G. Annealing behavior of severely deformed IF steel via the differential speed rolling method. *Mater. Sci. Eng. A* **2012**, *558*, 90–94.
13. Iro, A.; Zhou, Z.; Utsunomiya, H.; Sakai, T. Reduction in Planar Anisotropy of Pure Niobium Sheet by Asymmetric Rolling. *Tetsu Hagane J. Iron Steel Inst. Jpn.* **2011**, *97*, 572–577.
14. Jiang, J.; Ding, Y.; Zuo, F.; Shan, A. Mechanical properties and microstructures of ultrafine-grained pure aluminum by asymmetric rolling. *Scr. Mater.* **2009**, *60*, 905–908.
15. Polkowski, W.; Jozwik, P.; Polanski, M.; Bojar, Z. Microstructure and texture evolution of copper processed by differential speed rolling with various speed asymmetry coefficient. *Mater. Sci. Eng. A* **2013**, *564*, 289–297.
16. Cho, J.H.; Kim, H.W.; Kang, S.B.; Han, T.S. Bending behavior, and evolution of texture and microstructure during differential speed warm rolling of AZ31B magnesium alloys. *Acta Mater.* **2011**, *59*, 5638–5651.
17. Chang, L.; Kang, S.; Cho, J. Influence of strain path on the microstructure evolution and mechanical properties in AM31 magnesium alloy sheets processed by differential speed rolling. *Mater. Des.* **2013**, *44*, 144–148.
18. Lee, J.; Konno, T.; Jeong, H. Grain refinement and texture evolution in AZ31 Mg alloys sheet processed by differential speed rollingloo. *Mater. Sci. Eng. B* **2009**, *161*, 166–169.
19. Kim, W.; Yoo, S.; Jeong, H.; Kim, D.; Choe, B.; Lee, J. Effect of the speed ratio on grain refinement and texture development in pure Ti during differential speed rolling. *Scr. Mater.* **2011**, *64*, 49–52.
20. Kim, W.; Yoo, S.; Lee, J. Microstructure and mechanical properties of pure Ti processed by high-ratio differential speed rolling at room temperature. *Scr. Mater.* **2010**, *62*, 451–454.
21. Terada, D.; Inoue, S.; Tsuji, N. Microstructure and mechanical properties of commercial purity titanium severely deformed by ARB process. *J. Mater. Sci.* **2007**, *42*, 1673–1681.
22. Ng, H.; Przybilla, T.; Schmidt, C.; Lapovok, R.; Orlov, D.; Höppel, H.W.; Göken, M. Asymmetric accumulative roll bonding of aluminium-titanium composite sheets. *Mater. Sci. Eng. A* **2013**, *576*, 306–315.
23. Romberg, J.; Freudenberger, J.; Scharnweber, J.; Gaitzsch, U.; Marr, T.; Eschke, A.; Kühn, U.; Oertel, C.G.; Okulov, I.; Petters, R.; *et al.* Metallographic Preparation of Aluminium-Titanium Composites. *Pract. Metallogr.* **2013**, *50*, 739–753.
24. Hall, E. The deformation and ageing of mild steel: III. Discussion and results. *Proc. Phys. Soc. Sect. B* **1951**, *64*, 747–753.
25. Petch, N. The cleavage strength of polycrystalls. *J. Iron Steel Inst.* **1953**, *174*, 25–28.

5

Predicted Fracture Behavior of Shaft Steels with Improved Corrosion Resistance

Goran Vukelic [1,*] and Josip Brnic [2]

1 Department of Marine Engineering and Ship Power Systems, Faculty of Maritime Studies Rijeka, University of Rijeka, Rijeka 51000, Croatia
2 Department of Engineering Mechanics, Faculty of Engineering, University of Rijeka, Rijeka 51000, Croatia; brnic@riteh.hr
* Correspondence: gvukelic@pfri.hr

Academic Editor: Hugo F. Lopez

Abstract: One of the crucial steps in the shaft design process is the optimal selection of the material. Two types of shaft steels with improved corrosion resistances, 1.4305 and 1.7225, were investigated experimentally and numerically in this paper in order to determine some of the material characteristics important for material selection in the engineering design process. Ultimate tensile strength and yield strength have been experimentally obtained, proving that steel 1.4305 has higher values of both. In addition, J-integral is numerically determined as a measure of crack driving force for finite element models of standardized fracture specimens (single-edge notched bend and disc compact tension). Obtained J values are plotted *versus* specimen crack growth size (Δa) for different specimen geometries (a/W). Higher resulting values of J-integral for steel 1.4305 as opposed to 1.7225 can be noted. Results can be useful as a fracture parameter in fracture toughness assessment, although this procedure differs from experimental analysis.

Keywords: steel 1.4305; steel 1.7225; fracture

---

1. Introduction

Material selection is a crucial step in the process of engineering design. Optimal selection of the material can significantly reduce the possibility of failures, along with understanding the nature and stress intensity that occurs in a designed structure. Engineering practices usually distinguishes one or few causes of failure: excessive force and/or temperature-induced elastic deformation, yielding, fatigue, corrosion, creep, *etc.* Selection of improper materials may have a negative effect on operational lifetime cycle and result in flaw appearance, which can cause structural failure.

A successful material-selection process implies reconciling requirements, such as appropriate strength of a material, sufficient level of rigidity, heat resistance, *etc.* For structures susceptible to crack growth, it is necessary to ensure that the material has been selected on the basis of fracture mechanics parameters.

Considering shaft design, the fracture mechanics approach must be used in order to account for high stresses and harsh operating conditions. Implementation of the fracture mechanics approach has the benefit of reducing potential failures, such as the fatigue induced fracture presented in a study of marine main engine crankshaft failure [1]. The agitator steel shaft failed due to an inadequate design, which was incapable of withstanding torsional-bending fatigue during operation [2]. The gearbox shaft failure occurred due to high stress concentrations at the corners of the wobbler of the shaft, causing fatigue crack initiation [3]. Improved design and machining practice suggested that this would help to prolong service life of the component. A forklift collapsed due to failure of axle shaft, caused by material inclusions and poor heat treatment [4].

Most of the mentioned failures occurred on steels typically used in the manufacturing of shafts intended for use in harsh environments, where a higher corrosion resistance is necessary. To be able to properly choose a suitable material for such an environment, characterization of a material is essential.

Fracture mechanics parameters that define material resistance to crack propagation are usually determined through experimental investigations of the material under consideration. Fracture behavior is usually estimated using some of the well-established fracture parameters, such as stress intensity factor (K), J-integral, or crack tip opening displacement (CTOD). J-integral is appropriate for quantifying material resistance to crack extension when dealing with ductile fracture of metallic materials, which includes nucleation, growth, and coalescence of voids [5]. For a growing crack, J-integral values can be determined for a range of crack extensions (Δa) and can be presented in the form of the J-resistance curve. This curve is usually obtained experimentally following standardized procedures, but it can be successfully complemented or even substituted by numerical methods, e.g., the finite element (FE) method. Some of the recent articles on this topic include discussion on the accuracy of J-integral obtained by experiments, two-dimensional (2D) FE analysis, three-dimensional (3D) FE analysis, or the Electric Power Research Institute method [6]. J-integral and CTOD are related through plastic constraint factors evaluated using 3D FE analyses of a clamped, single-edge tension specimen [7]. Methodology to evaluate 3D J-integral for finite strain elastic-plastic solid using FE analysis is proposed [8]. Stress intensity factors and T-stress of 3D interface cracks and notches are computed using the scaled boundary FE method [9].

This paper presents a comparison of numerically predicted J-values taken from the measure of crack driving force for two types of steel commonly used in shaft manufacturing, steels 1.4305 and 1.7225. Obtained material data may help designers to find the best solution in appropriate material selection.

2. Materials and Methods

2.1. Considered Materials

The two materials compared are stainless steel 1.4305 (AISI 303) and alloy steel 1.7225 (AISI 4140). Steel 1.4305, commonly named chromium-nickel steel, is a derivative of a common grade stainless steel 1.4301, but with improved machinability. Chromium-molybdenum steel 1.7225 has an excellent strength to weight ratio, is readily machinable, and suitable for forging between 900 and 1200 °C.

The two materials differ substantially with respect to chemical composition (Table 1).

Table 1. Chemical composition of considered materials (wt%).

Material	C	Cr	Mn	Si	Mo	S	P
1.4305	0.047	17.4	2.0	0.584	–	0.252	0.0323
1.7225	0.45	1.06	0.74	0.32	0.17	0.018	0.014
Material	**Ni**	**Al**	**V**	**Nb**	**W**	**Cu**	**Rest**
1.4305	7.95	–	–	–	–	–	71.734
1.7225	0.04	0.02	0.01	0.02	0.02	0.04	97.078

Comparing the composition of steel 1.4305 to standard EN 10088-2:2005, it can be noted that the nickel percentage in the considered steel is just below the standard range (8%–10%), while manganese equals the maximum standard value (2%). As for steel 1.7225, comparing it to standard EN 10083-3:2006, it can be noted that carbon is equal to the maximum standard value (0.45%), while other elements are in standard ranges.

Both steels are widely considered for shafts intended to be used in marine, the petro or chemical industries, or as vehicular components, because both offer improved corrosion resistance, but with 1.4305 having significantly better resistance in a corrosive environment due to its elevated chromium content. Considering the differences in composition and corrosion resistance, it can be concluded that the two materials correspond to somewhat different ranges of specific applications. Steel 1.7225 can be found in bridge crane shafts [10], which are prone to fatigue failure, marine diesel engine crankshafts [11], where a material has to be adequate to severe working conditions, automotive applications [12], where axle shafts are sensitive to improper heat treatment, or in diesel engines of commercial vehicles [13], in which crankshafts need to be machined properly in order to avoid fatigue fractures. Chromium-nickel stainless steels can be found in agitator steel shafts [2] or mixer unit shafts [14], which prone to intergranular stress cracking at weld heat affected zones.

The equipment used to determine the mechanical properties of the materials was a computer-directed, materials testing machine (Zwick/Roell 400 kN, Zwick/Roell, Ulm, Germany) (Figure 1). Specimens were manufactured from rods made of the considered steels. Specimen geometry and uniaxial tensile test procedure were set, according to appropriate American Society for Testing and Materials (ASTM) standard [15]. Experimentally-obtained stress-strain curves are shown in Figure 2, and values of yield strength (σ_{YS}) and tensile strength (σ_{TS}) of considered materials are given in Table 1.

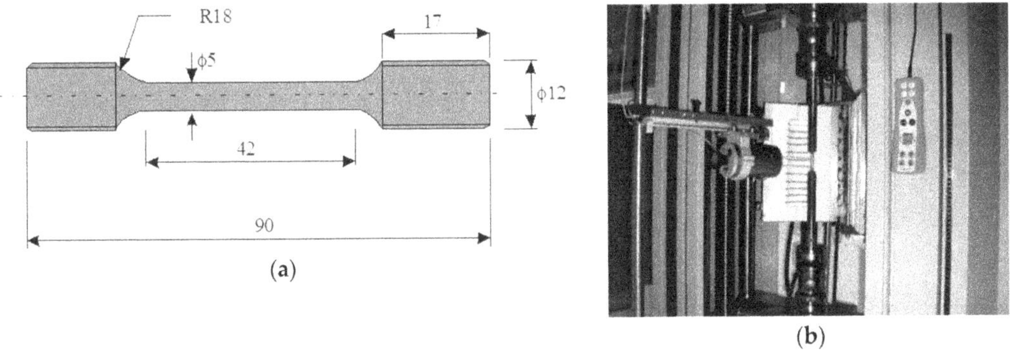

Figure 1. (**a**) Test specimen (dimension in mm); (**b**) testing machine.

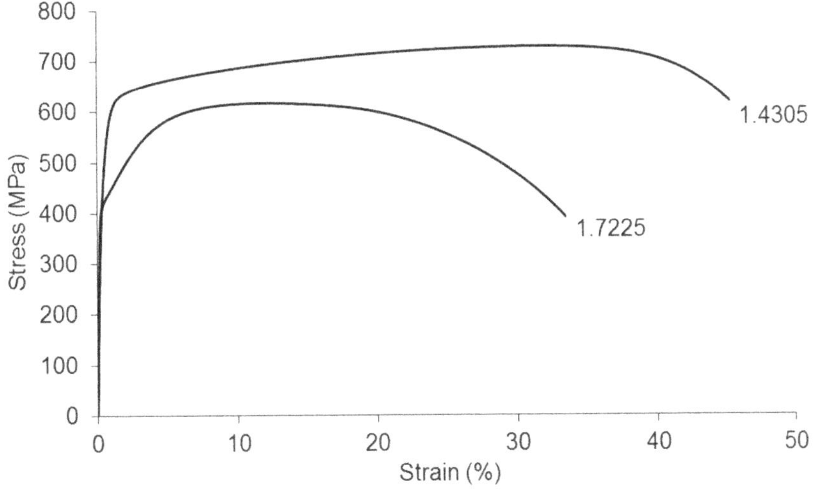

Figure 2. Uniaxial engineering stress-strain (σ-ε) diagrams for the considered materials [16,17].

When dealing with the fracture resistant design of structures, fracture toughness has an importance similar to yield strength when dealing with structure design against plastic deformation. The simple Charpy procedure for impact energy determination can be used to determine the fracture parameters. On the basis of known Charpy V-notch (CVN), impact energy correlation with fracture toughness can be made, e.g., with Roberts-Newton formula, independent of CVN energy range and temperature level [18]:

$$K_{Ic} = 8.47(\mathrm{CVN})^{0.63} \tag{1}$$

Some experimental data related to CVN energy is presented in Table 2. Charpy V-notch energy was also experimentally determined [16,17]. Experimentally-obtained results show that steel 1.4305 manifests a higher tensile strength and elongation than steel 1.7225, but CVN energy results show that the steel 1.4305 presents a lower CVN and lower fracture toughness than steel 1.7225, which can be attributed to the state of the as-received materials. Steel 1.7225 was a soft annealed as-received material (containing more percentage of carbon), compared to steel 1.4305, which was annealed and cold drawn (containing a lower percentage of carbon).

Table 2. Yield strength (σ_{YS}) and tensile strength (σ_{TS}) of the considered materials [16,17].

Material	σ_{YS} (Mpa)	σ_{TS} (Mpa)	CVN (J)	K_{Ic} (MPa· m$^{1/2}$)
1.4305	467	728	46	94.5
1.7225	415	617	166	212.1

In Figure 3, an optical micrograph of the as-received steel 1.4305 (soft annealed and cold drawn) is presented.

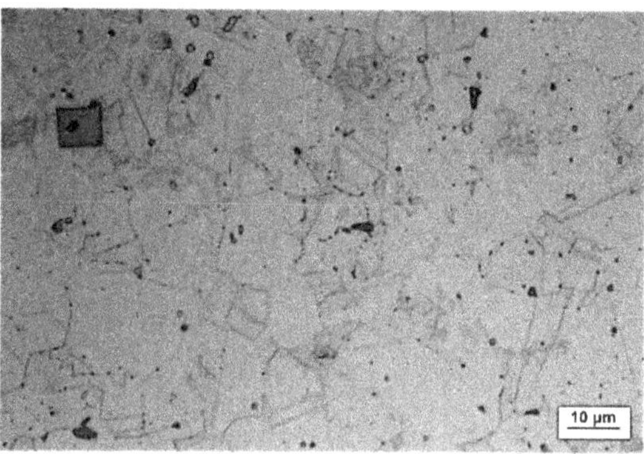

Figure 3. Optical micrograph of steel 1.4305; as-received material, soft annealed and cold drawn, cross-section of the specimen, aqua regia, 1000×.

In Figure 4, an optical micrograph of the as-received steel 1.7225 (soft annealed) is presented.

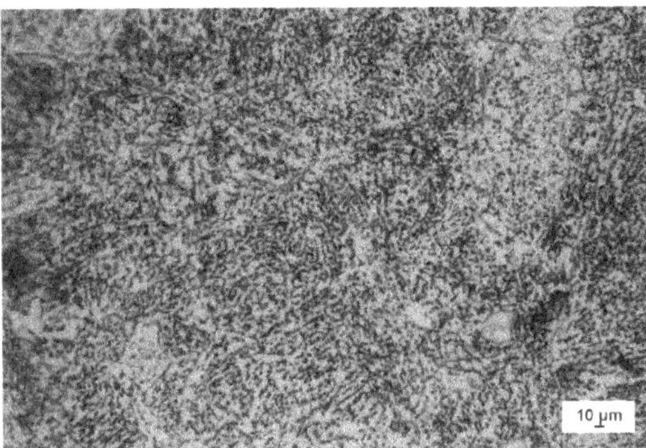

Figure 4. Optical micrograph of steel 1.7225; as-received material, soft annealed, cross-section of the specimen, 4% nital, 1000×.

Considering the microstructure of steel 1.4305, it can be said that the basic microstructure of the as-received material is austenite, but there is also a mixture of austenite and ferrite. Considering the microstructure of steel 1.7225, it can be said that its main phase (main structure) consists of a thin pearlitic microstructure, where also a few ferrite and some particles of cementite can be observed.

2.2. *Predicted Fracture Behavior of Considered Materials*

J-integral is used to numerically predict the fracture behavior of the considered materials. *J*-integral was introduced by Rice and Cherepanov [19,20], separately, as a path-independent integral, which can be drawn around the tip of a crack and viewed both as an energy release rate parameter and a stress intensity parameter. In a two-dimensional form, it can be written as:

$$J = \int_{\Gamma} \left(w\mathrm{d}y - T_i \frac{\partial u_i}{\partial x} \mathrm{d}s \right) \tag{2}$$

where $T_i = \sigma_{ij} n_j$ are components of the traction vector, u_i are the displacement vector components, and d*s* is an incremental length along the arbitrary contour path Γ enclosing the crack tip.

In order to predict fracture behavior of steels 1.4305 and 1.7225, an experimental single specimen test method [15] following an elastic unloading compliance technique was numerically simulated. This test method uses measured crack mouth opening displacement to estimate the growing crack size. Resulting *J*-integral values can be taken as a fracture toughness parameter and plotted *versus* crack extension. The first step of the numerical procedure is to conduct a structural stress analysis.

According to the appropriate ASTM standard [15], 2D FE models of the two types of specimens, single edge notched bend (SENB), and disc compact tensile (DCT), are defined and initial *a/W* (*W* = 50 mm) ratios of 0.25, 0.5 and 0.75 are taken (Figure 5). The material behavior is considered to be a multilinear isotropic hardening type. FE models of specimens are meshed with 8-node isoparamateric quadrilateral elements. The mesh is refined around the crack tip in order to capture high deformation gradients in the regions where yielding occurs. To simulate compliance procedures of single specimen test method, quasi-static load was imposed on specimen. Only half of the specimen needs to be modeled due to their symmetry. The node releasing technique was used to simulate crack propagation.

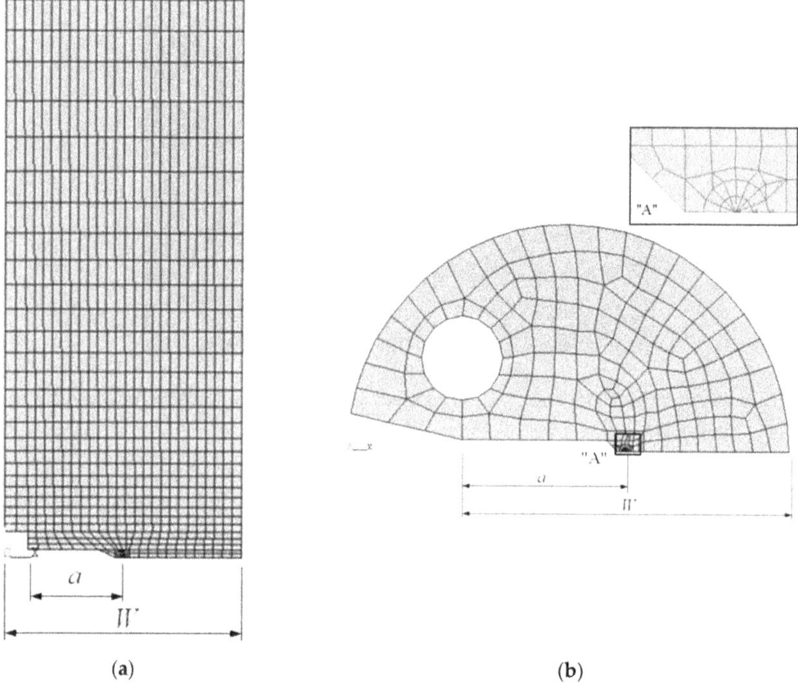

Figure 5. FE model of: (**a**) SENB specimen; (**b**) DCT specimen.

FE stress analysis results taken from the integration points of finite elements surrounding the crack tip were used to evaluate *J*-integral values using the following equation [21]:

$$J = \sum_{p=1}^{np} W_p G_p \left(\xi_p, \eta_p\right) \tag{3}$$

where W_p is the Gauss weighting factor, *np* is the number of integration points, and G_p is the integrand evaluated at each Gauss point *p*:

$$\begin{aligned} G_p = &\left\{ \frac{1}{2}\left[\sigma_{xx}\frac{\partial u_x}{\partial x} + \sigma_{xy}\left(\frac{\partial u_x}{\partial y} + \frac{\partial u_y}{\partial x}\right)\frac{\partial u_x}{\partial x} + \sigma_{yy}\frac{\partial u_y}{\partial y}\right]\frac{\partial y}{\partial \eta} \right. \\ &\left. - \left[\left(\sigma_{xx}n_1 + \sigma_{xy}n_2\right)\frac{\partial u_x}{\partial x} + \left(\sigma_{xy}n_1 + \sigma_{yy}n_2\right)\frac{\partial u_y}{\partial x}\right]\sqrt{\left(\frac{\partial x}{\partial \eta}\right)^2 + \left(\frac{\partial y}{\partial \eta}\right)^2} \right\}_g \end{aligned} \tag{4}$$

J values are summed along the path that encloses the crack tip, giving total value of *J*. Three different paths around the crack tip have been defined in each example, and the average value was taken as a final. Although *J*-integral is independent of the chosen path, this was done in order to account for any possible *J*-value variations in the vicinity and away from the crack tip. This procedure had already been verified in cases when numerically-obtained parameters have been found to be corresponding with the experimental values [22].

Since no fracture experimental results were available for steels 1.4305 and 1.7225 to verify the accuracy of the procedure, *J*-integral values were first determined for the SENB specimen, with an initial crack length of *a/W* = 0.25, 0.5, 0.75, made of 1.6310 steel. Numerically-obtained results were compared with available experimental data for the same specimen configuration and material [23] (Figure 6). Good correspondence between numerically-predicted and experimental results encouraged further application of the *J*-integral calculation method.

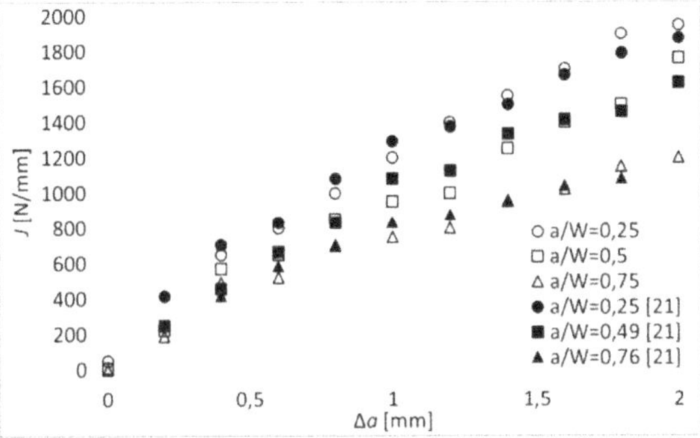

Figure 6. Comparison of numerically-predicted and experimentally-obtained J values for SENB specimens of 1.6310 steel.

3. Results

The entire numerical procedure described in Section 2.2 was performed for FE models of SENB and DCT specimens. With respect to the geometry of specimens, initial a/W (crack length/width ratio with W = 50 mm) ratios of 0.25, 0.5 and 0.75 are taken with Δa = 0...2 mm (Δa being crack advance). For every case, J-integral values, as a measure of crack driving force, are calculated and the final results are shown in Figure 7 (for steel 1.4305) and Figure 8 (for steel 1.7225).

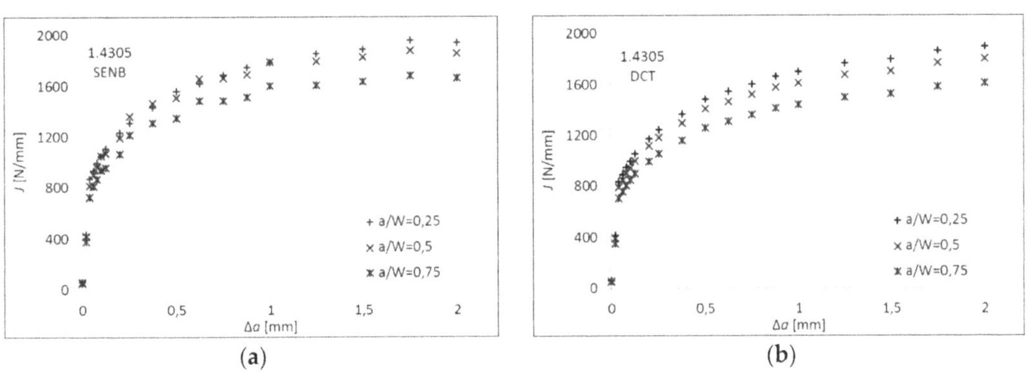

Figure 7. Numerically-predicted J values for steel 1.4305 using FE models of: (**a**) SENB specimen; (**b**) DCT specimen.

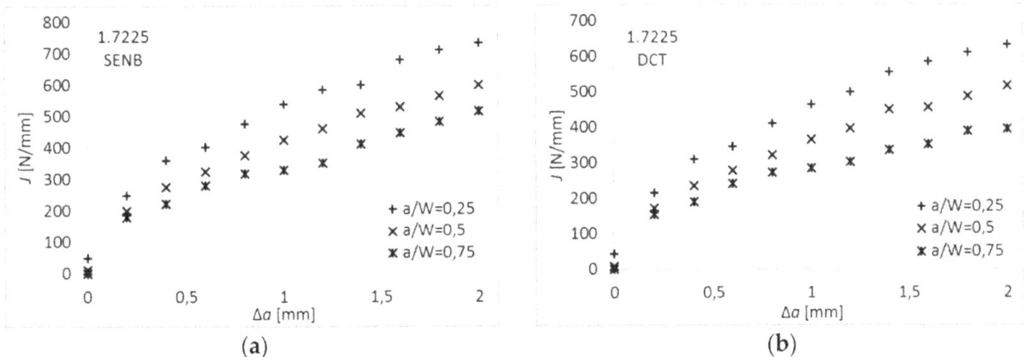

Figure 8. Numerically-predicted J values for steel 1.7225 using FE models of: (**a**) SENB specimen; (**b**) DCT specimen.

4. Discussion

The fracture behaviors of the mentioned materials are given in Figures 7 and 8 using *J*-integral as a measure of crack driving force. It can be noted that steel 1.7225 has higher resulting values of *J*-integral than steel 1.4305, making it more suitable for structures that need less susceptibility to fracture. The predicted difference in resistance to crack extension between steel 1.4305 and steel 1.7225 can be related to the different material properties and compositions (Tables 1 and 2) of the two materials. Steel 1.4305 has a significantly higher value of nickel, which can contribute to the noted behavior. Nickel is usually added at over 8% (here 7.95%) content to chromium-nickel stainless steels in order to increase strength, impact strength, and toughness. It also improves resistance to oxidization and corrosion along with chromium. Chromium is added at over 12% content in stainless steels to significantly improve corrosion resistance. Added benefits are also hardenability, strength, response to heat treatment, and wear resistance.

In numerical modeling, crack geometry (a/W ratios) was kept the same for both materials in relative specimens, so the geometry probably could not contribute to a difference in J values for the two steels.

J-integral values differ for $a/W = 0.75$ when compared with $a/W = 0.25$ and 0.5, which are quite similar in terms of values for steel 1.4305. Additionally, higher a/W ratios correspond to lower *J*-integral values of materials and *vice versa*. *J*-integral values obtained using a DCT specimen FE model give somewhat conservative results when compared with ones obtained using the SENB specimen FE model. Experimentally-obtained K_{Ic} values and numerically predicted J values cannot be easily compared because J values go beyond the elastic behaviour of the material, where K is appropriate parameter. Further, J is taken as a measure of crack driving force here. Although the mentioned numerical procedure does not give results that can be directly related to ones obtained experimentally, given results can be useful for the assessment of fracture toughness.

The numerically-obtained data, along with experimentally-obtained yield strength, tensile strength, and CVN energy for steels 1.4305 and 1.7225 presented in this paper may be of importance for designers of engineering structures when concerning material selection. In addition, numerical assessment of *J*-integral could be useful as a predictor of the possible fracture behaviour of a material. Altogether, experimentally- and numerically-obtained data can be used in the design process to assess the possible load capacity of a structure.

Acknowledgments: This work has been financially supported by the Croatian Science Foundation under the project 6876, and by the University of Rijeka under the projects 13.09.1.1.01 and 13.07.2.2.04. Funds for covering the costs to publish in open access publications have been provided.

Author Contributions: Goran Vukelic and Josip Brnic conceived the idea of the research; Josip Brnic performed the experiments; Goran Vukelic analyzed the data; Goran Vukelic performed numerical analysis; Goran Vukelic and Josip Brnic wrote the paper.

Conflicts of Interest: The authors declare no conflicts of interest. The founding sponsors had no role in the design of the study; in the collection, analyses, or interpretation of data; in the writing of the manuscript; or in the decision to publish the results.

Abbreviations

The following abbreviations are used in this manuscript:

FE	Finite Element
DCT	Disc Compact Tension
SENB	Single Edge Notched Bend
CVN	Charpy V-notch

References

1. Fonte, M.; de Freitas, M. Marine main engine crankshaft failure analysis: A case study. *Eng. Fail. Anal.* **2009**, *16*, 1940–1947. [CrossRef]

2. Zangeneh, Sh.; Ketabchi, M.; Kalaki, A. Fracture failure analysis of AISI 304L stainless steel shaft. *Eng. Fail. Anal.* **2014**, *36*, 155–165. [CrossRef]
3. Moolwan, C.; Netpu, S. Failure analysis of a two high gearbox shaft. *Procedia Soc. Behav. Sci.* **2013**, *88*, 154–163. [CrossRef]
4. Das, S.; Mukhopadhyay, G.; Bhattacharyya, S. Failure analysis of axle shaft of a fork lift. *Case Stud. Eng. Fail. Anal.* **2015**, *3*, 46–51. [CrossRef]
5. Kossakowski, P.G. Simulation of ductile fracture of S235JR steel using computational cells with microstructurally-based length scales. *J. Theor. Appl. Mech.* **2012**, *50*, 589–607.
6. Dai, Q.; Zhou, C.; Peng, J.; He, X. Experiment, finite element analysis and EPRI solution for *J*-integral of commercially pure titanium. *Rare Metal. Mat. Eng.* **2014**, *42*, 257–263.
7. Huang, Y.; Zhou, W. *J-CTOD* relationship for clamped SE(T) specimens based on three-dimensional finite element analyses. *Eng. Fract. Mech.* **2014**, *131*, 643–655. [CrossRef]
8. Koshima, T.; Okada, H. Three-dimensional *J*-integral evaluation for finite strain elastic-plastic solid using the quadratic tetrahedral finite element and automatic meshing methodology. *Eng. Fract. Mech.* **2015**, *135*, 34–63. [CrossRef]
9. Saputra, A.A.; Birk, C.; Song, C. Computation of three-dimensional fracture parameters at interface cracks and notches by the scaled boundary finite element method. *Eng. Fract. Mech.* **2015**, *148*, 213–242. [CrossRef]
10. Zambrano, O.A.; Coronado, J.J.; Rodríguez, S.A. Failure analysis of a bridge crane shaft. *Case Stud. Eng. Fail. Anal.* **2014**, *2*, 25–32. [CrossRef]
11. Fonte, M.; Duarte, P.; Anes, V.; Freitas, M.; Reis, L. On the assessment of fatigue life of marine diesel engine crankshafts. *Eng. Fail. Anal.* **2015**, *56*, 51–57. [CrossRef]
12. Tawancy, H.M.; Al-Hadhrami, L.M. Failure of a rear axle shaft of an automobile due to improper heat treatment. *J. Fail. Anal. Prev.* **2013**, *13*, 353–358. [CrossRef]
13. Bai, S.Z.; Hu, Y.P.; Zhang, H.L.; Zhou, S.W.; Jia, Y.J.; Li, G.X. Failure analysis of commercial vehicle crankshaft: A case study. *Appl. Mech. Mater.* **2012**, *192*, 78–82. [CrossRef]
14. Fuller, R.W.; Ehrgott, J.Q., Jr.; Heard, W.F.; Robert, S.D.; Stinson, R.D.; Solanki, K.; Horstemeyer, M.F. Failure analysis of AISI 304 stainless steel shaft. *Eng. Fail. Anal.* **2008**, *15*, 835–846. [CrossRef]
15. American Society for Testing and Materials (ASTM International). *Metals Test Methods and Analytical Procedures*; Annual BOOK of ASTM Standards; ASTM International: Baltimore, Maryland, MD, USA, 2005; Volume 03.01.
16. Brnic, J.; Turkalj, G.; Canadija, M.; Lanc, D.; Krscanski, S. Responses of austenitic stainless steel American iron and steel institute (AISI) 303 (1.4305) subjected to different environmental conditions. *J. Test. Eval.* **2012**, *40*, 319–328. [CrossRef]
17. Brnic, J.; Turkalj, G.; Canadija, M.; Lanc, D.; Brcic, M. Study of the effects of high temperatures on the engineering properties of steel 42CrMo4. *High Temp. Mater. Proc.* **2015**, *34*, 27–34. [CrossRef]
18. Roberts, R.; Newton, C. *Interpretive Report on Small Scale Test Correlations with K_{Ic} Data*; Welding Research Council Bulletin: New York, NY, USA, 1981; pp. 1–16.
19. Rice, J.R. A path independent integral and the approximate analysis of strain concentration by notches and cracks. *J. Appl. Mech.* **1968**, *35*, 379–386. [CrossRef]
20. Cherepanov, G.P. The propagation of cracks in a continuous medium. *J. Appl. Math. Mech.* **1967**, *31*, 503–512. [CrossRef]
21. Mohammadi, S. *Extended finite element method*; Blackwell Publishing: Singapore, 2008; pp. 56–58.
22. Vukelic, G.; Brnic, J. Prediction of fracture behavior of 20MnCr5 and S275JR steel based on numerical crack driving force assessment. *J. Mater. Civ. Eng.* **2015**, *27*, 14132–14132(6). [CrossRef]
23. Narasaiah, N.; Tarafder, S.; Sivaprasad, S. Effect of crack depth on fracture toughness of 20MnMoNi55 pressure vessel steel. *Mater. Sci. Eng. A* **2010**, *527*, 2408–2411. [CrossRef]

6

Influence of Loading Rate on the Hydrogen-Assisted Micro-Damage in Bluntly Notched Samples of Pearlitic Steel

Jesús Toribio *,†, Diego Vergara † and Miguel Lorenzo †

Academic Editor: Recep Avci

Fracture & Structural Integrity Research Group, University of Salamanca, 37008 Salamanca, Spain; diego.vergara@ucavila.es (D.V.); mlorenzo@usal.es (M.L.)

* Correspondence: toribio@usal.es

† These authors contributed equally to this work.

Abstract: The influence of loading rate (crosshead speed) on the fracture process of bluntly notched samples of pearlitic steel under hydrogen environment is analyzed in this paper. Results indicate that the location of the zone where fracture initiates (fracture process zone) in pearlitic steel samples with a blunt notch directly depends on the loading rate or crosshead speed. For slow testing rates, such a zone is placed in the specimen center due to hydrogen diffusion towards the prospective fracture places located in the central area of the section. On the other hand, in the case of high testing rates, the process of hydrogen-assisted fracture initiates near the sample periphery, *i.e.*, in the vicinity of the notch tip, because in such quick tests hydrogen does not have enough time to diffuse towards inner points of the specimen.

Keywords: pearlitic steel; notched samples; constraint; crosshead speed; hydrogen-assisted fracture; micro-fracture maps

1. Introduction

Prestressing steel wires are highly susceptible to hydrogen embrittlement (HE) or hydrogen assisted fracture (HAF) in cathodic environments [1–5]. Under such environmental conditions, hydrogen diffuses towards the internal regions of the material sample and can reach a critical concentration dependent on the stress-stain state at a given point and at a specific time [6].

With regard to the influence of the loading rate, previous analyses [7–10] showed the inverse relationship between HE susceptibility and strain rate, *i.e.*, the lower the loading rate, applied displacement rate or crosshead speed during the test, the higher the hydrogen degradation in the sample. In this paper, the influence of the loading rate is shown in the micro-fracture maps (MFMs) appearing after fracture in bluntly notched samples of high-strength pearlitic steel tested in a hydrogenating environment.

MFMs of notched samples of the same material were shown in a previous paper by the authors containing experimental results [11] for high loading rates (0.01 mm/min). This paper goes further in the analysis, including several new contributions: (i) a new MFM appearing after fracture after testing at low loading rates (0.001 mm/min); and (ii) a numerical explanation of this MFM by using stress-assisted hydrogen diffusion models.

Although the notch machining in the wires could affect the generation of dislocations in the material at the very local level (just at the close vicinity of the notch tip), the main dislocations presented in the material are due to cold drawing. According to [12], the local stress in the notch vicinity is not affected by screw dislocations but the position of edge dislocations influences the stress state distribution. Anyway, the key issue in the studies with notched geometries is the triaxial stress state generated in the material when an external loading is applied [13], which directly affects the distribution of stresses within the material, and thereby the hydrogen diffusion and fracture behaviour.

2. Experimental Procedure

Pearlitic steel samples were used in this work. Different degrees of cold drawing were chosen from the initial hot rolled bar up to the commercial prestressing steel wire (C 0.800%, Mn 0.690%, Si 0.230%, P 0.012%, S 0.009%, Al 0.004%, Cr 0.265%, V 0.060%). Cold drawing consists of the pass of the steel wire through a hard die, thus obtaining a new wire with a smaller transverse section [14]. To obtain commercial prestressing steels, it is necessary to carry out several progressive reductions, *i.e.*, cold drawing is a multi-pass process.

Cumulative straining by cold drawing induces material microstructure changes [15–19] in the form of progressive slenderizing and orientation of pearlitic colonies in the cold drawing direction [20,21], as well as densification and orientation of pearlite (ferrite/cementite) lamellae in that direction [22,23].

The aforesaid microstructural changes directly affect hydrogen diffusion in the steel, as shown in Figure 1 where three drawing levels appear: hot rolled bar (0 steps of cold drawing, Figure 1a), prestressing steel wire (6 steps of cold drawing, Figure 1c), and an intermediate steel in the drawing chain (3 steps of cold drawing, Figure 1b). A metallographic study with scanning electron microscopy (SEM, JEOL model JSM-SG 20LV, Tokyo, Japan) was carried out to obtain the micrographs observed in Figure 1. Previously, the material was polished and chemically etched with Picral during 4–5 s.

Apart from the microstructural orientation of pearlite, another factor affecting hydrogen diffusion in the different steels is the density of non-metallic inclusions due to the interactions between inclusions and hydrogen [24,25]. In the matter of mechanical properties, yield strength (σ_Y) and ultimate tensile strength (UTS) are shown in Figure 1 for the three selected drawing levels.

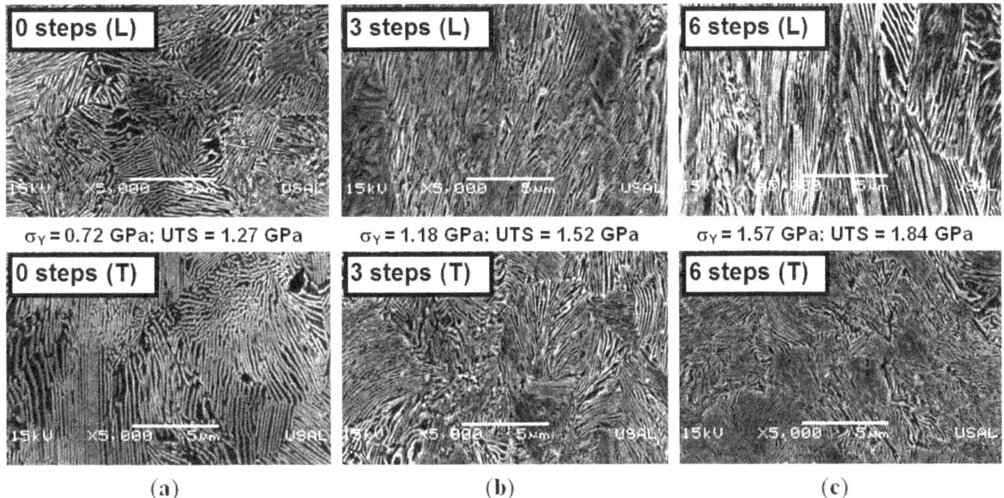

Figure 1. Micrographs of longitudinal (L) and transverse (T) sections of steels at different levels of cold drawing: (**a**) 0 steps, hot rolled bar; (**b**) 3 steps; (**c**) 6 steps, prestressing steel.

Specimens for testing were axisymmetric bluntly-notched bars (see Figure 2) with the relative dimensions $A/\text{Ø} = 0.30$ and $R/\text{Ø} = 0.40$, where A is the notch depth, R the notch radius and Ø the wire diameter (changing with cold drawing from 12 mm in the hot rolled bar to 7 mm in the commercial prestressing steel wire, 8.9 mm being the diameter on the intermediate wire). Notches were machined in each wire after cold drawing, maintaining in all cases the same relative dimensions (those given at the beginning of this paragraph).

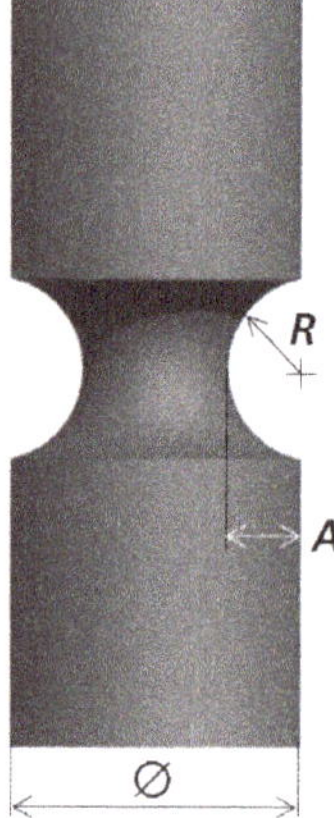

Figure 2. Round notched geometry used in the tests.

The specimens were subjected to constant extension rate tensile (CERT) tests in which the cathodic electrochemical conditions promoted HE. An aqueous solution of $Ca(OH)_2$ with 0.1 g/L of NaCl (pH 12.5) was used in a cell connected to a potentiostat applying a constant potential of −1200 mV SCE. In Figure 3 a scheme of the experimental setup is shown, including the three electrodes: saturated calomel electrode (SCE), working electrode (steel specimen) and auxiliary electrode (Pt). According to previous studies [26], these electrochemical conditions promote HE. Two crosshead speeds (or applied displacement rates) were chosen in this work: (i) 0.001 mm/min (relatively low speed: quoted as type 1), and (ii) 0.01 mm/min (relatively high speed: quoted as type 2).

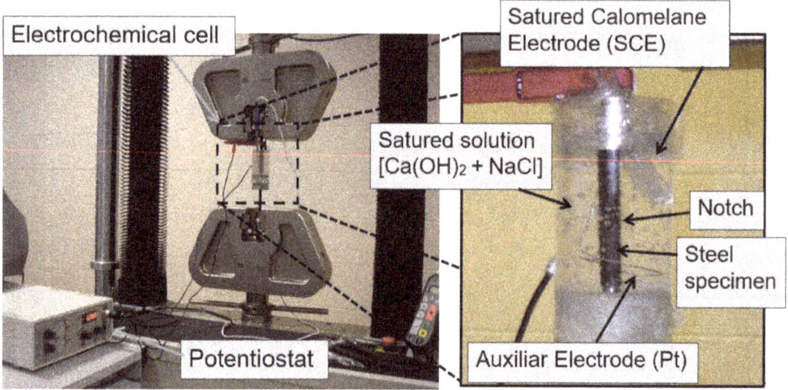

Figure 3. Constant extension rate tensile (CERT) test under cathodic electrochemical conditions including a detailed view of the electrochemical cell.

3. Fractography

In a previous paper by the authors [11], different fracture topographies were found in notched samples of progressively drawn steel under HE environmental conditions, namely: (i) the so-called tearing topography surface or TTS [27,28] associated with hydrogen-assisted micro-damage in pearlitic steels [29]; (ii) ductile microvoid coalescence (MVC); (iii) a non-conventional microscopic mode called quasi-MVC (and noted as MVC* throughout this paper), consisting of a partially ductile zone with particular appearance resembling MVC that could be seen as candidate to TTS in which hydrogenation is not enough [30,31] and (iv) brittle cleavage (C).

This paper goes further in the fractographic analysis to study the kinematic effects on the MFMs, *i.e.*, the influence of the loading rate (or crosshead speed) on the micromechanisms of fracture. Figures 4 and 5 show the fractographs for speeds type 1 and type 2, respectively. They were obtained by means of a SEM and using a low magnification factor, thereby providing enlarged views of the real fracture surfaces after the CERT tests. In all pictures the first digit represents the drawing degree (0, 3 and 6 indicating the number of drawing steps undergone by the steel) and the second one the crosshead speed (1 and 2 indicating respectively the low and high loading rates).

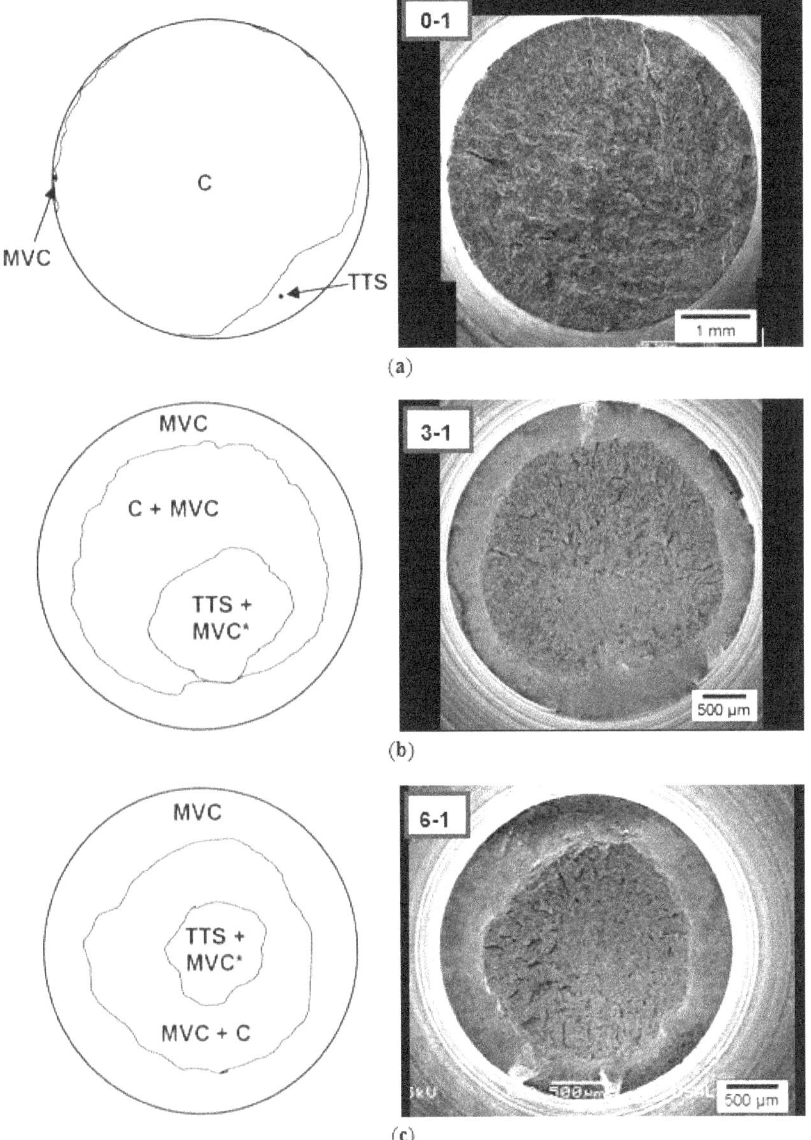

Figure 4. Micro-fracture maps (MFMs) for low loading rate (crosshead speed type 1 of 0.001 mm/min): (**a**) hot rolled steel (base material that is not cold drawn at all); (**b**) three drawing steps; (**c**) six drawing steps (heavily-drawn commercial prestressing steel).

The location of the MVC* region in the fracture surface depends on the crosshead speed. In the case of high speed (type 2, Figure 5), fracture initiation by TTS takes place at the sample periphery. Figure 6 shows the fractographic appearance of the TTS zone obtained by SEM for two drawing degrees (hot rolled bar and prestressing steel). In previous analyses [29,30] it was clearly demonstrated that the TTS region is associated with hydrogen-assisted micro-damage in pearlite at the finest microscopical level, its size being linked to the mechanical (geometry and type of loading) and electrochemical (pH and potential) characteristics of the HE tests [29]. Therefore, the TTS domain is an experimental evidence of the fracture process zone (FPZ) due to hydrogen effects, *i.e.*, such an area is precisely where fracture initiates by a HE mechanism. Later, subcritical cracking advances towards the inner points, creating MVC* regions, up to reaching a critical instant of fracture in which sudden final fracture happens by brittle cleavage (C) with isolated MVC areas. This fully agrees with the three MFM schemes described in previous research [11].

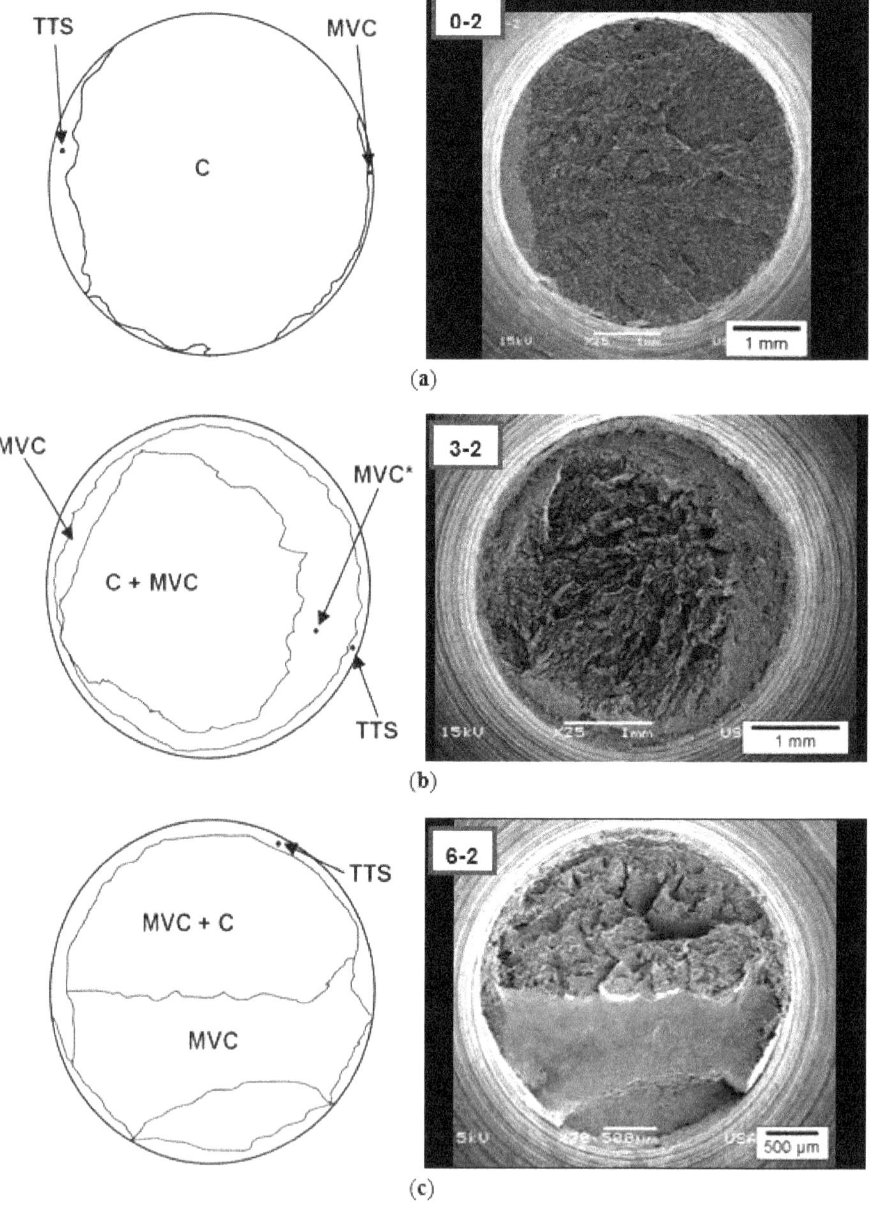

Figure 5. Micro-fracture maps (MFMs) for high loading rate (crosshead speed type 2 of 0.01 mm/min): (**a**) hot rolled steel (base material that is not cold drawn at all); (**b**) three drawing steps; (**c**) six drawing steps (heavily-drawn commercial prestressing steel).

On the other hand, in the case of cold drawn steels tested at low speed (type 1, Figure 4), the FPZ appears at the wire core (center of the sample) and progresses towards the periphery, *i.e.*, just the opposite of the MFM schemes described previously in [11]. Figure 4 shows how the FPZ is placed in the internal area, exhibiting a mixed fractography consisting of both TTS and MVC*. This experimental fact suggests that, in this case of slow loading rate, hydrogen diffuses towards the inner points and promotes fracture initiation in that area up to reaching the critical concentration. At the critical instant, final fracture takes place by a micromechanism similar to that obtained in standard tension tests, *i.e.*, cup and cone ductile fracture.

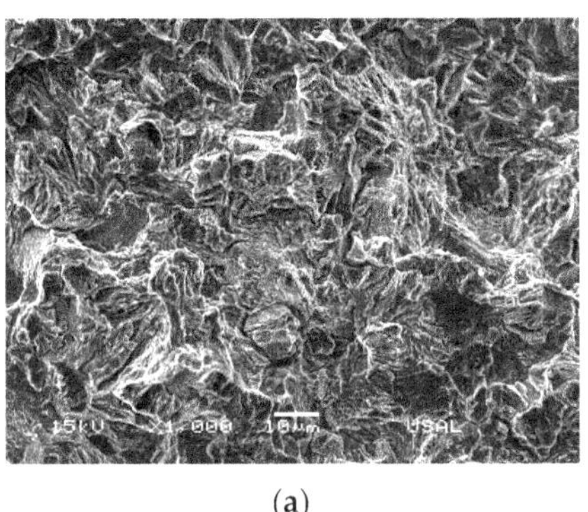

(a)

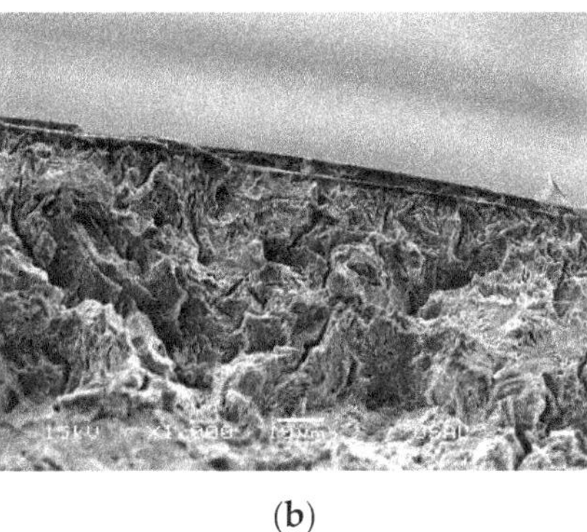

(b)

Figure 6. Hydrogen-assisted micro-damage (TTS) in wires after CERT tests: (**a**) hot rolled steel, speed type 1 (Figure 4a); (**b**) prestressing steel, speed type 2 (Figure 5c).

Figure 7 summarizes the four models of MFM found in the fractographic analysis. The first three are in agreement with previous research [11]: model I (Figure 7a), model II (Figure 7b) and model III (Figure 7c) and all of them are associated with surface (peripherical) initiation of HAF in the vicinity of the notch tip. In the fourth model (IV) represented in Figure 7d the HAF process initiates at the center of the cross-sectional area of the specimen and propagates radially towards the periphery.

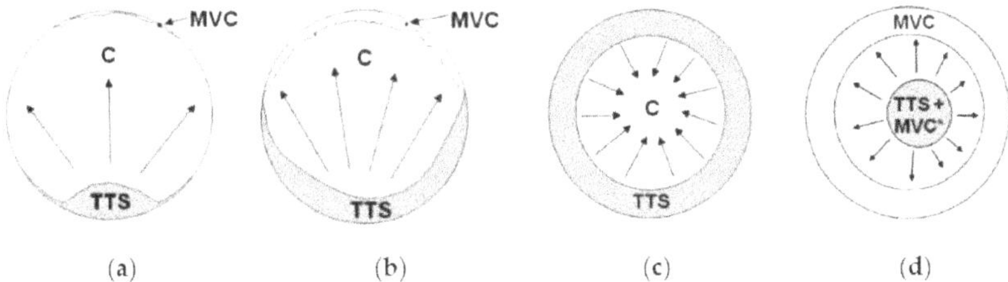

Figure 7. Schemes of the different models of micro-fracture map (MFM) in bluntly-notched samples of pearlitic steel under hydrogen embrittlement (HE) environmental conditions: (**a**) model I; (**b**) model II; (**c**) model III; (**d**) model IV.

The four models of MFM appearing in the different HE tests on bluntly-notched specimens of pearlitic steel are given in Table 1 for each loading rate (crosshead speed) and all drawing degrees (microstructural orientation). For hot rolled material and slightly drawn steels undergoing only one drawing step, the FPZ is localized in the form of surface flaw (model I of MFM) or shallow quasi-circumferential crack (model II of MFM).

Table 1. Models of MFM for each loading rate (crosshead speed) and drawing degree.

Loading Rate (Crosshead Speed)	Cold Drawing Step						
	0	1	2	3	4	5	6
Type 1 (0.001 mm/min)	I	II	IV	IV	IV	IV	IV
Type 2 (0.01 mm/min)	I	II	III	III	III	III	III

As the drawing degree increases, the MFMs evolve towards an axisymmetric shape due to a double effect [11]: (i) increase of material anisotropy produced by microstructural orientation (barrier effect created by cementite alignment tending to the wire axis or cold drawing direction); and (ii) more uniform microstructure as a consequence of the aforesaid orientation, thereby promoting an axisymmetric FPZ of the models III and IV.

For cold drawn steels undergoing at least two drawing steps and subjected to quick HE tests (type 2), the MFM corresponds to model III, *i.e.*, HAF initiates at the sample surface or periphery (notch tip), whereas the same drawn steels subjected to slow HE tests (type 1) exhibit central HAF initiation at the inner points (model IV) due to the fact that hydrogen does have enough time to diffuse towards such inner areas of the sample.

To evaluate the HE susceptibility of the analyzed steels, Figure 8 plots the variation with the drawing degree of the ratio F_{HE}/F_0 where F_{HE} is the critical force (maximum value; fracture instant) in cathodic environment and F_0 is the same in air, considering the two loading rates used in the tests (0.001 mm/min and 0.01 mm/min). The role of the loading rate in CERT tests is only noticeable after the fourth drawing step, with a clear effect on the final commercial prestressing steel wire. Thus, samples tested with the lower loading rate (0.001 mm/min) exhibit a higher HE susceptibility than those tested with higher loading rate (0.01 mm/min).

There is no clear relationship between the fracture surface morphologies (models I to IV, Table 1) and the degree of susceptibility to HE (Figure 8). In addition, the visual aspect of the fracture surface after testing similar specimens in an inert environment is quite similar to that appearing after CERT tests under low loading rates (0.001 mm/min) [32], *i.e.*, the fracture path follows a transversal plane of the wire (mode I in the fracture mechanics sense). On the other hand, a deflected fracture surface (mixed mode I-II) appears in heavily drawn wires during CERT test under high loading rate (0.01 mm/min). Thus the hydrogen concentration in the material is not the unique factor affecting the fracture surface in notched wires, as discussed in the next section of the paper.

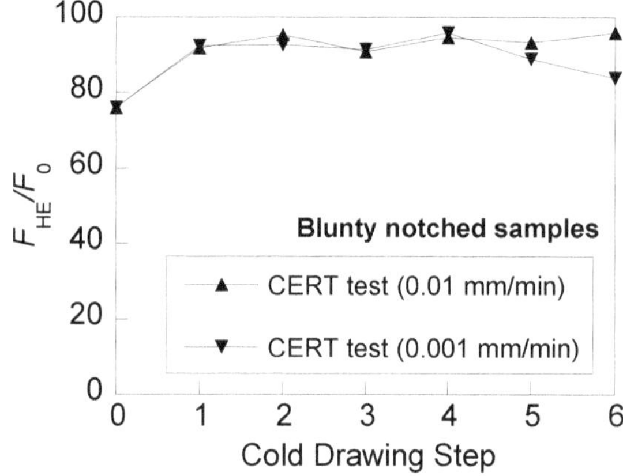

Figure 8. Variation with the cold drawing degree of the hydrogen embrittlement (HE) susceptibility of the progressively drawn steels for the two loading rates.

4. Hydrogen Diffusion

The key role of hydrostatic stress (σ) in hydrogen diffusion is well known [33–35]. The equation describing stress-assisted diffusion of hydrogen can be formulated as follows [36,37]:

$$\frac{\partial C}{\partial t} = D\left(\nabla^2 C - \frac{V_H}{RT}\nabla C \nabla \sigma - \frac{V_H}{RT} C \nabla^2 \sigma\right) \quad (1)$$

where C is the hydrogen concentration, D is the hydrogen diffusion coefficient, V_H the partial molar volume of hydrogen, R the universal gases constant and T the absolute temperature.

In this paper, the analysis of hydrogen diffusion was performed for a representative of the whole family of cold drawn steels: the wire undergoing three steps of drawing (steel number 3). The following values were used for the computations of hydrogen diffusion in the notched samples: $V_H = 2 \times 10^6$ m^3/mol [38]; $D = 3.21 \times 10^{-11}$ m^2/s [39], obtained after linear interpolation between the values for the hot rolled bar (not cold drawn at all, $D = 6.6 \times 10^{-11}$ m^2/s [40]) and the fully drawn wire (commercial prestressing steel, $D = 4.99 \times 10^{-12}$ m^2/s [41]).

In the bluntly-notched geometry analyzed in this paper (see Figure 1) the distribution of hydrostatic stress σ reaches its maximum value at the center of the cross sectional area of the sample, so that hydrogen will be "pumped" to such a location according to Equation (1). The hydrogen concentration C profile in the transverse section of the notched specimen at the final instant of the test $t = t_{HE}$ is represented in Figure 9 in dimensionless terms (C_r as the relative hydrogen concentration, *i.e.*, the ratio C/C_0 where C_0 is the initial equilibrium concentration for the stress-free metal).

It is seen that the hydrogen concentration profile (numerically computed) allows one to explain the experimental results of the HE tests, and particularly the effect of the loading rate (or crosshead speed) on the initiation of HAF.

In the case of slow tests (crosshead speed type 1, 0.001 mm/min) hydrogen does have enough time to diffuse towards the inner points located at the center of the cross-sectional area of the specimen, so that the maximum hydrogen concentration is achieved in that central region (specimen axis) due to the inwards positive gradient of hydrostatic stress driving (or "pumping") hydrogen towards such an area. This experimental evidence is fully consistent with a MFM of the model IV (Figure 7d), associated with HAF initiation in the core or central area (Figure 4b; steel 3; low crosshead speed).

On the other hand, in the case of quicker tests (crosshead speed type 2, 0.01 mm/min) hydrogen does not have enough time to reach sufficient concentration at the inner points located at the center of the cross-sectional area of the specimen, so that the maximum hydrogen concentration is achieved at the periphery (just at the notch tip), which is fully consistent with a MFM of the model III (Figure 7c), associated with HAF initiation with ring shape in the vicinity of the notch tip (Figure 5b; steel 3; high crosshead speed).

Local fracture event in the particular steel takes place at a certain locus (x) when hydrogen concentration, C, reaches a critical value which is itself time dependent: $C(x, t) = C_{cr}(\sigma(x, t))$ [6,36]. According to the model of hydrogen diffusion assisted by the stress field in the material, Equation (1), two driving forces govern hydrogen diffusion towards prospective damage places: the gradient of hydrogen concentration and the gradient of hydrostatic stress.

Thus, for low loading rates in CERT tests of wires undergoing three steps of cold drawing, the critical hydrogen concentration happens at the wire core, so that initiation of fracture is central (MFM model IV). However, for CERT tests under high loading rates applied on the same wire, the time is not high enough for hydrogen to diffuse towards the inner points and reach the central area, so that the maximum concentration is reached at the surface and the fracture initiation is located at the wire periphery (MFM model III). This is explained in Figure 9, where the radial distribution of relative hydrogen concentration (C_r) is shown at the instant of final failure during a CERT test ($t = t_{HE}$) of a steel undergoing three steps of cold drawing during manufacturing. Depth is defined in radial direction from the notch tip, and the relative hydrogen concentration is derived from Equation (1).

Figure 9 reveals a key issue: whereas in the fastest test (0.01 mm/min) the hydrogen concentration profile is decreasing (maximum at the surface), in the slowest one (0.001 mm/min) such a profile is increasing (reaching its maximum just at the specimen center). This way, in steels tested under low loading rate hydrogen has more time for diffusing until reaching the wire core before provoking final fracture of the specimen. This is demonstrated with the analysis of the fracture mechanisms shown in Figure 4, where it is shown that the TTS zone appears only at the sample centre in these slow CERT tests. On the contrary, in the steels tested under high loading rates the hydrogen-assisted micro-damage appears in the specimen surface, just at the notch tip (peripheral TTS) precisely where the hydrogen concentration reaches its maximum, so that hydrogen does not have enough time to diffuse towards the core region and the remaining material does not seem to be markedly affected by hydrogen ($C < C_{cr}$).

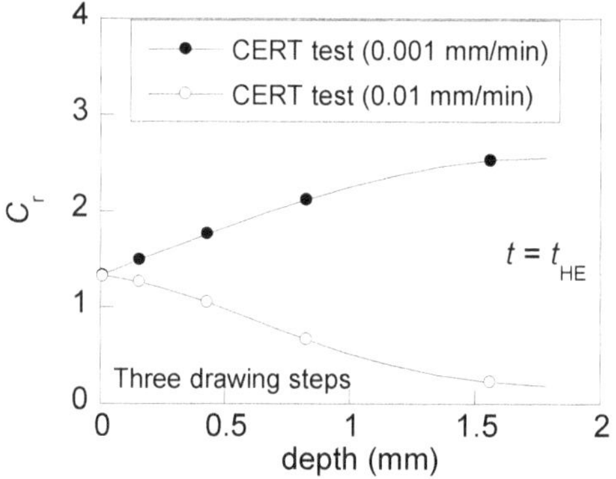

Figure 9. Radial distribution of relative hydrogen concentration for both loading rates at the instant of final failure during a CERT test ($t = t_{HE}$) of a steel undergoing three steps of cold drawing. Depth is defined in radial direction from the notch tip.

5. Conclusions

The following conclusions may be drawn on the basis of the analysis performed in this paper:

(i) The fracture surfaces in bluntly notched samples of pearlitic steel subjected to constant extension rate tensile (CERT) tests in an environment promoting hydrogen embrittlement (HE) may be classified into four schematic micro-fracture maps (MFMs).

(ii) In all MFMs the fracture process zone (FPZ) is associated either with tearing topography surface (TTS) or with an area resembling micro-void coalescence (MVC* or quasi-MVC), *i.e.*, a candidate to TTS that is not fully hydrogenated.

(iii) Cold drawing produces microstructural orientation in the pearlitic steel and therefore anisotropic fracture behavior under hydrogen environments, thus promoting axisymmetric shape of the FPZ (TTS alone or a mixture of TTS and MVC*).

(iv) For low loading rates (slow HE tests) a special MFM appears in which hydrogen-assisted fracture (HAF) initiates in the central core of the cross-sectional area of the bluntly notched specimen of pearlitic steel.

(v) For high loading rates (quick HE tests) all MFMs show experimental evidence of surface (peripherical) initiation of HAF due to the short time for hydrogen diffusion and penetration towards the inner zone.

Acknowledgments: The authors wish to acknowledge the financial support provided by the following Spanish Institutions: Ministry for Science and Technology (MCYT; Grant MAT2002-01831), Ministry for Education and Science (MEC; Grant BIA2005-08965), Ministry for Science and Innovation (MICINN; Grants BIA2008-06810 and BIA2011-27870), Junta de Castilla y León (JCyL; Grants SA067A05, SA111A07 and SA039A08), and the steel supplied by Emesa Trefilería (La Coruña, Spain).

Author Contributions: J.T. and D.V. conceived and designed the experiments; D.V. performed the experiments; D.V. and M.L. carried out the numerical simulations; J.T., D.V. and M.L. analyzed the data and wrote the paper.

Conflicts of Interest: The authors declare no conflict of interest.

References

1. Enos, D.G.; Williams, A.J.; Scully, J.R. Long-term effects of cathodic protection of prestressed concrete structures: Hydrogen embrittlement of prestressing steel. *Corrosion* **1997**, *53*, 891–908. [CrossRef]
2. Vehovar, L.; Kuhar, V.; Vehovar, A. Hydrogen-assisted stress-corrosion of prestressing wires in a motorway viaduct. *Eng. Fail. Anal.* **1998**, *5*, 21–27. [CrossRef]
3. Tkachov, V.I. Problems of hydrogen degradation of metals. *Mater. Sci.* **2000**, *36*, 481–488. [CrossRef]
4. Perrin, M.; Gaillet, L.; Tessier, C.; Idrissi, H. Hydrogen embrittlement of prestressing cables. *Corros. Sci.* **2010**, *52*, 1915–1926. [CrossRef]
5. Mallick, A.; Das, S.; Mathur, J.; Bhattacharyya, T.; Dey, A. Internal reversible hydrogen embrittlement leads to engineering failure of cold drawn wire. *Case Stud. Eng. Fail. Anal.* **2013**, *1*, 139–143. [CrossRef]
6. Toribio, J.; Kharin, V.; Vergara, D.; Lorenzo, M. Optimization of the simulation of stress-assisted hydrogen diffusion for studies of hydrogen embrittlement of notched bars. *Mater. Sci.* **2011**, *46*, 819–833. [CrossRef]
7. Toribio, J.; Elices, M. The role of local strain rate in the hydrogen embrittlement of round-notched samples. *Corros. Sci.* **1992**, *33*, 1387–1409. [CrossRef]
8. Wang, M.; Akiyama, E.; Tsuzaki, K. Crosshead speed dependence of the notch tensile strength of a high strength steel in the presence of hydrogen. *Scr. Mater.* **2005**, *53*, 713–718. [CrossRef]
9. Merson, E.D.; Krishtal, M.M.; Merson, D.L.; Eremichev, A.A.; Vinogradov, A. Effect of strain rate on acoustic emission during hydrogen assisted cracking in high carbon steel. *Mater. Sci. Eng. A* **2012**, *550*, 408–417. [CrossRef]
10. Raykar, N.R.; Raman, R.K.S.; Maiti, S.K.; Choudhary, L. Investigation of hydrogen assisted cracking of a high strength steel using circumferentially notched tensile test. *Mater. Sci. Eng. A* **2012**, *547*, 86–92. [CrossRef]
11. Toribio, J.; Vergara, D. Role of microstructural anisotropy in the hydrogen-assisted fracture of pearlitic steel notched bars. *Int. J. Fract.* **2013**, *182*, 149–156. [CrossRef]
12. Smith, E. The elastic stress distribution near a circular cylindrical notch due to external dislocations. *Int. J. Eng. Sci.* **2004**, *42*, 1841–1846. [CrossRef]
13. Sanyal, G.; Das, A.; Singh, J.B.; Chakravartty, J.K. Effect of notch geometry on fracture features. *Mater. Sci. Eng. A* **2015**, *641*, 210–214. [CrossRef]
14. Toribio, J.; Lorenzo, M.; Vergara, D.; Kharin, V. Hydrogen degradation of cold-drawn wires: A numerical analysis of drawing-induced residual stresses and strains. *Corrosion* **2011**, *67*, 075001:1–075001:8. [CrossRef]
15. Nam, W.J.; Bae, C.M. Void initiation and microstructural changes during wire drawing of pearlitic steels. *Mater. Sci. Eng. A* **1995**, *203*, 278–285. [CrossRef]
16. Zelin, M. Microstructure evolution in pearlitic steels during wire drawing. *Acta. Mater.* **2002**, *50*, 4431–4447. [CrossRef]
17. Sauvage, X.; Guelton, N.; Blavette, D. Microstructure evolutions during drawing of a pearlitic steel containing 0.7 at. % copper. *Scri. Mater.* **2002**, *46*, 459–464. [CrossRef]
18. Guo, N.; Luan, B.; Wang, B.; Liu, Q. Microstructure and texture evolution in fully pearlitic steel during wire drawing. *Sci. China Technol. Sci.* **2013**, *56*, 1139–1146. [CrossRef]
19. Guo, N.; Luan, B.; Wang, B.; Liu, Q. Deformation bands in fully pearlitic steel during wire drawing. *Sci. China Technol. Sci.* **2014**, *57*, 796–803. [CrossRef]
20. Toribio, J.; Ovejero, E. Microstructure evolution in a pearlitic steel subjected to progressive plastic deformation. *Mater. Sci. Eng. A* **1997**, *234–236*, 579–582. [CrossRef]
21. Toribio, J.; Ovejero, E. Microstructure orientation in a pearlitic steel subjected to progressive plastic deformation. *J. Mater. Sci. Lett.* **1998**, *17*, 1037–1040. [CrossRef]
22. Toribio, J.; Ovejero, E. Effect of cumulative cold drawing on the pearlite interlamellar spacing in eutectoid steel. *Scr. Mater.* **1998**, *39*, 323–328. [CrossRef]
23. Toribio, J.; Ovejero, E. Effect of cold drawing on microstructure and corrosion performance of high-strength steel. *Mech. Time-Depend. Mater.* **1998**, *1*, 307–319. [CrossRef]

24. Dua, X.S.; Caoa, W.B.; Wanga, C.D.; Lib, S.J.; Zhao, J.Y.; Sun, Y.F. Effect of microstructures and inclusions on hydrogen-induced cracking and blistering of A537 steel. *Mater. Sci. Eng. A* **2015**, *646*, 181–186. [CrossRef]
25. Inés, M.N.; Asmus, C.A.; Mansilla, G.A. Influence of total strain amplitude on hydrogen embrittlement of high strength steel. *Procedia Mater. Sci.* **2015**, *8*, 1039–1046. [CrossRef]
26. Parkins, R.N.; Elices, M.; Sánchez-Gálvez, V.; Caballero, L. Environment sensitive cracking of pre-stressing steels. *Corros. Sci.* **1982**, *22*, 379–405. [CrossRef]
27. Thompson, A.W.; Chesnutt, J.C. Identification of a fracture mode: the tearing topography surface. *Metall. Trans.* **1979**, *10A*, 1193–1196. [CrossRef]
28. Costa, J.E.; Thompson, A.W. Hydrogen cracking in nominally pearlitic 1045 steel. *Metall. Trans.* **1982**, *13A*, 1315–1318. [CrossRef]
29. Toribio, J.; Lancha, A.M.; Elices, M. The tearing topography surface as the zone associated with hydrogen embrittlement processes in pearlitic steel. *Metall. Trans.* **1992**, *23A*, 1573–1584. [CrossRef]
30. Toribio, J.; Vasseur, E. Hydrogen-assisted micro-damage evolution in pearlitic steel. *J. Mater. Sci. Lett.* **1997**, *16*, 1345–1348. [CrossRef]
31. Gamboa, E.; Atrens, A. Environmental influence on the stress corrosion cracking of rock bolts. *Eng. Fail. Anal.* **2003**, *10*, 521–558. [CrossRef]
32. Toribio, J.; Vergara, D.; Lorenzo, M. Role of microstructural anisotropy of prestressing steel on the fractographic appearance of hydrogen-assisted micro-damage. In Proceedings of the Corrosion 2011, Houston, TX, USA, 13–17 March 2011; NACE International: Houston, TX, USA; pp. 2759–2767.
33. Toribio, J. Role of hydrostatic stress in hydrogen diffusion in pearlitic steel. *J. Mater. Sci.* **1993**, *28*, 2289–2298. [CrossRef]
34. Panasyuk, V.V.; Ivanyts'kyi, Y.L.; Hembara, O.V.; Boiko, V.M. Influence of the stress-strain state on the distribution of hydrogen concentration in the process zone. *Mater. Sci.* **2014**, *50*, 315–323. [CrossRef]
35. Raykar, N.R.; Maiti, S.K.; Raman, R.K.S. Influence of hydrostatic stress distribution on the modelling of hydrogen assisted stress corrosion crack growth. *Blucher Mech. Eng. Proc.* **2014**, *1*, 1–15.
36. Toribio, J.; Kharin, V.; Vergara, D.; Lorenzo, M. Two-dimensional numerical modelling of hydrogen diffusion in metals assisted by both stress and strain. *Adv. Mater. Res.* **2010**, *138*, 117–126. [CrossRef]
37. Toribio, J.; Kharin, V.; Vergara, D.; Lorenzo, M. Hydrogen diffusion in metals assisted by stress: 2D Numercal Modelling and Analysis of Directionality. *Solid State Phenom.* **2015**, *225*, 33–38. [CrossRef]
38. Hirth, J.P. Effects of hydrogen on the properties of iron and steel. *Metall. Trans.* **1980**, *11A*, 861–890. [CrossRef]
39. Toribio, J.; Kharin, V.; Lorenzo, M.; Vergara, D. Role of drawing-induced residual stresses and strains in the hydrogen embrittlement susceptibility of prestressing steels. *Corros. Sci.* **2011**, *53*, 3346–3355. [CrossRef]
40. Lillard, R.S.; Enos, D.G.; Scully, J.R. Calcium hydroxide as a promoter of hydrogen absorption in 99.5% Fe and a fully pearlitic 0.8% C steel during electrochemical reduction of water. *Corros. Sci.* **2000**, *56*, 1119–1132. [CrossRef]
41. Toribio, J.; Elices, M. Influence of residual stresses on hydrogen embrittlement susceptibility of prestressing steels. *Inter. J. Solids Struct.* **1991**, *25*, 791–803. [CrossRef]

7

The Influence of Processing Conditions on Microchemistry and the Softening Behavior of Cold Rolled Al-Mn-Fe-Si Alloys

Ning Wang [1,2], Ke Huang [2,3], Yanjun Li [2] and Knut Marthinsen [2,*]

1. Gränges Technology AB, 612 33 Finspång, Sweden; ning.wang@granges.com
2. Department of Materials Science and Engineering, NTNU, Trondheim NO-7491, Norway; ke.huang@epfl.ch (K.H.); yanjun.li@ntnu.no (Y.L.)
3. Thermomechanical Metallurgy Laboratory—PX Group Chair, Ecole Polytechnique Fédérale de Lausanne (EPFL), CH-2002 Neuchâtel, Switzerland

* Correspondence: knut.marthinsen@ntnu.no

Academic Editor: Hugo F. Lopez

Abstract: Using different homogenization treatments, different initial microchemistry conditions in terms of solid solution levels of Mn, and number densities and sizes of constituents and dispersoids were achieved in an Al-Mn-Fe-Si model alloy. For each homogenized condition, the microchemistry and microstructure, which further change both during deformation and subsequent annealing, were quantitatively characterized. The influence of the different microchemistries, with special focus on different particle structures (constituents and dispersoids), on the softening behavior during annealing after cold rolling and the final grain structure has been systematically studied. Time-Temperature-Transformation diagrams with respect to precipitation and recrystallization as a basis for analysis of the degree of concurrent precipitation during back-annealing have been established. Densely distributed fine pre-existing dispersoids and/or conditions of significant concurrent precipitation strongly slows down recrystallization kinetics and lead to a grain structure of coarse and strongly elongated grains. At the lowest annealing temperatures, recrystallization may even be completely suppressed. In conditions of low number density and coarse pre-existing dispersoids, and limited additional concurrent precipitation, recrystallization generally results in an even, fine and equi-axed grain structure. Rough calculations of recrystallized grain size, assuming particle stimulated nucleation as the main nucleation mechanism, compare well with experimentally measured grain sizes.

Keywords: Al-Mn-alloys; homogenization; dispersoids; cold rolling; back annealing; recrystallization; concurrent precipitation; TTT-diagrams

---

1. Introduction

AA3xxx alloys that contain Mn as a main alloying element have a wide range of applications, e.g., in the building sector, in packaging industry and in equipment for heating and cooling. AA3xxx alloy sheets are typically processed via hot-rolling followed by cold rolling and back-annealing where the final annealing step gives the grain structure and texture of the sheet material before further possible forming operations (e.g., deep drawing). The microstructure and properties in this condition are a result of the whole thermo-mechanical history of the material and the corresponding closely coupled evolution in microchemistry (solid solution level of alloying elements, volume fraction and size of constituents and dispersoids), grain structure, and texture. After casting, the supersaturated Mn in solid solution will precipitate from the matrix in subsequent heat treatment processes to form different types of second-phase particles (dispersoids), which may have a significant influence on the deformation behavior and subsequent softening behavior after annealing [1–4].

Numerous investigations have been carried out related to the influence of second-phase particles, including both large constituent particles [5,6] and fine dispersoids [7–14], on the recrystallization behavior and texture in particular (e.g., [7,11,13,14]) of aluminum alloys. It is well known that large particles (larger than about 1 μm) promote recrystallization due to the activation of particle-stimulated nucleation (PSN), and that fine densely distributed dispersoids retard and may even inhibit recrystallization due to the effect of Zener pinning both of low- and high-angle grain boundary motion [5,13,15]. However, these complex interactions and their influence on the microstructure and texture evolution are still not fully understood and, in particular, adequate quantitative descriptions are still largely missing. This is a general challenge with respect to process optimization and alloy design for "tailor-made" properties of these alloys, and even more so with the introduction of more recycled aluminum where alloying elements like Mn, Fe and Si will accumulate in secondary alloys. More generally the role of dispersoids is of vital importance in the framework of recrystallization induced by severe plastic deformations and the stability of the refined micro- and nanostructure obtained by cold working, with the aim of creating stable grain refined metals with improved properties [16,17].

A particular focus of the current study has been to investigate how different levels of Mn in solid solution (and thus the potential for concurrent precipitation) and/or how different pre-existing particle structures (in terms of size and the number density of constituents and dispersoids) influence the deformation and subsequent back-annealing behavior. While some related papers have given special attention to the grain structure and texture evolution during back-annealing [18–20], the present paper provides a more detailed characterization of the microchemistry and microstructural evolution during homogenization and deformation as well as a quantitative analysis of the final recrystallized grain structures. The key to understand the recovery and recrystallization behavior, is the prior sub-structure and microchemistry state in terms of constituents particles, solid solution levels and dispersoids. A detailed characterization of the as-homogenized and deformed (cold-rolled) states is therefore included. To discuss the degree of concurrent precipitation and its influence on the back-annealing behavior, Time-Temperature-Transformation (TTT) diagrams with respect to precipitation and recrystallization have been established. The precipitation behavior during back-annealing of selected conditions is followed in detail for a more quantitative analysis of the dispersoid effects. The recrystallized grain sizes are discussed in view of model concepts assuming particle stimulated nucleation of recrystallization to be the main nucleation mechanisms and estimates for the stored energy (driving force for recrystallization) and Zener drag obtained from the experiments.

2. Experimental Section

2.1. Material Preparation and Thermomechanical Treatments

The material investigated in this work was a model direct chill (DC) cast aluminum extrusion billet with a diameter of 228 mm and 1 m in length supplied by Hydro Aluminum (Sunndalsøra, Norway). The chemical composition of the alloy was (in wt. %): Mn 0.97, Fe 0.53, and Si 0.15. Samples were machined from the central region of the as-cast billet. In addition to the as-cast state, three distinctively different homogenization procedures (see Table 1 for details) were used in order to achieve different levels of Mn in solid solution and obtain different volume fractions and size distributions of constituent particles and dispersoids. The different material conditions were labeled C2-X (X = 0–3), to be consistent with related work on the same alloys [18–20], as shown in Table 1. The different dispersoid density levels indicated in Table 1 will be confirmed and discussed later in the paper. The homogenization treatments were performed in an air circulation furnace with a temperature accuracy of ±2 °C, starting from room temperature (about 20 °C) with a heating rate of 50 °C/h, followed by different annealing schedules as specified in Table 1. Materials were water quenched to room temperature at the end of the homogenization procedure.

Table 1. Four homogenization procedures and resulting different concentration levels of Mn in solid solution.

Sample	Mn_{ss} (wt. %)	Dispersoids Density	Homogenization Procedure
C2-0	0.69	-	As-cast condition
C2-1	0.35	Low	50 °C/h up to 600 °C + 24 h@600 °C + quenching
C2-2	0.23	High	50°C/h up to 450 °C + 4 h@450 °C + quenching
C2-3	0.21	Medium	50 °C/h up to 600 °C + 4 h@600 °C + 25 °C/h down to 500 °C + 4 h@500 °C + quenching

Cold rolling was performed in order to obtain a deformed material for subsequent investigation of softening behavior of materials in different homogenized states. Heavily lubricated rolls and maximum roll velocity, which minimizes the shear zone in the surface region of the sheet, were used in order to obtain a microstructure similar to industrial processed materials. The homogenized materials, in the form of cuboid pieces of length 200 mm, width 80 mm and an initial thickness of 15 mm and 30 mm were cold rolled to a thickness of about 1.5 mm, *i.e.*, thickness reductions of 90% and 95%, corresponding to true strains of $\varepsilon = 1.6$ and 3.0, respectively. The rolled sheets were subsequently isothermally back-annealed in a salt bath at different temperatures in the range 300–500 °C and times in the range 5–10^5 s, followed by water quenching to room temperature. The specimens were stirred after they were immersed in the salt bath in order to minimize the heating time, an aspect that is especially important for short annealing times and for thick specimens [21]. The softening and precipitation behavior during annealing were followed by Vickers hardness (VHN) and electrical conductivity (EC) measurements performed on the RD-TD (RD: rolling direction; TD: transverse direction) plane of sheets [18,19]. Following the procedure used in [13], TTT-diagrams with respect to precipitation and recrystallization were established, as a basis for analyzing the degree of recrystallization and concurrent precipitation, where a 25% drop in hardness from the deformed condition to the fully recrystallized condition was used to indicate the onset of recrystallization and a 2.5% increase in electrical conductivity was defined as the start of precipitation, assuming changes of Mn in solid solution (and thus precipitation of Mn-containing dispersoids) to be the main contribution to changes in EC [13,19].

2.2. Microstructure Characterization

The microstructure of the samples was analyzed in a Zeiss Ultra 55-Limited Edition field emission scanning electron microscope (FESEM, Oberkochen, Germany), equipped with an energy dispersive spectrometer (EDS, Bruker, Berlin, Germany) for chemical composition analysis, a back-scattered electron detector (BSE, Zeiss, Germany) for the sub-grain structure and particle characterization, and electron back-scattered detector (EBSD, Nordif, Trondheim, Norway) for orientation imaging mapping (OIM). Characteristic size parameters of primary particles as measured in two-dimensional (2D) cross-sections, including circle equivalent diameter, mean diameter, area fraction and number density, were measured by the image analysis software IMAGE-PRO (Media Cybernetics, Rockville, MD, US). To transpose the measured 2D distributions into a 3D-distribution a Johnson-Saltykov analysis has been used [22–24]. The basis for this analysis is that the particles are assumed to belong to different size classes, and where these size classes scale with the logarithm of the circle equivalent diameter, $d_k = 10^{0.1k}$. For simplicity the particles in the present work are all assumed spherical. Corrections for non-spherical shaped particles and a finite information depth may be performed [23], although such corrections have not been included in the present work.

Both light optical microscopy (LOM, Leica Microsystems, Wetzlar, Germany) and SEM-EBSD orientation mapping was used to image the grain structure and measure grain sizes of specimens in different conditions. Grain size was measured by using the line intercept method. The measurements were performed both in the rolling direction (RD) and the normal direction (ND) and grain size was determined by the average of more than 300 grains.

Vickers hardness was measured with a Matsuzawa DVK-1S hardness tester (Toshima, Tokyo, japan) with a load of 1 kg, loading speed 100 μm/s and 15 s loading time. The measurements were performed by mounting the contact probe on a clean, planar sample surface, after grinding on a SiC paper with a surface finish of 1200 Mesh. The hardness was the average of at least 5 measurements on the rolled surfaces. Electrical conductivity was measured by a Sigmascope EX 8 (Sindelfingen, Germany) at room temperature of about 293 K (20 °C). The electrical conductivity values were averaged by 3 measurements on each sample. An accuracy of ±0.05 MS/m can be achieved by the experimental setup.

3. Results

3.1. Initial Microstructure before Rolling

The manganese levels in solid solution, Mn_{SS}, of the different initial conditions, *i.e.*, the as-cast state and after the three different homogenization procedures were estimated based on Thermo-Electrical Power (TEP) measurements (see e.g., [25] for details) and the results are shown in Table 1 above. The Mn content in solid solution in the as-cast condition (C2-0) is 0.69%. As compared to the non-homogenized condition C2-0, it can be seen that the Mn content following the three homogenization procedures has been much reduced, and in particular for the two last variants (C2-2 and C2-3). Although quite different, the two latter homogenization procedures have quite efficiently promoted the decomposition of the Mn supersaturated solid solution from the as-cast condition, where the C2-2 condition (heating 50 °C/h to 450 °C + 4 h@450 °C) has only slightly higher Mn content in solid solution than the second homogenization procedure involving longer holding times at higher temperatures (C2-3; involving 4 h@600 °C + 25 °C/h down to 500 °C + 4 h@500 °C + quenching). The reduction of Mn in solid solution can mainly be attributed to precipitation of Mn-bearing dispersoids, and also to the growth of Mn-containing constituent particles [26,27]. The low value of Mn remaining in solid solution in the two last variants indicates that these alloy conditions have a limited potential for concurrent precipitation during a subsequent annealing stage. However, the different dispersoid structures (see details below) in terms of the size and number density will still have a very different influence on the softening behavior, as will be demonstrated later in the paper.

SEM-BSE micrographs of the particle structures of the four conditions after homogenization are illustrated in Figure 1. The solidification microstructure (as-cast) is shown in Figure 1a. The eutectic primary particles with rod and plate like shape are distributed in the interdendritic regions and at grain boundaries. The area fraction of the primary particles is measured to be 2.1%, and mean diameter of primary particles is estimated to be 0.8 μm. The microstructure of the C2-1 material after homogenization (24 h@600 °C) is shown in Figure 1b. Most of the eutectic networks are broken up, and primary particles have become coarsened and spheroidized. The area fraction and the mean diameter of the primary particles have increased, and are measured to be 2.9% and 1.5 μm, respectively. Furthermore, only a few small precipitated dispersoids are seen to be distributed within the interdendritic regions of the homogenized sample. The number density is measured to be 0.17×10^6 ($1/mm^2$), and the mean diameter is 170 nm. These observations indicate that the decreased content of Mn in solid solution in the C2-1 material mainly contributes to the growth of the primary particles, and only to a small amount to the precipitation of dispersoids.

For the C2-2 material which is homogenized at a lower temperature (450 °C), there is only a slight coarsening of the constituent particles as compared to the as-cast state due to the low diffusion rate of Mn in the Al matrix [28]. Moreover, most of the eutectic constituent particles keep their rod-like or plate-like shapes, as in the as-cast condition. The area fraction of constituent particles in C2-2 is measured to be 1.9%, and mean diameter of the constituent particles is measured to be 1.0 μm. During heating and holding at 450 °C for the C2-2 sample, the decomposition of the supersaturated solid solution is mainly through precipitation of dispersoids. For the more elaborate homogenization of the C2-3 material, the area fraction of constituent particles is measured to be 2.7%, and mean diameter of constituent particles is measured to be 1.5 μm. The fact that the area fraction of constituent particles in the C2-3 condition is larger than that of the C2-2 sample is due to coarsening of the large constituent particles on behalf of the fine dispersoids, during heating and holding at the higher temperature (>530 °C), which makes the constituent particles growing. In addition, for the C2-3 condition, the homogenization procedure, which combines both a high and medium holding temperature, also results in general growth of large constituent particles (600 °C) as well as coarsening of existing dispersoids (at 500 °C).

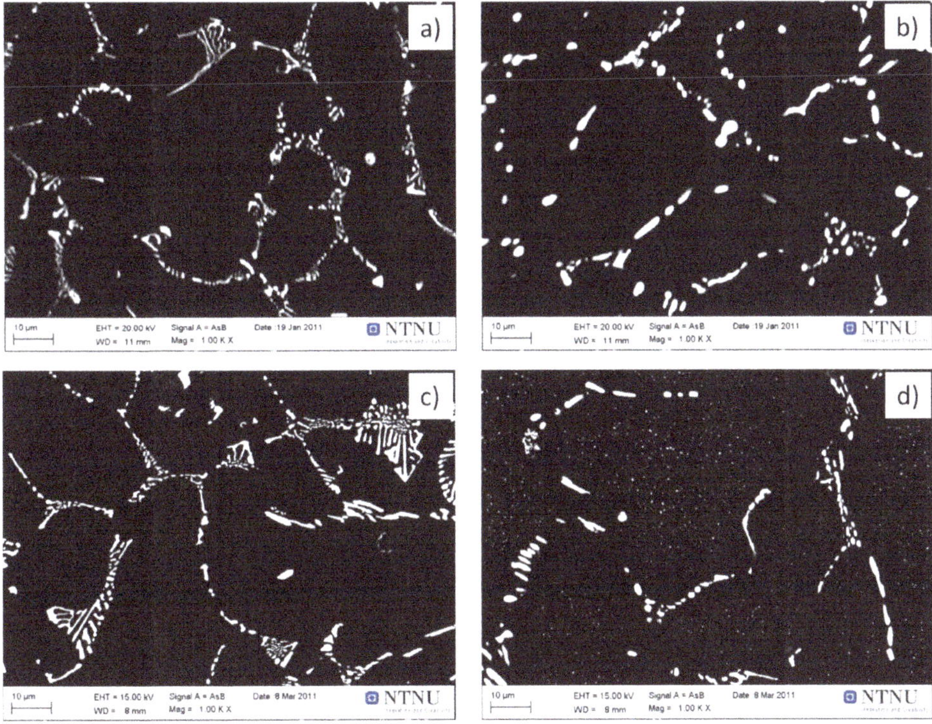

Figure 1. Back scatter electron scanning electron microscopy (SEM) images of constituent particles after the different homogenization procedures: (**a**) C2-0 (non-homogenized); (**b**) C2-1; (**c**) C2-2; and (**d**) C2-3.

In order to show more clearly the morphology and distribution of the dispersoids precipitated after homogenization in the C2-2 and C2-3 samples, high magnification electron backscattered images were recorded, as illustrated in Figure 2, where a significant difference in dispersoid structures can be seen, in terms of number density and size. The diameter of the dispersoids in C2-2 and C2-3 were quantitatively measured and the size distributions are shown in Figure 3 (the fully drawn lines represent smooth curves fitted to the experimental values). The number density and mean diameter of dispersoids are listed in Table 2, together with corresponding values for the C2-1 alloy. The chemical compositions of the dispersoids measured by SEM-EDS show that the dispersoids are mainly of α-Al(Mn,Fe)Si type. C2-2 contains a large percentage of small dispersoids with a maximum in the size distribution in the range 100–200 nm. No dispersoids larger than 400 nm could be detected. The number density is measured to be 2.8×10^6 mm^{-2} and the mean diameter estimated to be 105 nm. C2-3 contains more large particles with the peak position shifted to ~230 nm along with a broader distribution and a lower peak value. The number density is measured to be 0.9×10^6 mm^{-2} and mean diameter estimated to be 156 nm. A large part of the dispersoids precipitated during heating has dissolved during holding at 600 °C for 4 h. The solid solution supersaturated with Mn decomposes further during cooling from 600 °C to 500 °C, however, there is limited precipitation of new dispersoids, and the result is mainly further growth of the pre-precipitated dispersoids and large constituent particles [26,27]. Nucleation of dispersoids in 3xxx alloys mainly occurs in the low temperature range (300–450 °C) during heating, while at higher temperatures the evolution of dispersoids is mainly controlled by coarsening and dissolution [26,27].

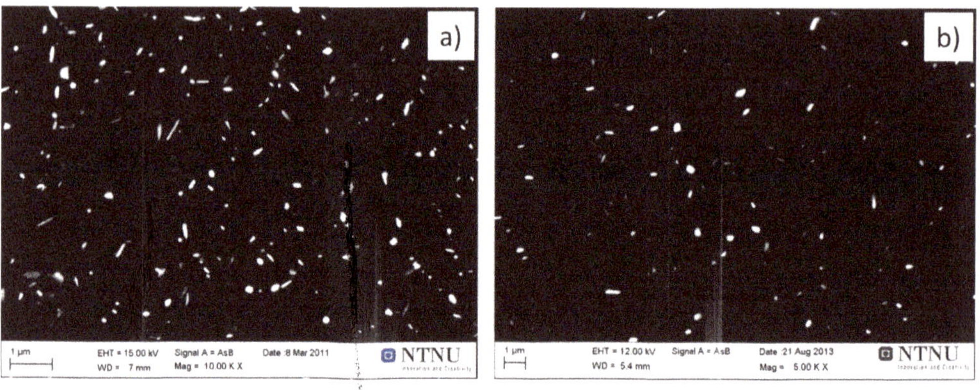

Figure 2. Back scatter electron SEM images of dispersoids after homogenization: (**a**) C2-2; and (**b**) C2-3.

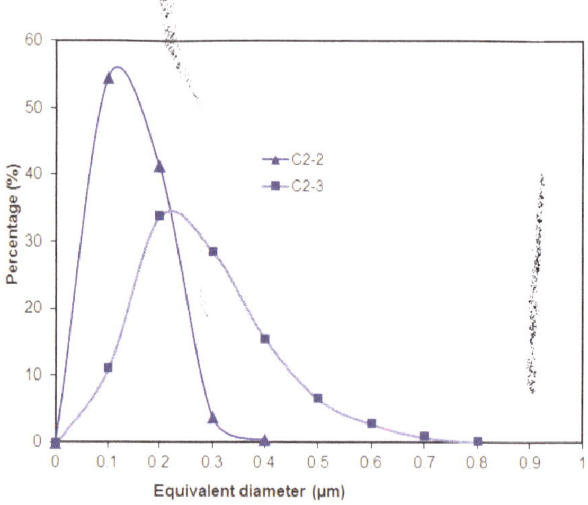

Figure 3. Distribution of mean diameter of precipitated particles after homogenization in the C2-2 and C2-3 materials.

Table 2. The number density and mean diameter of dispersoids in materials in the as-homogenized conditions.

Sample	Number Density (mm^{-2})	Mean Diameter (nm)
C2-1	0.17×10^6	170
C2-2	2.8×10^6	105
C2-3	0.9×10^6	156

3.2. Deformation Microstructure

Characteristics of the deformed state may have a significant influence on the subsequent softening behavior. Therefore, a thorough characterization of the deformation structure has been carried out, including the evolution in primary particle structures with increasing cold deformation. The morphology of constituent particles of the four material condition imaged by SEM-BSE after cold rolling to a strain of $\varepsilon = 1.6$ are illustrated in Figure 4. As compared to the as-homogenized conditions the network of constituent particles, after deformation, has been compressed and some broken up into smaller particles with rod and plate like shape and distributed more densely and uniformly in the matrix. Moreover there is a clear tendency of the particles to be rotated and aligned into strings along the RD direction. There are also some distinct differences between the different conditions. Both the non-homogenized material, C2-0, and the C2-2 and C2-3 conditions, have a higher number density of particles after deformation as compared with the C2-1 condition. For the latter condition the particles are coarser and the number density is smaller. The shape and size of the primary particles after deformation is mainly inherited from the as-cast and as-homogenized conditions, e.g., in the C2-1 condition the particles are generally larger and have a more rounded shape than those in the other conditions, in particular C2-0 and C2-2.

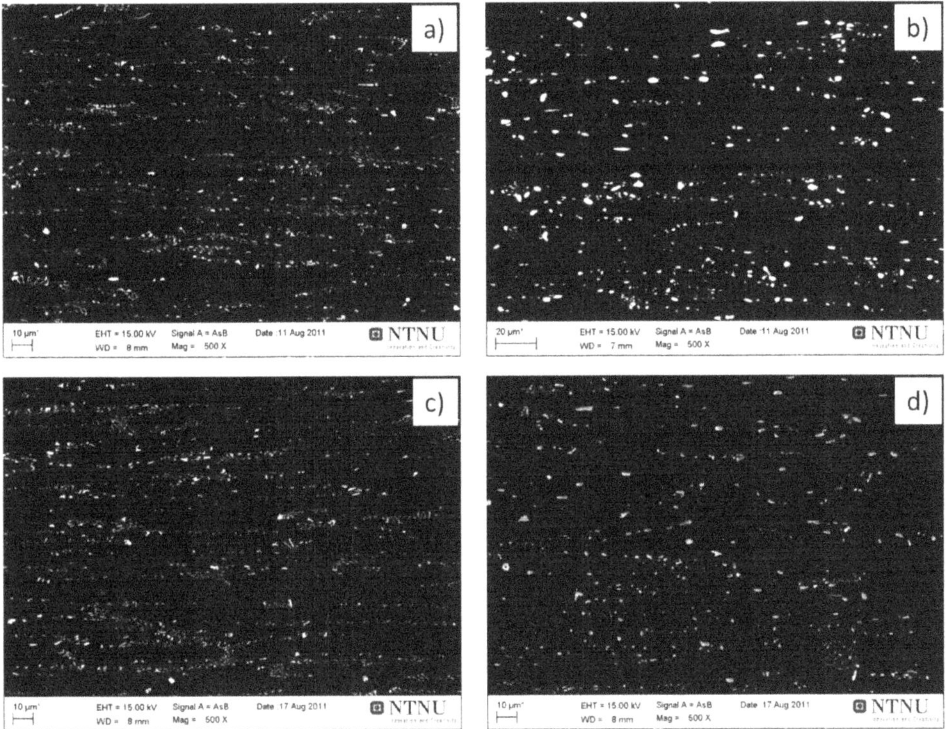

Figure 4. Back scatter electron SEM images of constituent particles after cold rolling strain of $\varepsilon = 1.6$: (**a**) C2-0; (**b**) C2-1; (**c**) C2-2; and (**d**) C2-3. RD-ND (vertical) section.

The sizes of the particles in the cold-rolled conditions were also measured quantitatively in terms of 2D size-distributions. Following a Johnson-Saltykov analysis [22,23], they are presented as 3D cumulative size distributions for the different materials before deformation (as-homogenized) and after true strains of $\varepsilon = 1.6$ and 3.0 in Figure 5. All the 3D cumulative size distributions of constituent particles are fitted to an equation of the following form [29]:

$$F(\eta) = N_0 \exp(-L \cdot \eta) \quad (1)$$

where η is the volume equivalent particle diameter, and N_0 and L are characteristic size distribution parameters. The resulting values for N_0 and L in the different conditions are listed in Table 3.

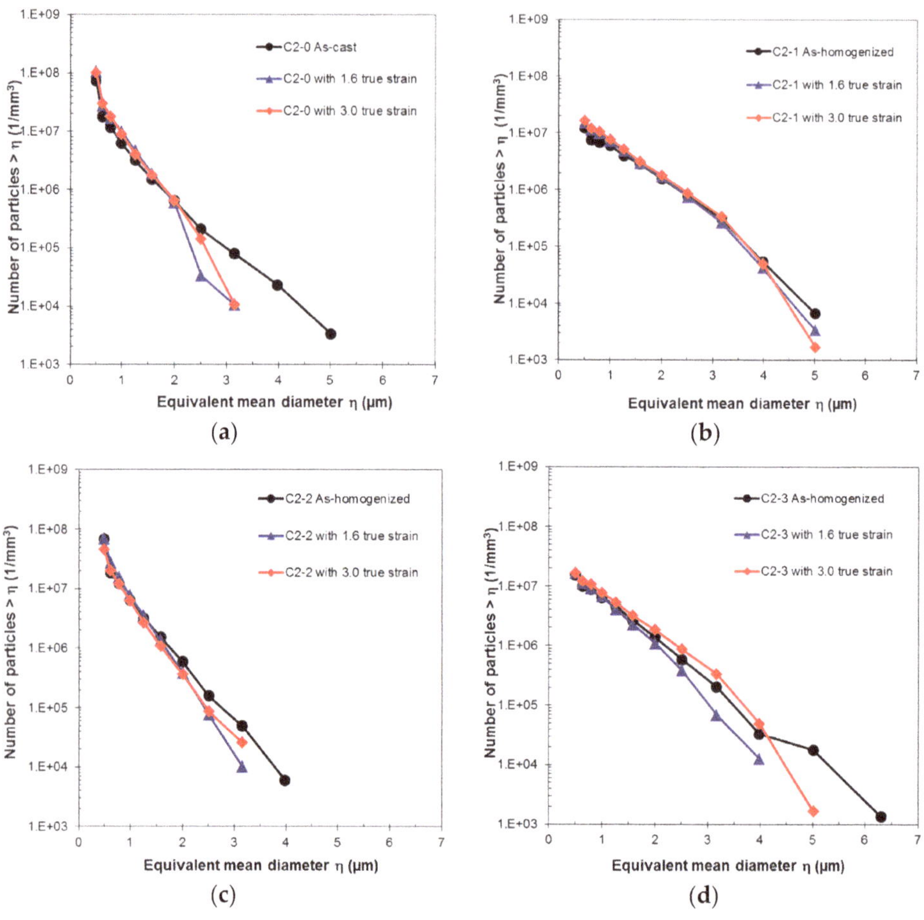

Figure 5. 3D cumulative size distribution of constituent particles for the different material conditions before and after rolling to strains of $\varepsilon = 1.6$ and 3.0: (**a**) C2-0; (**b**) C2-1; (**c**) C2-2; and (**d**) C2-3.

Table 3. The values of the constituent size distribution parameters, N_0 and L, for the different material conditions (*cf.* Figure 5).

Materials	True Strain ε	N_0	L (μm^{-1})
C2-0	0 (as-cast)	1.13×10^8	2.54
	1.6	3.24×10^8	3.37
	3.0	2.67×10^8	3.14
C2-1	0 (as-homogenized)	2.91×10^7	1.57
	1.6	4.55×10^7	1.78
	3.0	5.64×10^7	1.87
C2-2	0 (as-homogenized)	9.72×10^7	2.48
	1.6	2.21×10^8	3.19
	3.0	1.12×10^8	2.78
C2-3	0 (as-homogenized)	3.10×10^7	1.58
	1.6	4.97×10^7	2.04
	3.0	5.00×10^7	2.02

As compared to the as-homogenized condition, the as-deformed samples all contain fewer particles with size larger than 1 μm. This can be attributed to particle break-up taking place during deformation which changes the size distribution and increases the number density of particles smaller than 1 μm at the cost of the larger ones. The effect of break-up is best seen for the largest particles, *i.e.*, generally for those larger than ~3–4 μm. The size distributions in Figure 5 clearly confirm the qualitative considerations of the particle size differences between the different conditions made above with reference to the micrographs in Figure 4. The C2-0 material contains a considerably larger number density of particles with size smaller than 1 μm as compared to both the C2-1 and C2-3 materials. Moreover, no particles with equivalent diameter larger than 4 μm were found in the C2-0 material. The same applies for the C2-2 materials and for both more than 90% of the particles at a strain of $\varepsilon = 3$ are less than 1 μm. The C2-1 material on the other hand contains more coarse particles, and the equivalent diameter of the largest particles was measured to be ~5 μm. In this material, more than 30% of the particles at a strain of $\varepsilon = 3$ are larger than 1 μm. For C2-3 almost 20% of the particles at a strain of $\varepsilon = 3$ are larger than 1 μm. These large differences of the particle structure do lead to differences in the PSN activity during subsequent annealing and softening, as discussed later in the paper.

In order to show the influence of different microchemistry (in terms of constituents and dispersoid structures) on the sub-structure evolution during deformation, the sub-grain size after the different cold-rolling strains of the different material variants were measured. The sub-grain sizes were measured from SEM-BSE images (RD-ND section; *cf.* Figure 6a,b below, unrecrystallized) using the linear intercept method in the length direction (RD; δ_{length}) and normal direction (ND; δ_{width}), and the results are presented in Table 4. The sub-grain sizes typically vary between ~1.0 μm for the lowest strain in RD direction to less than 0.3 μm for the largest strain in the ND-direction. As expected, the sub-grain sizes are decreasing with increasing strain. Except for the C2-1 alloy, where the sub-grain sizes are generally larger, there is no significant influence of the different particle structures on the sub-grain structures. The sub-grain sizes in the deformed condition are of importance during the subsequent annealing treatment as they are the main factor that determines the stored energy of the as-deformed conditions, *i.e.*, the driving pressure for recovery and recrystallization.

Table 4. Sub-grain sizes measured from SEM-BSE micrographs, in terms of length and width, for the different material variants after strains of $\varepsilon = 1.6$ and 3.0.

Materials	True Strain ε	δ_{length} (μm)	δ_{width} (μm)
C2-0	1.6	0.92	0.46
	3.0	0.58	0.24
C2-1	1.6	1.37	0.59
	3.0	0.98	0.32
C2-2	1.6	0.85	0.31
	3.0	0.67	0.25
C2-3	1.6	0.92	0.42
	3.0	0.72	0.39

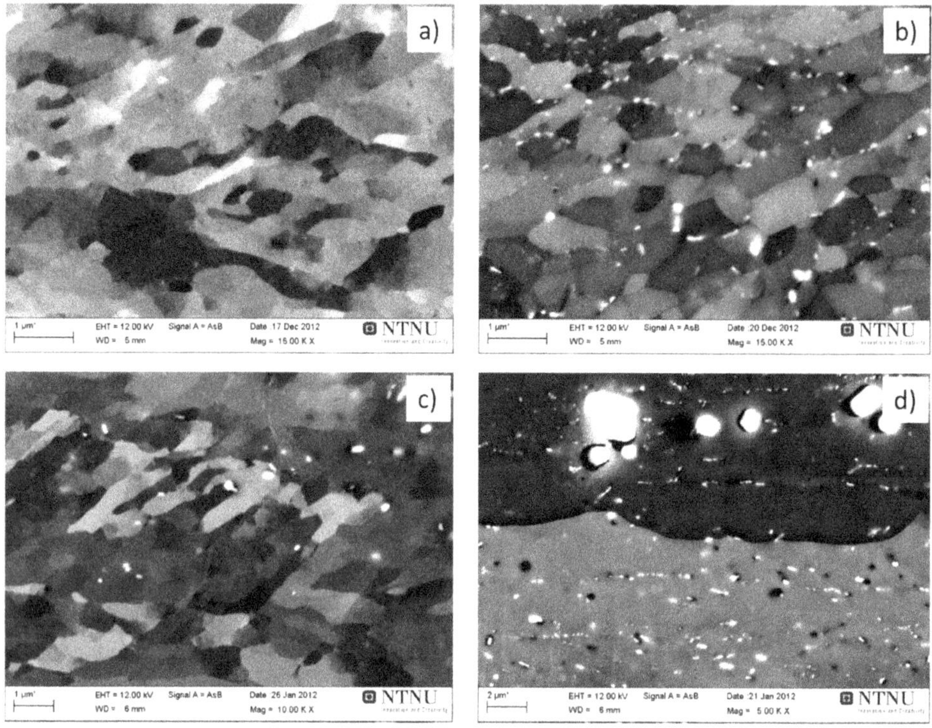

Figure 6. SEM-BSE images exemplifying of the microstructure and dispersoid structure evolution during different annealing times at 350 °C for the C2-0 and C2-1 materials, respectively, after strain of ε = 1.6: (**a**) C2-0, annealing of 5 s; (**b**) C2-0, annealing of 10^4 s; (**c**) C2-1, annealing of 5 s; and (**d**) * C2-1, annealing of 10^4 s. * [30] Copyright 2012 by The Minerals, Metals and Materials Society, reprinted with permission.

3.3. *Softening Behavior*

The softening behavior of the differently homogenized materials (*i.e.*, with different microchemistry) after a strain of 1.6, in view of Vickers hardness (VHN) and electrical conductivity (EC) are shown in Figure 7a–d, where the behavior at four different annealing temperatures are compared, *i.e.*, 350 °C, 400 °C, 450 °C and 500 °C. The corresponding TTT-diagrams are shown in Figure 7e–h.

As shown, the softening behavior (kinetics) is very different for the different materials and the temperature dependence for each of them is also very different. It is observed that both for the C2-0 variant and the C2-2 material the kinetics is quite sluggish except for at the highest annealing temperature. At the two lowest annealing temperatures, 350 and 400 °C, neither material fully recrystallizes even within 10^5 s of annealing. This has been confirmed (not shown) by light optical microscopy which show only partial recrystallization even after this long annealing times. For the C2-0 alloy the slow kinetics is connected with quite considerable concurrent precipitation of Mn (into Mn-containing dispersoids) as indicated by the considerable increase in EC during annealing and also illustrated by the TTT diagram in Figure 7b. Only at the highest temperature (500 °C) recrystallization is so fast that the influence of precipitation is limited. For the C2-2 material the very slow kinetics, especially at the two lowest annealing temperatures, are associated with a considerable amount of pre-existing fine dispersoids. However, although less than for C2-0, some additional (concurrent) precipitation is also observed for this material (note the scale difference), and which at the two lowest temperatures mainly take place prior to or simultaneous with the recrystallization reaction.

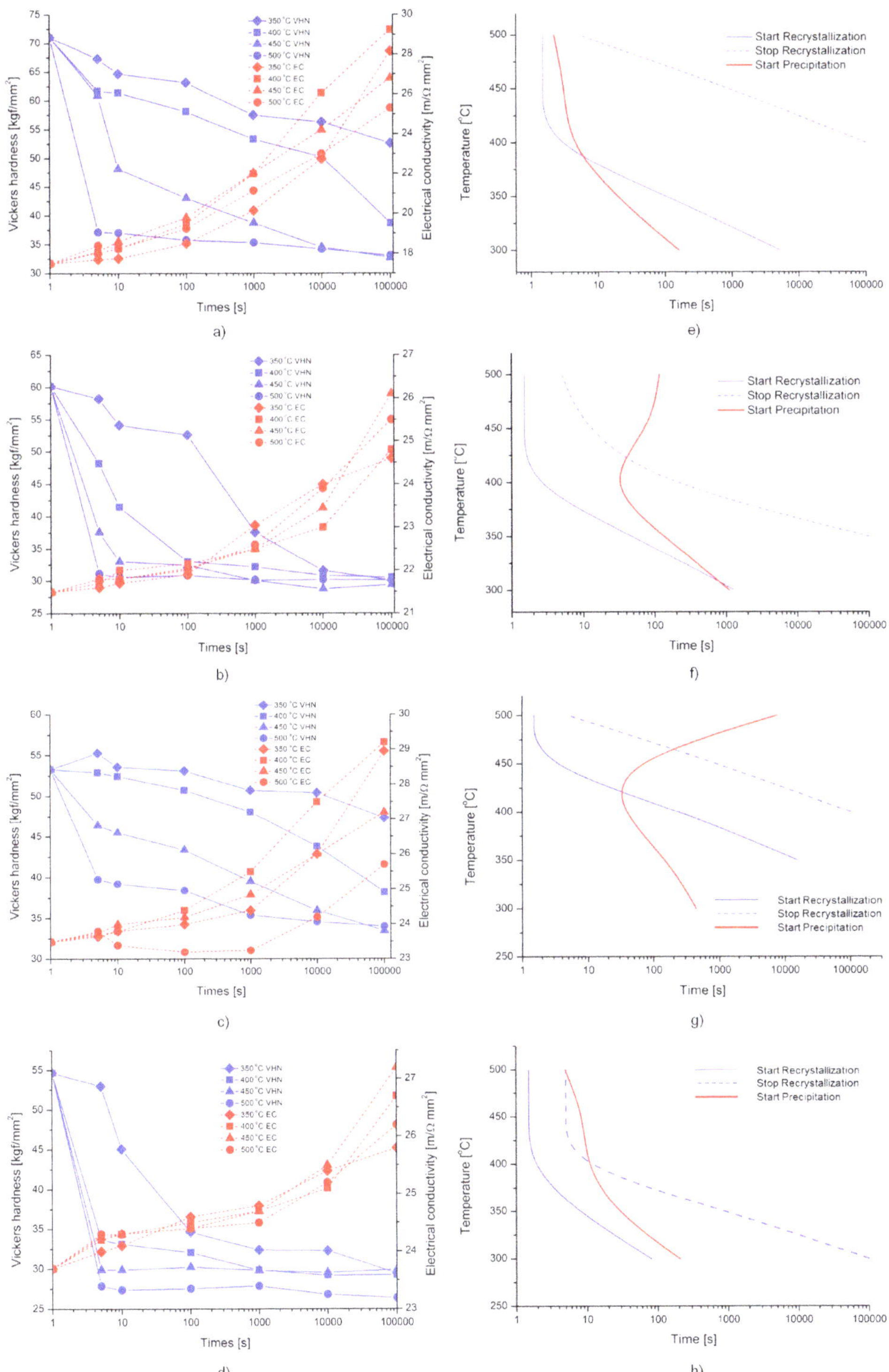

Figure 7. (**a**) Vickers hardness and electrical conductivity (EC) for the different materials material with the true strain of ε = 1.6 during isothermal back annealing at 350 °C, 400 °C, 450 °C and 500 °C for different times (**a**–**d**) and TTT-diagrams (**e**–**g**) for: C2-0 (**a**,**e**); C2-1 (**b** *,**f**); C2-2 (**c** *,**g**); and C2-3 (**d**,**h**). * [30] Copyright 2012 by The Minerals, Metals and Materials Society, reprinted with permission.

For the C2-1 material the softening reaction is fairly fast for the three higher temperatures, fully recrystallized conditions are reached within 100 s (confirmed by LOM) even at 400 °C. For the lowest temperature (350 °C), however, the softening is strongly delayed and the material is barely recrystallized even after 10^4 s of annealing. As for the C2-0 alloy this is related to considerable concurrent precipitation at this temperature, while at 400 °C and above, recrystallization is mainly finished before any significant precipitation takes place (Figure 7f). For the C2-3 alloy, kinetics is generally faster, and only at the lowest temperature any influence of precipitation can be observed (Figure 7c,g). As compared to the other alloy conditions, precipitation in this alloy is both slower and much less, with limited influence on the recrystallization behavior.

The softening behavior upon annealing of the same four differently homogenized variants after a prior rolling strain of $\varepsilon = 3.0$ has also been monitored. However, the qualitative differences between the different material conditions (and temperatures) follow the same pattern as for the same materials after a strain of $\varepsilon = 1.6$ (Figure 7). The corresponding softening curves, the evolution in electrical conductivity (as a measure of precipitation) and associated TTT-diagrams are therefore not shown here, but are included in the Figure S1.

The measured grain sizes in the fully recrystallized conditions (when achieved), of the different materials are summarized in Table 5. Although very different for the different materials, the general trend is that the grain size decreases with increasing annealing temperature, while the effect of increasing rolling strain is not equally obvious. However, the apparent increase in grain size with strain for some conditions is most probably due to the difference in concurrent precipitation where generally stronger concurrent precipitation is observed with larger cold deformation. Also poor statistics/measuring uncertainty might have contributed. The most obvious and pronounced differences are reflecting the differences in microchemistry and in particular the influence of fine dispersoids on the recrystallization behavior, whether pre-existing (as for C2-2) and/or as a result of significant concurrent precipitation. In both cases highly elongated coarse grain structures result.

Table 5. Recrystallized grain sizes of the different material conditions cold rolled to strains of $\varepsilon = 1.6$ and 3.0, respectively, after annealing at different temperatures for 10^5 s.

Alloys	True Strain ε	Recrystallized Grain Size in RD × ND Direction (μm)			
		350 °C	400 °C	450 °C	500 °C
C2-0	1.6	-	-	132 × 33	37 × 18
	3.0	-	285 × 37	81 × 21	52 × 19
C2-1	1.6	40 × 20	19 × 16	22 × 17	21 × 16
	3.0	86 × 25	27 × 15	18 × 15	21 × 16
C2-2	1.6	-	119 × 36	114 × 40	86 × 29
	3.0	-	182 × 30	156 × 30	109 × 29
C2-3	1.6	27 × 18	25 × 16	29 × 19	26 × 19
	3.0	26 × 16	21 × 13	27 × 17	21 × 17

3.4. Precipitation Behavior

To more closely correlate the microchemistry state and in particular quantify the influence of dispersoids and concurrent precipitation (*i.e.*, the Zener pressure) on the microstructure evolution during annealing and the associated softening behavior, one or two (for C2-0) annealing conditions for each of the four homogenization variants have been chosen for a further analysis in view of their precipitation behavior, *i.e.*, all materials after deformation to a strain of $\varepsilon = 1.6$ and subsequent annealing at 350 °C, 400 °C , and 450 °C. The number density of dispersoids and their mean diameter as a function of annealing time have been measured, and the results are presented in Table 6. The measurements were carried out by a manual 2D-counting procedure based on high magnification BSE-micrographs. Areas without constituent particles were selected to perform these measurements, and the measured data refer to selected areas of the samples which may be characterized as areas of "maximum density" of dispersoids.

Table 6. Dispersoid number density and mean diameter as a function of annealing time measured for the different alloy conditions at selected annealing temperatures in the range 350 °C to 450 °C after cold rolling to strain of ε = 1.6.

Material	Back Annealing Time (s)						
	0	5	10	10^2	10^3	10^4	10^5
	Number Density × 10^4 (mm^{-2})						
C2-0 (350 °C)	-	-	59	185	597	809	1247
C2-0 (450 °C)	-	171	208	189	421	577	688
C2-1 (350 °C)	26	29	30	37	75	82	64
C2-2 (400 °C)	177	181	206	216	302	287	178
C2-3 (350 °C)	73	75	72	76	74	78	55
	Mean Diameter, *d* [nm]						
C2-0 (350 °C)	-	-	38	34	37	54	56
C2-0 (450 °C)	-	39	41	34	57	55	61
C2-1 (350 °C)	145	164	168	171	142	154	177
C2-2 (400 °C)	90	98	94	106	108	114	148
C2-3 (350 °C)	125	128	147	139	128	137	151

The dispersoid structure evolution for the differently homogenized structures is largely consistent with the evolution in electrical conductivity as presented in Figure 7. The C2-0 material starts out with no dispersoids, however, after 10 s of annealing a measurable number density of fairly small dispersoids (diameter of about 40 nm) is obtained, which increases with a factor of 20× during the annealing period, while the mean size increases only moderately. For the C2-1 material a certain number density of much coarser dispersoids (~150 nm) is present already in the deformed state (formed during homogenization). During annealing, the number density clearly increases, although much less than for C2-0 (factor of 3), while the mean size also in this case increases only moderately.

The different dispersoid structures of the C2-0 and C2-1 material are exemplified in Figure 6, where the microstructures as observed by SEM-BSE micrographs after 5 s and 10^4 s of annealing are shown. As seen no dispersoids can be observed in the C2-0 material after only 5 s of annealing while a high density in present in the one annealed for 10^4 s. We also see that even at this long annealing time this is still mainly an unrecrystallized sub-grain structure (which has only experienced some recovery/sub-grain growth from the initially deformed state), consistent with previous results (Figure 7). As observed the high density of small dispersoids decorate the sub-grain boundaries which thus are pinned by these, preventing recrystallization. In C2-1 on the other hand a few coarser dispersoids are present already after 5 s of annealing, and some more after annealing for 10^4 s (Figure 6d) which are mainly observed inside large recrystallized grains (note the differences in scale bar on the different micrographs). The distinctively larger particles at the upper half in Figure 6d are primary (constituent) particles.

For the C2-2 material, a fairly high density of medium sized dispersoids (~100 nm) is present already at the beginning the annealing process. During the first 1000 s of annealing, the number density distinctively increases, while there are only marginal changes in the mean diameter. This observation indicates that some new dispersoids precipitate from the matrix, which is also in accordance with an increase in EC during this stage as demonstrated by the EC evolution in Figure 7a. After longer annealing times, the number density decreases again and the mean diameter more clearly increases, *i.e.*, mainly coarsening of the dispersoid structure takes place. A similar observation is also made for the C2-1 material at the longest annealing time. The "removal" of Mn from solid solution is still strong during this period, as seen from the increase in EC in Figure 7, so Mn in solid solution mainly diffuses into the pre-existing dispersoids and contributes to their growth.

For the C2-3 material a considerable amount of coarse dispersoids is present from the beginning of annealing and they remain mainly constant both in number density and mean size throughout annealing. These dispersoid number density and size values are comparable to those of the C2-1 material. The somewhat stronger influence on the softening behavior of the latter may indicate that the new dispersoids forming during the initial stages of annealing of the C2-1 material more effectively prevent and/or delay recrystallization than those already present in the deformed structure of the C2-3 material.

4. Discussion

This paper was mainly aimed to investigate the effects of different microchemistries (constituents, solid solution levels and dispersoids) on the substructure evolution during rolling to different strains and the subsequent softening behavior during back-annealing. Particular focus has been given to the influence of different dispersoid structures, whether pre-existing or mainly formed during back-annealing, in terms of the number density and size, on softening kinetics and final recrystallized grain size.

At low to medium temperatures of homogenization, the precipitation of dispersoids is controlled by nucleation and growth due to the relatively low diffusion rate of Mn in Al. At a homogenization temperature of 450 °C, the number density of dispersoids can reach a maximum and the spatial distance of dispersoids reaches a minimum. When increasing the temperature further, the diffusion speed of Mn in the matrix is increased significantly, and for annealing at 600 °C, dissolution of small and coarsening of large dispersoids as well as constituent particles becomes the prominent process due to the increased solubility and the fact that long distance diffusion becomes possible. This explains the difference in solid solution level and dispersoid structures of the two conditions C2-1 and C2-2. The slow and elaborate cooling procedure of condition C2-3 ensures additional precipitation of Mn into constituents and dispersoids while preserving a fairly coarse dispersoid structure and bringing the solid solution level even below the one in C2-2 (*cf.* Table 1).

In terms of constituents both the C2-1 and C2-3 conditions contain fewer and larger particles, than the non-homogenized condition, due to coarsening during homogenization. The C2-2 material, which was homogenized at a relatively low holding temperature (450 °C), contains smaller, but considerably more particles, due to less coarsening. Moreover, in the C2-2 material the constituent particles are mainly thin plates which can break up more easily during deformation, which increases the number density and decreases the size of constituent particles during rolling. On the other side, both C2-1 and C2-3 contains more particles with sizes larger than 1 μm, which is typically a critical size for successful PSN in aluminum alloys. The microchemistry states observed here are in accordance with earlier work on the evolution of particles during homogenization of similar alloys [26,27].

The present work has clearly confirmed that the different microchemistries following the different homogenization procedures have a profound effect on the back-annealing behavior, in line with previous findings in similar alloys (e.g., [13,18–20]). Generally, conditions which are mainly unaffected by dispersoid effects give a fairly fine grained structure. This applies for conditions without dispersoids or when the dispersoids are coarse and/or with a low number density and in condition with no or limited concurrent precipitation. Faster kinetics and slightly refined grain structures are generally obtained with increased annealing temperature and a larger cold rolling strain. On the other hand, in conditions with either a considerable density of fine pre-existing dispersoids and/or significant concurrent precipitation of fine dispersoids, the influence can be quite pronounced with much slower kinetics and strongly modified, generally much coarser, grain structures. In some cases of low stored energy and/or low annealing temperatures recrystallization can be completely prevented and only extended recovery takes place. Although only indicative with respect to start and stop of precipitation and recrystallization, respectively, the respective TTT-diagrams are quite useful in discussing the interaction of these two reactions and in particular the influence of concurrent precipitation on the recrystallization behavior. Nonetheless, the present work have also shown that a high density of fine dispersoids whether pre-existing or formed during annealing (concurrent precipitation), both have a strong and comparable effect on kinetics and recrystallized microstructure, the latter in the form of large and highly elongated grains. The recrystallization textures for most of the conditions considered here have been reported elsewhere [18–20]. In the absence of dispersoids the texture is found to be generally weak, indicating that PSN plays a prominent role as nucleation mechanism for recrystallization in these alloys. In the presence of dispersoids, on the other, the textures become dominated by a distinct P-texture and a ND-rotated cube component [13,14,18–20]. The two latter are also generally associated with large particles, which indicates that PSN also plays a dominant role in these conditions.

The effect of rolling strain, increasing from $\varepsilon = 1.6$ to 3.0, on the softening behavior, is in general much less pronounced than the microchemistry effects. Higher strain gives a higher stored energy (and thus larger driving force) for recrystallization, consistent with slightly higher as-deformed hardness at the higher strain (compare Figure 7a–d and Figure S1). Accordingly the kinetics is consistently faster for all material conditions. In principle this should also promote nucleation of recrystallization, and then a smaller grain size, which is not obvious from the present experimental results (*cf.* Table 5), where even some counter-intuitive results are observed with increasing grain size with increasing strain. As mentioned above these results may be the result of different concurrent precipitation behavior and/or an artifact due to poor statistics/measurement uncertainties in the measured results. Moreover, the directional homogeneities introduced by the rolling process where characteristic length scales (between e.g., dispersoid layers and constituents (PSN sites) are shortened along ND and extended along RD with increasing strain may also play a role.

Although the shape effect discussed above cannot be completely accounted for, a rough estimate of grain size in selected cases can now be made, assuming PSN to be the dominating nucleation mechanism for recrystallization for all conditions. In doing so we make use of ideas from the ALSOFT model [29], and the available experimental data for the deformation sub-structure, constituent particles and dispersoids (*cf.* Tables 3, 4 and 6). Following Vatne *et al.* [29], it is assumed that the critical step for nucleation from large constituent particles is growth out the deformation zone around these particles, *i.e.*, the critical particle size for a successful PSN event is given by the Gibbs-Thompson equation, *i.e.*, $\eta^* = 4\gamma_{GB}/P_{eff}$ where γ_{GB} is the particle-matrix interfacial energy between the nucleus and the deformed aluminum matrix. Here, it is assumed that the size of the deformation zone scales with the size of the particle itself.

The quantity P_{eff} is effective driving force for recrystallization and is given by the following expression:

$$P_{eff} = P_D - P_Z = \frac{Gb^2}{2}\rho_i + \frac{3\gamma_{SB}}{\delta} - \frac{3F_V\gamma_{GB}}{2r} \tag{2}$$

Here, P_D is the driving pressure for recrystallization/stored energy in terms of sub-grain size δ (mean diameter) and the cell interior dislocation density ρ_i. P_Z (last term in Equation (2)) is the Zener pinning pressure due to a particle volume fraction F_V of radius r [4,15], and γ_{SB} and γ_{GB} are the boundary energies for sub-grains and high angle grain boundaries, respectively. G is shear modulus and b is Burgers vector. It should be noted that the Zener-pressure expression in Equation (2) is based on the assumption of spherical equi-axed particles with a random spatial distribution. The latter assumption is generally not valid in our case, as the dispersoids tend to be located at the grain boundaries where they may more effectively pin the boundaries (a higher effective Zener drag [2,4,10,31]). However, for qualitative purposes Equation (2) may still be used.

Making use of Equation (1) for the size distribution of constituent particles, the following relationship for the number density of PSN nuclei can then be derived [29]:

$$N_{PSN} = C_{PSN}N_0\exp\left(-C_{PE}\frac{4\gamma_{GB}L}{P_D - P_Z}\right) \tag{3}$$

where N_0 and L are defined from Equation (1) f, and C_{PSN} and C_{PE} are model constants of the order ~1. On the assumption that site-saturation nucleation kinetics applies, the recrystallized grain size then becomes:

$$D_{RX} = 1/\sqrt[3]{N_{PSN}} \tag{4}$$

In estimating P_D and P_Z, respectively, according to Equation (2), we use $\gamma_{GB} = 0.3\ \text{Jm}^{-2}$ [32] and $\gamma_{SB} = \frac{Gb\theta}{4\pi(1-v)} \ln\left(e\frac{\theta_c}{\theta}\right) = 0.15\ \text{Jm}^{-2}$ (assuming an average misorientation angle of $\theta \sim 4°$ and $\theta_c \sim 15°$), $G = 26.5$ GPa, $b = 2.86$ Å together with relevant data from Tables 4 and 6 with $F_V = N_V \frac{\pi}{6} d_{3D}^3$; $N_V = N_A/d_{3D}$, where d_{3D} is the equivalent spherical particle diameter (3D) derived from the mean cross sectional diameter in Table 6. Moreover the equivalent circle diameter (CED) is used for the sub-grain size. Following [32], the contribution from cell interior dislocations can be ignored as it represents only a small contribution to the stored energy at the onset of recrystallization (typically 1%–2% [32]). The results are presented in Table 7, where values marked with an asterisk refer to artificially changed input parameter values to highlight how the stored energy and the Zener drag may influence the result.

Table 7. Estimated values for initial (t = 0) stored energy, P_D, and Zener drag, P_Z, for C2-2 and C2-3 and resulting calculated grain sizes (assuming PSN only; see text below) for annealing at 400 °C and 350 °C, for 10^5 s, respectively, after different strains (based on Equations (3) and (4)).

Alloys	True Strain ε	δ_{CED} (μm)	P_D (MPa)	P_Z (MPa)	Calculated Recrystallized Grain Size (μm)		
					350 °C	400 °C	450 °C
C2-0	1.6		0.53	0.15			176
	1.6		0.60 *	0.15			76
	3.0		0.92	0.15			20
	3.0		0.92	0.3 *			30 *
C2-1	1.6		0.38	0.03	55		
	1.6		0.5 *	0.03	30 *		
	3.0		0.57	0.03	24		
	3.0		0.65 *	0.03	30 *		
C2-2	1.6	0.58	0.64	0.12	54		
	1.6			0.25 *		148	
	3.0	0.46	0.81	0.12	25		
	3.0			0.25 *		38	
C2-3	1.6	0.60	0.62	0.067	16		
	1.6						
	3.0	0.47	0.80	0.067	27		
	3.0						

* Conditions, for which the stored energy (P_D) or the Zener drag (P_Z)values is artificially changed as compared to the experimentally based estimates.

Ignoring the shape effect and that some of the experimentally measured grain sizes are counter-intuitive (in view of the strain dependence), the calculated results are grossly in agreement with the experimental results (*cf.* Table 5). This result supports the assumption of PSN being the dominant nucleation mechanism. For the C2-3 alloy the results actually compares quite well with the experiments, although the variation with strain is not consistent with the experiments. For some conditions a better agreement, is obtained by changing either the stored energy or the Zener drag values (data marked with *). Concerning the stored energy, the results clearly indicate that for some conditions the stored energy is underestimated, *i.e.*, the respective measured sub-grain size is too large and/or neglecting the dislocation contribution may have contributed to the underestimation in these cases. Especially for the C2-1 alloy this seems reasonable as the measured sub-grain size seems somewhat high. The indicated underestimations of the Zener drag is supported by more recent comprehensive model predictions with the ALSOFT model [33]. Both for the C2-0 alloy after a strain of ε = 3, and the C2-2 material, the grain sizes obtained with nominal input for the Zener drag, is much too low, indicating that the actual Zener drag effect on nucleation (determining the grain size), as estimated from the classical Zener expression, Equation (3), is much too low. In fact, increasing the nominal Zener drag with a factor of 2, gives results which are more in line with the experiments. It should also be noted that the actual Zener drag also increases during annealing, due to (additional) concurrent precipitation (*cf.* Table 6; [18]). With a long incubation for onset of recrystallization (possible slow time-dependent nucleation), and possibly also boundary-dispersoid correlations the increased Zener drag may actually be closer to reality, also during nucleation, than the lower nominal one.

5. Conclusions

Four different material conditions, following different homogenization procedures including the as-cast non-homogenized condition of an AlMnFeSi model alloy (close to a commercial AA3103 in composition) giving quite different initial microchemistries in terms of content of Mn in solid solution (potential for concurrent precipitation during subsequent processing), size and number density of constituent particles as well as dispersoids have been investigated during subsequent cold rolling and back-annealing.

It is shown that the softening behavior upon annealing after cold rolling is strongly influenced by the initial microchemistry. In particular, the amount of Mn in solid solution (potential for concurrent precipitation) and/or the density of fine pre-existing dispersoids are important factors, where the actual effect is strongly dependent on temperature and to a much less extent on the deformation strain. Both a well homogenized material (reduced Mn solid solution level) and high annealing temperatures reduces the influence of dispersoids giving relatively fast softening kinetics at adequate temperatures (even faster the higher the annealing temperature is) and the formation of a homogeneous fine grained structure. An increased rolling strain has generally the same effect, but much less pronounced. On the other hand, conditions strongly influenced by pre-existing dispersoids and/or concurrent precipitation, in the form of a high density of small dispersoids, have a significant influence on the softening behavior in general and the recrystallization behavior in particular. Such dispersoid structures strongly suppress nucleation and/or retard recrystallization (through a dynamic Zener pinning pressure) resulting in a sluggish recrystallization reaction and an inhomogeneous grain structure of coarse elongated grains. The recrystallization behavior also depends on the microchemistry state in view of size and number density of primary constituents particles. A large fraction of relatively large constituents (>1 μm) promote recrystallization by PSN, which generally speeds up the kinetics and gives smaller recrystallized grains.

Rough grain size estimates based on the assumption that PSN is the main nucleation mechanism for recrystallization with appropriate experimental input values for the stored energy and the Zener drag are qualitatively comparable with experimental results. However, the calculations also indicate that for some conditions the stored energy/driving force for recrystallization is underestimated while the actual Zener drag effect on nucleation seems underestimated by the classical Zener drag expression.

Acknowledgments: This research work has been supported by a KMB project (project number: 193179/I40), in Norway. The financial support by the Research Council of Norway and the industrial partners, Hydro Aluminium and Sapa Technology is gratefully acknowledged. Qinglong Zhao is gratefully acknowledged for providing some of the sub-grain size data in Table 5.

Author Contributions: Ning Wang performed all the experimental work and drafted the paper. Ke Huang contributed to analyses and discussions of the results and to writing of the paper. Yanjun Li and Knut Marthinsen conceived and designed the experimental program, contributed to analyses and discussions of the results and to writing of the paper.

Conflicts of Interest: The authors declare no conflict of interest.

References

1. Hutchinson, W.B.; Oscarsson, A.; Karlsson, A. Control of microstructure and earing behavior in aluminum-alloy AA3004 hot bands. *Mater. Sci. Technol.* **1989**, *5*, 1118–1127. [CrossRef]

2. Daaland, O.; Nes, E. Recrystallization texture development in commercial Al-Mn-Mg alloys. *Acta Mater.* **1996**, *44*, 1413–1435. [CrossRef]
3. Engler, O.; Yang, P.; Kong, X.W. On the formation of recrystallization textures in binary Al-1.3% Mn investigated by means of local texture analysis. *Acta Mater.* **1996**, *44*, 3349–3369. [CrossRef]
4. Humphreys, F.J.; Hatherly, M. *Recrystallization and Related Annealing Phenomena*; Elsevier: Oxford, UK, 2004.
5. Humphreys, F.J. Nucleation of recrystallization at 2nd phase particles in deformed aluminum. *Acta Metall.* **1977**, *25*, 1323–1344. [CrossRef]
6. Humphreys, F.J. Local lattice rotations at 2nd phase particles in deformed metals. *Acta Metall.* **1979**, *27*, 1801–1814. [CrossRef]
7. Engler, O.; Kong, X.W.; Lucke, K. Recrystallisation textures of particle-containing Al-Cu and Al-Mn single crystals. *Acta Mater.* **2001**, *49*, 1701–1715. [CrossRef]
8. Nes, E.; Embury, J.D. Influence of a fine particle dispersion on recrystallization behavior of a 2 phase aluminum-alloy. *Z. Metallkunde* **1975**, *66*, 589–593.
9. Nes, E. Effect of a fine particle dispersion on heterogeneous recrystallization. *Acta Metall.* **1976**, *24*, 391–398. [CrossRef]
10. Vatne, H.E.; Engler, O.; Nes, E. Influence of particles on recrystallisation textures and microstructures of aluminium alloy 3103. *Mater. Sci. Technol.* **1997**, *13*, 93–102. [CrossRef]
11. Ryu, J.H.; Lee, D.N. The effect of precipitation on the evolution of recrystallization texture in AA8011 aluminum alloy sheet. *Mater. Sci. Eng. A Struct. Mater. Prop. Microstruct. Process.* **2002**, *336*, 225–232. [CrossRef]
12. Tangen, S.; Sjolstad, K.; Nes, E.; Furu, T.; Marthinsen, K. The effect of precipitation on the recrystallization behavior of a supersaturated, cold rolled AA3103 aluminium alloy. *Mater. Sci. Forum* **2002**, *396–402*, 469–474. [CrossRef]
13. Tangen, S.; Sjolstad, K.; Furu, T.; Nes, E. Effect of concurrent precipitation on recrystallization and evolution of the *p*-texture component in a commercial Al-Mn alloy. *Metall. Mater. Trans. A Phys. Metall. Mater. Sci.* **2010**, *41A*, 2970–2983. [CrossRef]
14. Schafer, C.; Gottstein, G. The origin and development of the P{011}<111> orientation during recrystallization of particle-containing alloys. *Int. J. Mater. Res.* **2011**, *102*, 1106–1114. [CrossRef]
15. Nes, E.; Ryum, N.; Hunderi, O. On the zener drag. *Acta Metall.* **1985**, *33*, 11–22. [CrossRef]
16. Shankar, M.R.; King, A.H.; Compton, W.D. Microstructure and stability of nanocrystalline aluminum 6061 created by large strain machining. *Acta Mater.* **2005**, *53*, 4781–4793. [CrossRef]
17. Bacca, M.; Hayhurst, D.R.; McMeeking, R.M. Continuous dynamic recrystallization during severe plastic deformation. *Mech. Mater.* **2015**, *90*, 148–156. [CrossRef]
18. Huang, K.; Wang, N.; Li, Y.J.; Marthinsen, K. The influence of microchemistry on the softening behaviour of two cold-rolled Al-Mn-Fe-Si alloys. *Mater. Sci. Eng. A Struct. Mater. Prop. Microstruct. Process.* **2014**, *601*, 86–96. [CrossRef]
19. Huang, K.; Zhao, Q.L.; Li, Y.J.; Marthinsen, K. Two-stage annealing of a cold-rolled Al-Mn-Fe-Si alloy with different microchemistry states. *J. Mater. Process. Technol.* **2015**, *221*, 87–99. [CrossRef]
20. Huang, K.; Engler, O.; Li, Y.J.; Marthinsen, K. Evolution in microstructure and properties during non-isothermal annealing of a cold-rolled Al-Mn-Fe-Si alloy with different microchemistry states. *Mater. Sci. Eng. A Struct. Mater. Prop. Microstruct. Process.* **2015**, *628*, 216–229. [CrossRef]
21. Furu, T.; Orsund, R.; Nes, E. Subgrain growth in heavily deformed aluminum-experimental investigation and modeling treatment. *Acta Metall. Mater.* **1995**, *43*, 2209–2232. [CrossRef]
22. DeHoff, R.T.; Rhines, F.N. *Quantitative Microscopy*; McGraw-Hill: New York, NY, USA, 1968.
23. Ekstrøm, H.-E.; Østensen, L.; Hagstrøm, J. *Dispersoids and Constituent Distribution in 2.2 mm Thick Hot Rolled Bands of AA3104*; BE96–3364; Gränges Technology AB: Finspång, Sweden, 1998.
24. Ekstrom, H.E.; Hagstrom, J.; Ostensson, L. Particle size distributions in a DC-cast and rolled AA3104 alloy. *Mater. Sci. Forum* **2000**, *331–337*, 179–184. [CrossRef]
25. Engler, O.; Laptyeva, G.; Wang, N. Impact of homogenization on microchemistry and recrystallization of the Al-Fe-Mn alloy AA 8006. *Mater. Charact.* **2013**, *79*, 60–75. [CrossRef]
26. Li, Y.J.; Arnberg, L. Quantitative study on the precipitation behavior of dispersoids in DC-cast AA3003 alloy during heating and homogenization. *Acta Mater.* **2003**, *51*, 3415–3428. [CrossRef]

27. Li, Y.J.; Arnberg, L. Evolution of eutectic intermetallic particles in DC-cast AA3003 alloy during heating and homogenization. *Mater. Sci. Eng. A Struct. Mater. Prop. Microstruct. Process.* **2003**, *347*, 130–135. [CrossRef]
28. Altenpohl, D. *Aluminium und Aluminiumlegierungen*; Springer-Verlag: Berlin, Germany, 1965.
29. Vatne, H.E.; Furu, T.; Orsund, R.; Nes, E. Modelling recrystallization after hot deformation of aluminium. *Acta Mater.* **1996**, *44*, 4463–4473. [CrossRef]
30. Wang, N.; Flatoy, J.E.; Li, Y.J.; Marthinsen, K. Evolution in microchemistry and its effects on deformation and annealing behavior of an AlMnFeSi alloy. In Proceedings of the 13th International Conference on Aluminum Alloys (ICAA13), Pittsburgh, PA, USA, 3–7 June 2012; pp. 1837–1842.
31. Somerday, M.; Humphreys, F.J. Recrystallisation behaviour of supersaturated Al-Mn alloys—Part 1—Al-1.3 wt. %-Mn. *Mater. Sci. Technol.* **2003**, *19*, 20–29. [CrossRef]
32. Benum, S.; Nes, E. Effect of precipitation on the evolution of cube recrystallisation texture. *Acta Mater.* **1997**, *45*, 4593–4602. [CrossRef]
33. Marthinsen, K.; Wang, N.; Huang, K. Modelling microstructure and properties during annealing of coldrolled Al-Mn-Fe-Si-alloys with different microchemistries. *Mater. Sci. Forum* **2014**, *783–786*, 57–62. [CrossRef]

8

Influence of the Hardfacing Welds Structure on Their Wear Resistance

Janette Brezinová *, Dagmar Draganovská, Anna Guzanová, Peter Balog and Ján Viňáš

Department of Mechanical Technology and Materials, Technical University of Košice, Mäsiarska 74, 040 01 Košice, Slovakia; dagmar.draganovska@tuke.sk (D.D.); anna.guzanova@tuke.sk (A.G.); peter.balog@tuke.sk (P.B.); jan.vinas@tuke.sk (J.V.)
* Correspondence: janette.brezinova@tuke.sk

Academic Editor: Hugo F. Lopez

Abstract: The contribution presents the research results of hardfacing metals' resistance in conditions of abrasive wear. Two types of hardfacing electrodes with a different chemical composition were used in the creation of three layers of hardfacing metals. The chemical composition of electrodes determines the difference in a hardface deposit structure. We have investigated the influence of mixing the base metal and a filler metal and the influence of hardfacing welds structure on the resistance against abrasive wear. The results of the experiments have showed that the intensity of wear is very dependent on the parameters of wear as well as the morphology structure of hardfacing metals.

Keywords: hardfacing welds; structure; abrasive wear; wear resistance

---

1. Introduction

The wear of mechanical parts of machinery is still a current scientific, engineering and economic issue. The analysis of machine parts and structure faults shows they are often caused by tribological processes that take place at the functional surfaces. The interaction between functional surfaces in their relative motion causes adverse effects in the surface layers, leading to their deterioration. An external manifestation of this process is the removal or relocation of the functional surface particles. This process can be measured by a change in the size, weight or mechanical and physico-chemical properties. This process can be alleviated by improving the structural arrangement of nodes, the appropriate choice of materials or the creation of new surface layers [1–5]. When creating the surface layers for tribological use, we often encounter hardfacing technology which enables not only the restoration of the worn surface geometry but gives it new, often better properties than the properties of the original material, which leads to a prolongation of their life [6–8].

Welding consumables can be divided into groups according to their properties and wear resistance. The iron-based alloys include martensitic alloys, austenitic alloys and alloys with a high content of carbides [9]. Martensitic alloys are used for the restoration of shape and for hardfacing. Their main advantages include good resistance to wear of metal-metal type, good impact resistance and acceptable abrasion resistance.

Austenitic claddings are suitable for the restoration of shape and are characterized by an excellent impact resistance and an acceptable abrasion resistance.

The advantage of alloys with high carbide content is both the excellent abrasion resistance, good heat resistance, and acceptable corrosion resistance; however, the disadvantage is a weak impact resistance [10].

Cobalt- and nickel-based alloys are resistant to most types of wear. Due to their high cost, they are only used in cases when their properties are economically justified, such as in high temperature applications when the iron-based alloys with a high carbide content lack sufficient resistance. Nickel alloys are a slightly cheaper alternative [11].

The most progressive additive materials for welding and thermally sprayed coatings for the conditions of abrasive, erosive, corrosive and combined stresses in mining, energy and building industries include nanostructural claddings [12]. Characteristics of these materials not containing cobalt or nickel are based on the high content of carbide-forming and boride-forming metals Cr, Mo, W and Nb, carbon (up to 5%) and boron (5%–10%). They have an amorphous metallic glass character. Very hard and very fine nanostructures with a grain size of 2–75 nm are formed during their devitrification. A hardness of 900–1230 HV, 54–74 HRC, respectively, is reached by the resulting cladding. The base element in these alloys is iron even though, in the maximum content of alloying elements, it does not reach 50%. The claddings are carried out by MIG (Metal Inert Gas) hardfacing, and coatings are carried out by HVOF (High Velocity Oxygen Fuel) and plasma spraying technology.

Essential scientific knowledge of cladding tribology is detailed in [13,14] and others. These imply that one of the determining factors for the wear intensity in both the abrasive and erosive wear of claddings is their hardness. The hardness of cladding is a function of its chemical composition, the welding heat mode during hardfacing and its structure.

In the current analysis of abrasion resistance, the issue of cladding structure and substructure and their influence on abrasion resistance is not analyzed in detail. So far there is no consensus on the most appropriate type of structure in terms of resistance to abrasive wear. Most authors coincide in finding that the resistance to abrasive wear of alloys with the same hardness with a different chemical and structural composition is not the same. This depends on the hardness, the amount, the shape, the size and the distribution of structural components. Some authors consider austenitic-carbidic the most advantageous structure, others prioritize a martensitic-carbidic structure. This results from the diversity of the abrasive wear process and a wide range of real operating conditions [6,9,15–17].

Structural and substructural properties can directly affect material removal. Its intensity is conditioned by the strength and cohesive properties of the submicroscopic particles of structure that are loaded by wear and abrasion. Every structural component in the process of exploitation determines the level of resistance of the whole metal with its share. In abrasive wear, two crucial stages must be distinguished. The first one is the process of forcing the abrasive into the surface where the limiting factor is the indentation hardness. The second stage is the process of surface disintegration, where the crucial role belongs to the strength of interatomic bonds and the strength of the structural components' connection to each other at the grain boundaries [13].

In real conditions, the failure of the material surface layers by a high-cycle contact fatigue process (abrasive particles do not notch into the surface, causing only elastic deformation), a low-cycle contact fatigue process (a plastic deformation when notching abrasive particles into the surface) and the grooving together with the segregation of worn material particles may occur at various locations of the tough material-worn surface. At a high speed of abrasive particles' relative motion against the worn surface, it is necessary to also consider other degradation mechanisms: the heat affection of the material (tempering of steel, softening of the polymeric material), adsorption failure (reactions of the worn material with surface active agents that reduce surface hardness) and tribochemical reactions of the worn material with the environment. Although in abrasive wear of brittle materials plastic deformation occurs, the brittle fracture often determines the rate of wear. In tough materials this fault occurs probably right after the abrasive particle, due to tensile stress, acts here. The removal of the material by abrasion in brittle materials occurs by a brittle fracture rather than by the plastic deformation. Also, in the abrasive wear of heterogeneous materials which contain tough and brittle phases, infringement by brittle fracture may occur. The predominant mechanism of material removal depends on the characteristics of individual phases and their volume fractions [14].

The stated information implies that due to the complexity of the wear process, it is necessary to design a type of cladding based on the tribological analysis and the conditions of the surface stress. This analysis, however, is for many practical cases impracticable due to the imperfection of developed procedures and, therefore, operational and laboratory tests have recently been used for the evaluation of cladding properties.

The aim of the paper is to carry out a tribological and metallographical analysis for two types of hardfacing layers. The analysis can contribute to clarification of the relationship between the microstructure and wear resistance of layers.

2. Materials and Methods

The base material for production of hardfacing samples also used as a gauge in the tests of wear was steel of grade S235JRH. Dimensions of the test samples were 20 × 20 × 8 mm. This is a structural carbon steel with 0.22% C and excellent weldability.

For cladding formation hardfacing electrodes E 508 B and E 518 B (Welding Research Institute, Industrial Institute of Slovakia, Bratislava, Slovakia) were used, whose chemical composition meets the requirements for heterogeneity of structural constitution of claddings. Chemical composition of electrodes is shown in Table 1. Technological and use properties are as follows:

- E 508 B—thick flux-covered electrode. Recommended preheating: without preheating or preheating up to 250 °C, depending on the quality and dimensions of the hardfacing component. It is used for hardfacing of working parts of agricultural and forestry machines exposed to abrasion. The cladding is resistant to mild impacts, it is not thermally treated. If any treatment is needed, it shall be spheroidizing annealed and subsequently thermally treated to optimal hardness of approximately 600 HV (Vickers hardness).
- E 518 B—thick flux-covered electrode. Cladding preferably alloyed from the flux cover. Preheating for hardfacing is 400 °C with following cooling in wrap or in the oven. Cladding is not thermally treated. It is used for the parts of mining and earth-moving machinery, metallurgical equipment exposed to abrasive and erosive wear combined with impacts. Hardness of the cladding is 600–660 HV.

Table 1. Chemical composition and hardness of welding electrodes in wt. %.

Electrode	C	Mn	Si	Cr	Mo	HV
E 508 B	0.5	0.7	0.5	6.0	0.6	580
E 518 B	3.4	0.5	0.8	29	-	660

Cladding samples for the studies of structural condition and the fracture surfaces were made from one to three layers. After cooling the cladding was broken off by three-point bending. The surfaces obtained this way were used to study the failure mechanism and for the structural analysis. Abrasion resistance was determined with an assistance of a test device based on the principle of relative movement of test samples submerged into free abrasive [16]. For this test abrasive based on Al_2O_3 with grain size of 1.2 mm was used. Incidence angle between abrasive and test sample was 75° and speed of relative sample motion was 2.25 $m \cdot s^{-1}$. The weight losses were investigated using sensitive digital scale weight with a measurement accuracy of 10^{-4} g.

To determine the microhardness of individual structural phases, the microhardness HV0.05 was measured, with a load of 0.5 N and dwell time 10 s. The measurements were carried out using Hannemann head (Carl Zeiss AG, Feldbach, Switzerland). The course of the hardness in the cross-section of claddings was realized by the Vickers hardness pyramid (HPO 250, Liepzig, Germany) with a load of 294.3 N and dwell time 10 s.

Metallographic study and a study of fracture surfaces were realized using light microscope Olympus BXFM (Olympus Deutschland GmbH, Hamburg, Germany) and a scanning electron microscope TESLA-BS-301 (Tesla, Brno, Czech Republic).

3. Results and Discussion

The course of the hardness of claddings E 508 B showed a strong effect of mixing the weld and base metal, as shown in Figure 1. The hardness of the cladding in its first layer in the area of melting rises sharply to a value of 550 HV30. The second layer reaches an average value of 615 HV30, while in three-layer cladding the hardness reaches around 630 units. Theoretical and practical assumptions that the effect of mixing the base material and weld metal does not disappear until the third layer were confirmed [13]. The transition of the hardness curves from the minimal values of the base material to the maximum values in the cladding showed a step-change.

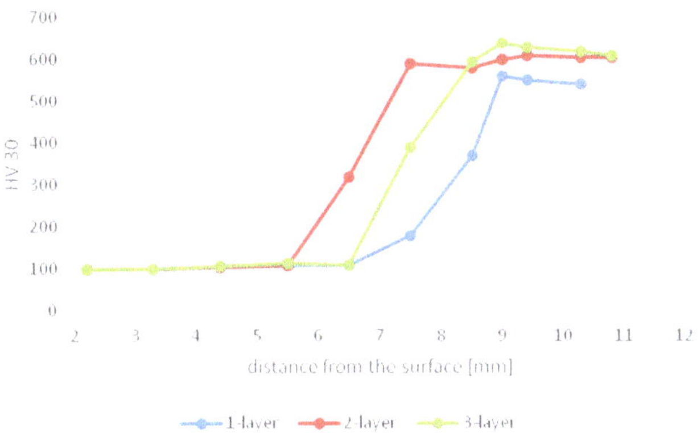

Figure 1. Course of hardness in cladding E 508 B with various numbers of layers.

A similar course of the hardness was shown by cladding E 518 B, except that the hardness value stated by the manufacturer, which is 660 HV30, was reached already in the first layer. Maximum values obtained in individual layers did not differ significantly and reached up to 730 HV. Due to the high content of C and Cr, the mixing with the base material was significantly less.

The course of three-layer cladding hardness is shown in Figure 2. In terms of achieved hardness, higher wear resistance can be assumed in claddings realized by the E 518 B electrode.

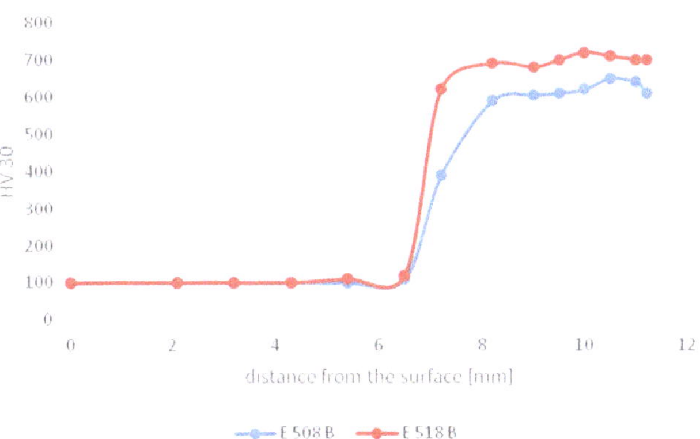

Figure 2. Course of hardness in three-layer hard-faced materials.

The results of abrasion resistance tests of evaluated hardfaced materials are graphically presented in Figures 3 and 4. From these figures it can be seen that, in electrode E 508 B, single-layer claddings show higher mass loss due to the mixing of the weld metal with low-carbon steel. Despite the fact that for electrode E 518 B small differences of hardness values were reached in individual layers, there is a big difference in mass loss between single-layer and multiple-layer claddings. Compared to electrode E 508 B, it reaches higher values. The results of mass loss values of tested claddings cannot be evaluated just by a chemical composition and measured values of hardness. Most authors explain the theory of material removal in wear by a contact interaction where a plastic deformation occurs [6]. This deformation is caused by the redistribution and dissipation of energy. The energy redistribution, followed by implementation of plastic deformation, causes a movement of anchored and free dislocations. The dislocation state of metals depends mainly on the structural state. The real structural state is therefore another factor influencing the wear resistance [13,15].

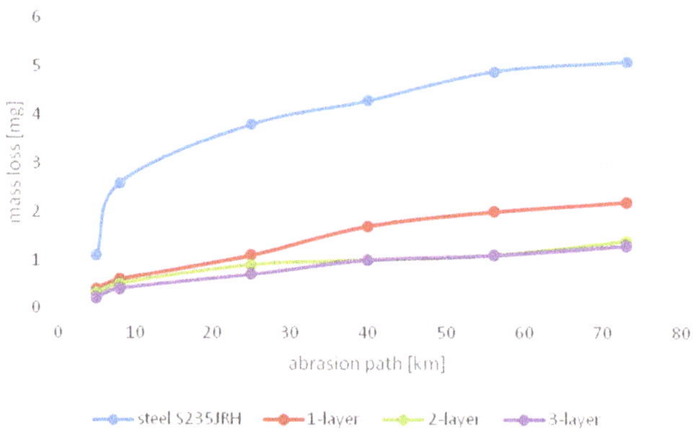

Figure 3. Course of wear for hardfacing material E 508 B.

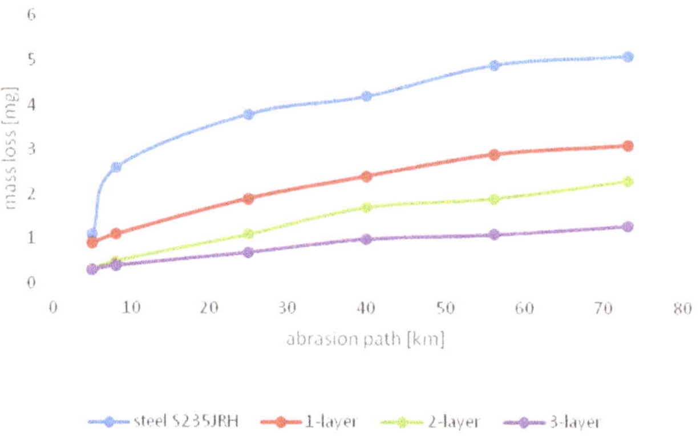

Figure 4. Course of wear for hardfacing material E 518 B.

As mentioned above, the investigated claddings vary significantly in internal structure. Electrode E 508 B represents cladding based on mild steel with a content of C up to 0.4%. The relatively high content of carbide-forming elements Cr and Mo determines the formation of the bainitic matrix structure and, by bonding with carbon, creates disperse carbides reinforcing the very fine bainitic matrix. Since claddings are multilayered and multipass in a layer, the mutual thermal influence causes very heterogeneous formation of the phases. In Figure 5 there is a structure of the first layer in the area of the mixing of the weld metal with the base material which was not thermally affected. It is a very fine bainitic structure with a ferrite net which retains the oriented nature. The microhardness of this layer reaches 424 HV 0.05. In the area of thermal affection, the grains are finer and the distinctive casting character is disappearing, as shown in Figure 6. The similar nature of the structure is also present in the second layer, where the hardness reaches 489 HV 0.05. In the third layer of the cladding, which was not thermally affected, the distinctive casting structure is retained, as shown in Figure 7 (538 HV 0.05). The cladding structure in the third layer in the thermally affected zone is shown in Figure 8. In the bainitic matrix, the very dispersely formed carbidic phase can be observed. Claddings are thus characterized by a relatively tough matrix in which the very disperse carbidic phase ensures sufficiently high hardness and, therefore, abrasion resistance.

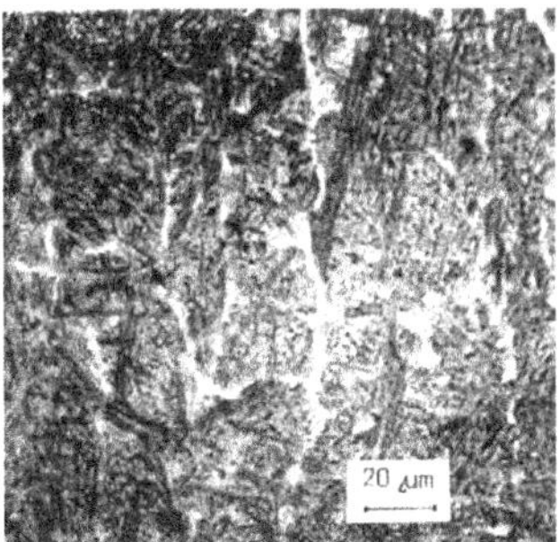

Figure 5. Microstructure of hardfacing weld E 508 B.

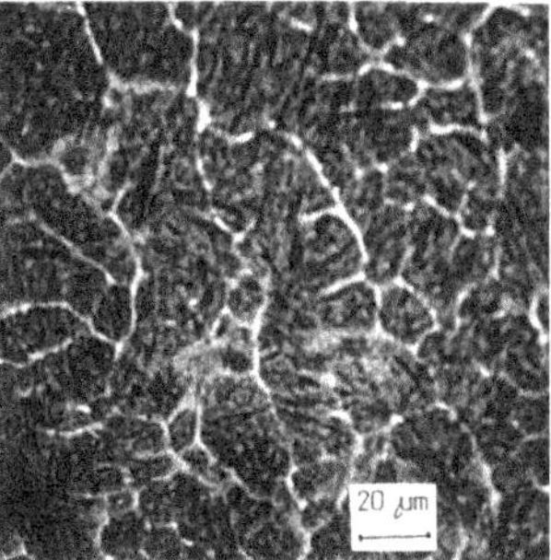

Figure 6. Thermally affected structure of the cladding E 508 B.

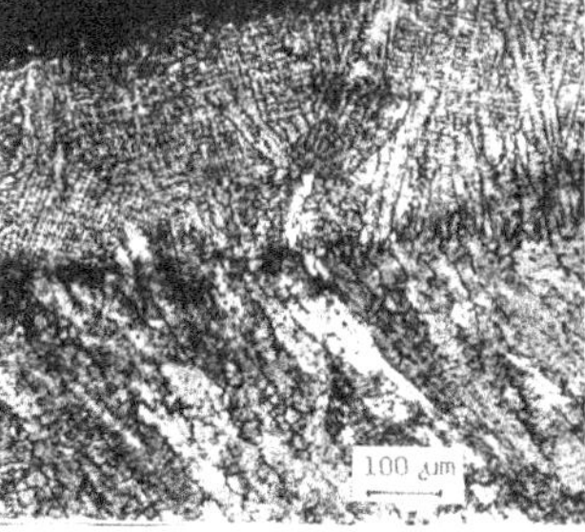

Figure 7. Cast structure of hardfacing weld E 508 B.

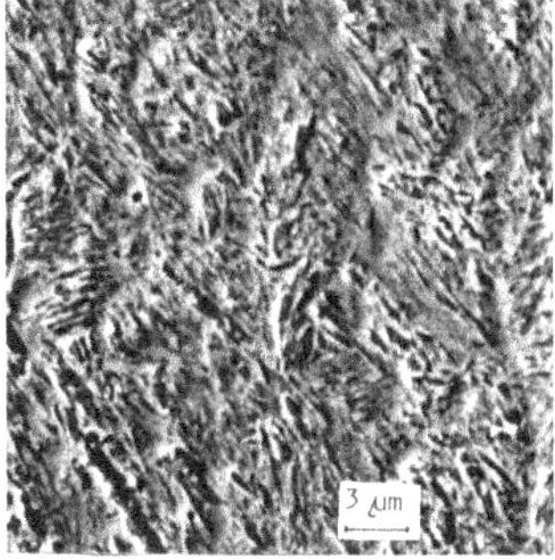

Figure 8. Carbidic phase in bainitic matrix of the weld E 508 B.

The structural composition of the claddings realized by electrode E 518 B is shifted into the area of cast iron, which is manifested by the significant dendritic character of the cast structure. The high content of C up to 3.5% and chrome up to 27.5% after the re-melting in the cladding gives a high-alloyed cast iron of ledeburitic type with a significant presence of primary and secondary carbides. The structure in the transition from the base material to the cladding is documented in Figure 9. Figure 9 shows that even in the short term of the re-melting, a very thin carburized layer was formed in the base material, demonstrated by large pearlitic grains. In the areas of the cladding where the cast structure was thermally affected, a very fine, oriented needle structure (Figure 10) or a transformed structure with large carbidic needles (Figure 11—TMP14) was formed due to the structural transformation. The structure in the cover layers of the cladding is a casting structure with a distinct dendritic composition (Figure 12). The average hardness of the ledeburitic matrix was of 562 HV 0.05. Large carbidic needles reached a hardness of 1630 HV 0.05 while small needles reached up to 2350 HV 0.05.

Figure 9. Cast structure of hardfacing weld E 518 B.

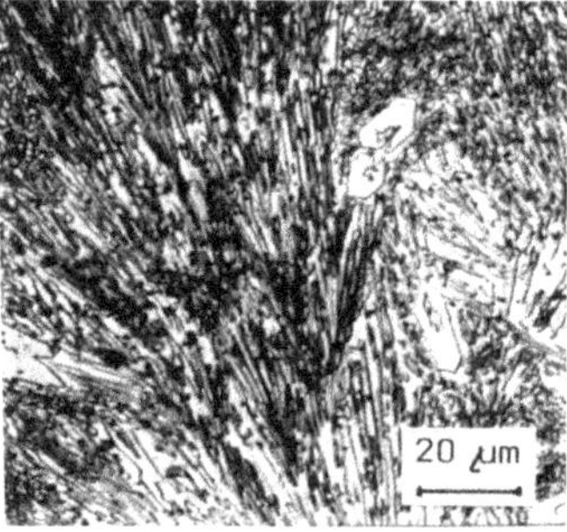

Figure 10. Detail of fine-grain structure of cladding E 518 B.

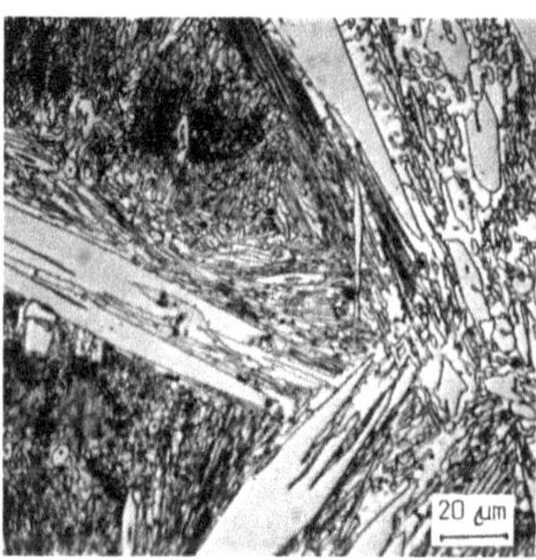

Figure 11. Detail of massive carbidic particles E 518 B.

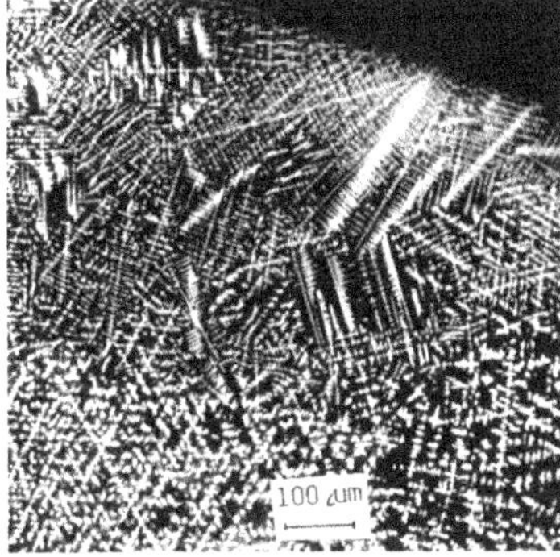

Figure 12. Cast structure of hardfacing weld E 518 B.

For the creation of the fracture surface, energy is demanded and, thus, the character of the fracture, whether it is cleaving, ductile failure or decohesion along the phase boundary, also provides the resistance to separation of microparticles under surface abrasion. Therefore, the examination of these surfaces makes it possible to predict the use properties of metallic materials.

Depending on the heat affection of the structural composition, the bulk of the fracture surfaces of E 508 B under bending stress was formed by decohesive failure at the border of crystallites or dendrites, as shown in Figure 13. Part of the surface failed by decohesion with a low power consumption—smooth facets and part of the surface failed in these claddings by pit morphology whose creation is energy-demanding and indicates the certain resistance to a brittle fracture, as shown in Figure 14.

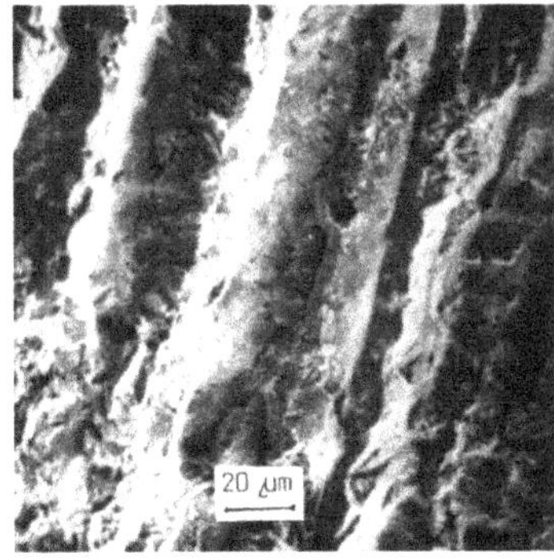

Figure 13. Decohesive failure at the dendrite borders.

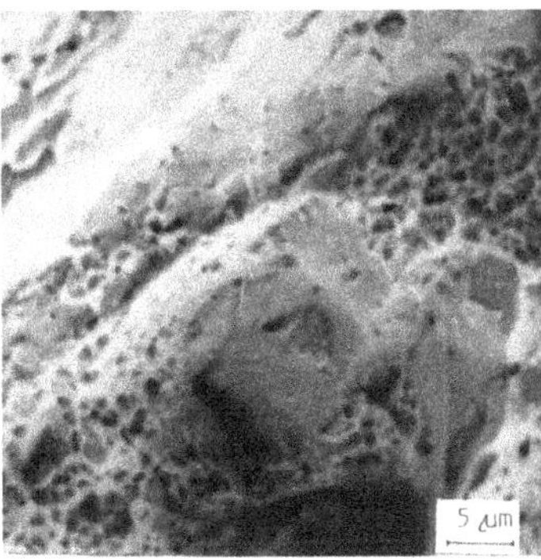

Figure 14. Morphology of E 508 B cladding failure.

The fracture surfaces of claddings of E 518 B obtained for the purpose of studying the mechanism of their formation confirm that the energy demand for their creation is very low. The view of the fracture surface is shown in Figure 15. Part of the fracture surface consists of decohesive facets without morphological signs of cleaving or ductile failure. Their formation has occurred by the simple separation of the surfaces at the grain borders with minimum energy consumption. A large part of the fracture, however, consists of facets of cleavage decohesion. These are facets formed by the separation of phase boundaries, but the mechanism of their failure is the cleavage, *i.e.*, low energy. These surfaces are shown in Figure 16 and are characteristic for the cast state of the cladding without thermal treatment.

Figure 15. Fracture surface of cladding E 518 B.

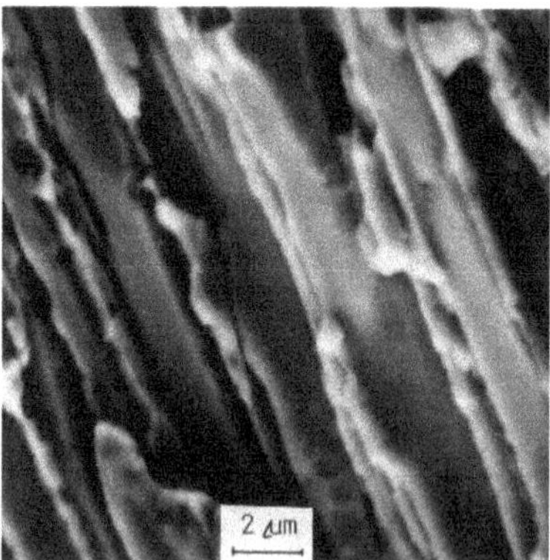

Figure 16. Detail of decohesive failure.

The stated information implies that claddings made by electrode E 508 B have a tough bainitic matrix dispersely reinforced by a carbidic phase, as shown in Figure 17. This symbiosis of the matrix and disperse carbidic reinforcement optimizes the abrasive wear resistance. The structure has a high density of dislocations whose movement is intensively stopped by the disperse phase, as shown in Figure 18. Plastic deformation thus requires high stress for the movement of the dislocations. It is known that the bainitic structures with the same chemical compositions optimize mechanical properties, *i.e.*, at high levels of strength (hardness) they retain adequate ductility.

Figure 17. Bainitic matrix containing carbides.

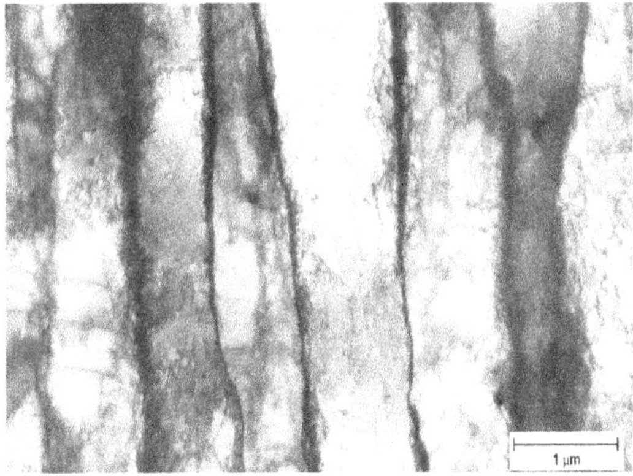

Figure 18. Dislocations in bainitic matrix.

The structure of claddings made by electrode E 518 B is ledeburitic due to the high content of carbon and the high content of chrome. There are primary as well as secondary carbides present in the structure, morphologically formed massively as well as dispersely, a shown in Figure 19. The dislocation state is high, represented mainly by anchored dislocations.

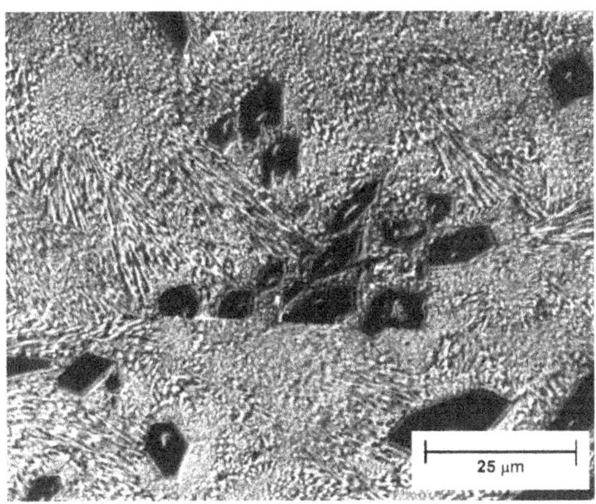

Figure 19. Carbides in ledeburitic structure.

Although the material is high strength (hard), it has a small reserve of plastic deformation, and therefore it is brittle. The low strength of the grain boundaries is also added to these properties. It can therefore be concluded that the cohesive strength of the boundaries and subboundaries is significantly lower than the strength of the matrix. The material separation is realized by the decohesion of particles and the separation by the cleaving mechanism.

4. Conclusions

Based on the achieved results of the study of the structural composition of claddings, the measurements of their hardness and their abrasive wear resistance, the following can be concluded:

1. The hardness and wear values confirm the correctness of the principle that the optimal properties are not achieved by hardfaced materials until the third layer. For practical application it is necessary to execute the welding process with minimal melting of the base material. In the cladding by electrode E 518 B, due to its chemical composition, this effect was less pronounced.
2. In set conditions of wear, the claddings made by electrode E 508 B seem more favorable. The tough bainitic matrix dispersely reinforced by a carbidic phase provides good resistance to abrasive wear. In terms of chemical composition and manufacturing technology, these claddings appear economical since they can be realized without a special regime of hardfacing. Cladding has excellent properties without a heat treatment.
3. Claddings made by electrode E 518 B, in terms of hardness values, appear to be of high quality. The formation of the ledeburitic structure with plenty of carbidic phase gives a high-strength though brittle cladding. Claddings therefore have little resistance to abrasive wear in combination with mild impacts.
4. Based on the obtained results, it can be concluded that in the optimized chemical composition of weld materials, it is necessary to also take the structural constitution of the cladding into consideration. Better results are obtained in two-component structural phases where the matrix has sufficient hardness with a good supply of plastic properties and the reinforcing disperse component is based on a carbidic or carbonitridic basis.

Acknowledgments: This work was supported by the Slovak Research and Development Agency under the contract No. SK-UA-2013-0013 and project VEGA No. 1/0600/13.

Author Contributions: Realization of hardfacing, three-point bending, abrasive tests—Ján Viňáš, Peter Balog. Metallography analysis, analysis of fracture, hardness testing—Janette Brezinová, Dagmar Draganovská, Anna Guzanová.

Conflicts of Interest: The authors declare no conflict of interest.

References

1. Viňáš, J.; Brezinová, J.; Guzanová, A.; Svetlík, J. Degradation of renovation layers deposited on continuous steel casting rollers by submerged arc welding. *Proc. Inst. Mech. Eng. B* **2013**, *227*, 1841–1848. [CrossRef]
2. Viňáš, J.; Brezinová, J.; Guzanová, A. Analysis of the quality renovated continuous steel casting roller. *Sadhana* **2013**, *38*, 477–490. [CrossRef]
3. Viňáš, J.; Brezinová, J.; Guzanová, A.; Balog, P. Evaluation of the quality of cladding deposited on continuous steel casting rolls. *Int. J. Mater. Res.* **2013**, *104*, 183–191. [CrossRef]
4. Correaa, E.O.; Alcântara, N.G.; Valeriano, L.C.; Barbedo, N.D.; Chaves, R.R. The effect of microstructure on abrasive wear of a Fe-Cr-C-Nb hardfacing alloy deposited by the open arc welding process. *Surf. Coat. Technol.* **2015**, *276*, 479–484. [CrossRef]
5. Jankauskas, V.; Antonov, M.; Varnauskas, V.; Skirkus, R.; Goljandin, D. Effect of WC grain size and content on low stress abrasive wear of manual arc welded hardfacings with low-carbon or stainless steel matrix. *Wear* **2015**, *328–329*, 378–390. [CrossRef]
6. Blaškovič, P.; Čomaj, M. *Renovácia Naváraním a Žiarovým Striekaním*; Alfa: Bratislava, Slovakia, 1991.
7. Dwivedi, D.K. Abrasive wear behavious of iron based hard serfacing alloy coatings developed by welding. *Surf. Eng.* **2004**, *20*, 87–92. [CrossRef]

8. Senthilkumar, B.; Kannan, T. Effect of flux cored arc welding process parameters on bead geometry in super duplex stainless steel claddings. *Measurement* **2015**, *62*, 127–136. [CrossRef]
9. Czichos, M. *Tribology*; Elsevier: Oxford, UK; New York, NY, USA, 1978.
10. Marinescu, I.D.; Hitchiner, M.P.; Uhlmann, E.; Rowe, W.B.; Inasaki, I. *Handbook of Machining with Grinding Wheels*; CRC Press: New York, NY, USA, 2007; p. 593.
11. Malkin, S.; Guo, C. *Grinding Technology: Theory and Applications of Machining with Abrasives*; Industrial Press: New York, NY, USA, 2008; p. 372.
12. Abušinov, A. Nanostrukturní návary a žárové nástřiky. *MM Průmyslové Spektrum* **2011**, *4*, 22–24.
13. Brožek, M. Výsledky zkoušek abrazivního opotřebení vrstev. In Proceedings of the Conference Renop 92, Trnava, Slovakia, 17–19 March 1992; pp. 139–142.
14. Hawk, J.A.; Wilson, R.D. *Abrasive Wear Failures*; ASM Handbook; ASM International: Materials Park, OH, USA, 2002; Volume 11, pp. 906–921.
15. Adamka, J.; Petríková, G. Vplyv štruktúry návarov na odolnosť proti abrazívnemu opotrebeniu. In Proceedings of the Conference Intertribo 93, Bratislava, Slovakia, 26–28 August 1993; pp. 70–76.
16. Jankura, D. Tribologické vlastnosti viacvrstvových tvrdonávarov. In Proceedings of the Conference Funkčné Povrchy, Trenčín, Slovakia, 14–16 June 2001; pp. 96–102.
17. Jankura, D. Hodnotenie vlastností viacvrstvových tvrdonávarov. *Mechanika* **1996**, *47*, 59–66.

9

Effective Synthesis and Recovery of Silver Nanowires Prepared by Tapered Continuous Flow Reactor for Flexible and Transparent Conducting Electrode

Hyung Duk Yun [1], Duck Min Seo [1,†], Min Yoeb Lee [2,†], Soon Yong Kwon [1,†] and Lee Soon Park [1,*]

Academic Editor: Hugo F. Lopez

[1] School of Material Science and Engineering, Ulsan National Institute of Science and Technology (UNIST), Ulsan 44919, Korea; yun2985@gmail.com (H.D.Y.); seodm@unist.ac.kr (D.M.S.); sykwon@unist.ac.kr (S.Y.K.)
[2] Department of Polymer Science and Engineering, Kyungpook National University, Daegu 41566, Korea; youp1999@naver.com
* Correspondence: parkls@unist.ac.kr
† These authors contributed equally to this work.

Abstract: Silver nanowires (AgNWs) with high aspect ratio were obtained utilizing a tapered tubular reactor by the polyol process. The tapered tubular type flow reactor allowed us to obtain nanowires in high yield without defects that is generally encountered in a closed reactor due to excessive shearing for a long time. After reaction the AgNWs were precipitated in the aqueous solution with the aid of a hydrogen bond breaker and were recovered effectively without using a high-cost centrifugation process. Dispersion of the AgNWs were used to prepare transparent conducting electrode (TCE) films by a spray coating method, which showed 86% transmittance and 90 Ωsq^{-1} sheet resistance.

Keywords: silver nanowires; transparent conducting electrode; continuous flow reactor; touch screen panel

---

1. Introduction

Transparent conducting electrodes (TCE) are widely used in touch screen panels(TSP), thin-film solar cells, and transparent/flexible displays. A sputtered film of indium tin oxide (ITO) shows high transmittance (95%T) at low sheet resistance (50 Ωsq^{-1}) [1]. However, ITO films lack mechanical rigidity especially under bending stress and have limitation in application to large area TSP (over 25 inches) due to the trade-off in high transmittance and low sheet resistance properties. Therefore many new materials are under development including carbon nanotubes [2,3], graphene [4,5], conductive polymers [6,7], and metal nanowires [8–10]. Among these materials one dimension metal nanowires have received great attention due to their potentials in fabricating large area TSPs, thin film solar cells, flexible OLEDs, and transparent displays. Silver nanowires (AgNWs) have been extensively studied during past years. They exhibit high conductivity(~105 Scm^{-1}) without any detrimental effects to their high transparency (>90%) [11]. Regarding the synthesis of silver nanowires several methods have been reported including hard-template [12,13] and soft template synthesis [14]. Among these synthetic routes, solution phase synthesis by polyol reduction has been the most intensively studied method [15,16]. However high-aspect ratio and rapid synthesis of AgNWs have not been achieved yet. Furthermore the cost of recovery and purification is also very high hindering the wide applications of AgNWs. Herein, we report a novel facile and high-concentration synthesis of AgNW utilizing tapered tubular type continuous flow reactors and a new method for the recovery of AgNW by precipitation method.

2. Experimental Section

2.1. Materials

Ethylene glycol (EG), $AgNO_3$, NaBr, poly(vinylpyrrolidone) (PVP, M_W ~ 10,000–40,000) and sodium dodecyl sulfate (SDS) were purchased from Sigma Aldrich (St. Louis, Mo, USA). All chemicals were used as received without further purification.

2.2. Synthetic Procedure and Recovery of AgNWs

Silver nanowires were synthesized be polyol process utilizing different type of reactors in this research. In bath type closed system, the AgNWs were synthesized in a three-neck round-bottomed flask by dropping method. First poly(vinylpyrrolidone) (PVP, M_W ~ 40,000) and sodium dodecyl sulfate (SDS) were dissolved in ethylene glycol (EG) at 75 °C for 12 h. In another beaker was dissolved $AgNO_3$ powder in EG at room temperature for 6 h. The PVP and SDS in EG solution was poured in a three-neck flask and heated to 160 °C, and $AgNO_3$ in EG solution was added dropwise for 5 min by using a dropping funnel while stirring with a magnetic bar. For the scale-up of AgNWs synthesis another type of batch reactor, the resin kettle reactor, was used with mechanical stirrer utilizing the same dropwise addition method.

For the high-concentration rapid synthesis of AgNWs the tapered tubular type reactor was used with a mechanical stirrer. Here the PVP/SDS in EG solution was added first in the tapered tubular reactor and then $AgNO_3$ in EG solution was added dropwise separately, while taking out the AgNW reaction mixture through the bottom outlet of tapered tubular reactor. This type of continuous flow reactor can minimize the destruction or damage of synthesized AgNWs due to high shearing force induced in the closed reactors of three-neck with magnetic stirrer or resin kettle type with mechanical stirrer, enabling high concentration and rapid synthesis of AgNWs.

In the small scale synthesis of AgNWs in three-neck flask and resin kettle, the recovery of AgNWs were carried out by centrifuge. The reaction mixture of AgNWs solution was diluted with a five-fold volume of ethanol and stirred for 30 min and then subjected to centrifuge for 30–60 min at 3000 rpm. After removing the supernatant by siphon, five-fold deionized water was added to the precipitate of AgNWs mixture and subjected to centrifuge to purify the obtained AgNWs.

2.3. AgNWs-PET Film and Property Measurements

The purified AgNWs (0.05 wt. %) were added into ethanol solvent with 0.25 wt. % ethyl cellulose to make AgNW spray coating solution. The AgNW on polyethylene terephthalate (AgNW-PET) film was obtained by 20 times spray coating of the AgNW coating solution. The size of the AgNWs was measured with SEM (Hitach High-Technologies, Tokyo, Japan), whereas the sheet resistance and transmittance of the AgNW-PET film was analyzed by four-point probe and UV-VIS spectrophotometer (Craic, San Dimas, CA, USA). The crystal structure of AgNW was studied by XRD (Rigaku, Tokyo, Japan) and the stability of AgNW-PET film under flexure stress was examined by a bending test.

3. Results and Discussion

Over the past few years, the polyol process has been the most promising method for preparing AgNWs by using poly(vinylpyrrolidone) (PVP) as the capping agent and ethylene glycol (EG) as the solvent and reducing agent to reduce $AgNO_3$ into Ag nanowires. In the polyol process, many factors have always affected the yield and morphology of AgNWs, such as additive agents, temperature, stirring speed, ratio of chemicals, reaction times, and injection speed of chemicals.

Although the synthesis of silver nanowires has been studied for many years and a variety of silver nanowires have been synthesized by different methods, continuous flow-type synthesis of high-concentration silver nanowires has rarely been reported. In this work we first examined the optimum chemical and physical condition for the synthesis of AgNWs by the polyol process utilizing closed batch type reactors. Based on this conditions the scale-up of AgNW synthesis was studied and applied to the synthesis of AgNWs by continuous flow reactions. A new method of recovery and purification of AgNWs was also developed.

3.1. Reaction Parameters of AgNWs Synthesis in Batch Type Closed Reactors

The synthesis of AgNWs is strongly affected by reaction conditions including chemical parameters such as the mole ratio of Ag^+ ion, repeating units/molecular weight of PVP, amount of sodium dodecyl sulfate (SDS) and various seed materials as well as physical parameters such as total solid content and method of addition of reactants, reaction time, method of stirring, structure and speed (rpm) of stirrer, and recovery/purification of AgNW product.

These special features of AgNW synthesis are due to the complex process of AgNW formation including the reduction of Ag^+ ion to Ag metal, seed formation, growth of silver nanometals (both nanowires and nanoparticles) under the capping of PVP, and other soft template materials in addition to the physical factors such as association and sedimentation of silver nanowires under high shearing forces.

In an attempt to separate these complex factors affecting AgNW formation, we first examined the synthesis of AgNWs in the closed reactor system. Here, we used the chemical composition of reactants which could promote a rapid and high-concentration synthesis of AgNW with high aspect ratio and high yield of AgNWs over Ag nanoparticles through the wide search of previously reported papers. In a typical synthesis of AgNWs, 1875 g PVP (M_W ~ 40,000) and 0.115 g SDS were dissolved in 20 mL EG by heating at 75 °C for 12 h under moderate stirring with magnetic stirrer. After PVP was dissolved completely, this solution was added in a three-neck flask and heated to 160 °C. Into the PVP/SDS solution was added $AgNO_3$/EG solution (which was made by dissolving 0.709 g $AgNO_3$ in 5 mL EG by stirring at room temperature for 6 h) by using dropping funnel over a period of 5 min and then stirred at 250 rpm with magnetic stirrer for certain period of time. After reaction the AgNWs were recovered and purified by centrifugation and subjected to SEM examination.

As for the effect of Ag^+ ion to PVP repeat unit, the mole ratio of 1:4 was found to give high yield of AgNWs as shown in Table 1 and Figure 1. From Figure 1 it was noted that large amount of amorphous Ag particles were observed below the Ag^+: PVP repeat unit mole ratio of 1:4 due to insufficient capping of PVP on the growing direction of AgNWs. This result may also be attributed to the change of the total solid content (sum of $AgNO_3$, PVP and SDS) in EG solution, showing that the larger the amount of $AgNO_3$ than the amount of PVP the larger the diameter of the AgNWs synthesized.

The most interesting part of this work was the effect of SDS on the AgNW synthesis and morphology. Previously, many surfactants have been used for the synthesis of AgNWs by the polyol process, for example, anionic surfactant sodium dodecylsulfonate (SDSN) by Tian [17] and cationic surfactant cetyltrimethylammonium bromide (CTAB) by Murphy [18]. They were used to generate either seed for AgNWs or soft capping agent. In our study the PVP was employed as main capping agent and SDS as soft capping agent. However, the small amount of SDS (*ca.* 0.3 mM) increased the reaction rate of Ag^+ reduction substantially; thus, it is suggested that SDS may play the role of phase transfer catalyst to supply Ag^+ ion effectively for reduction and incorporation into AgNW crystal growth.

From Table 1 it is noted that with small amount (0.3 mm) of SDS, the synthesis of AgNWs was almost completed in 5 min after addition of $AgNO_3$ solution into PVP/SDS solution. This may be the shortest reaction time ever reported in the synthesis of AgNWs so far. The effect of SDS concentration in AgNW synthesis is shown in Table 1 and Figure 2. At 0.24 mM SDS the aggregates of amorphous AgNWs were observed. When concentration of SDS was raised to 0.30 mM high yield of AgNWs was detected. Similarly, at 0.36 mM concentration of SDS the aspect ratio of AgNWs became significantly higher with the rapid supply of Ag^+ ions to the growing AgNW crystals. At higher concentration (0.42–0.48 mM) of SDS an increasing number of amorphous Ag particles were observed due to too rapid a supply of Ag^+ ions to be regularly packed into the AgNW crystal sites.

Table 1. Synthesis of AgNWs in batch type closed reactors.

Sample No.	$AgNO_3$ Solution		PVP/SDS Solution				SDS (mM)	Stirrer Speed (rpm)	Reaction Time after Edition of PVP/SDS (min)	Ag:PVP (Mole Ratio)
	$AgNO_3$ (g)	EG (mL)	PVP (g)	PVP (M_W)	EG (mL)	SDS (g)				
AgNW-P1	1.875	5	1.875	40,000	20	0.115	0.30	250	5	1:1.5
AgNW-P2	0.937	5	1.875	40,000	20	0.115	0.30	250	5	1:3.0
AgNW-P3	0.703	5	1.875	40,000	20	0.115	0.30	250	5	1:4.0
AgNW-P4	0.469	5	1.875	40,000	20	0.115	0.30	250	5	1:6.0
AgNW-S1	0.703	5	1.875	40,000	20	0.092	0.24	250	5	1:4.0
AgNW-P3	0.703	5	1.875	40,000	20	0.115	0.30	250	5	1:4.0
AgNW-S2	0.703	5	1.875	40,000	20	0.138	0.36	250	5	1:4.0
AgNW-S3	0.703	5	1.875	40,000	20	0.161	0.42	250	5	1:4.0
AgNW-S4	0.703	5	1.875	40,000	20	0.184	0.48	250	5	1:4.0
AgNW-R1	0.703	5	1.875	40,000	20	0.115	0.30	80	5	1:4.0
AgNW-P3	0.703	5	1.875	40,000	20	0.115	0.30	250	5	1:4.0
AgNW-R2	0.703	5	1.875	40,000	20	0.115	0.30	500	5	1:4.0
AgNW-P3	0.703	5	1.875	40,000	20	0.115	0.30	250	5	1:4.0
AgNW-T1	0.703	5	1.875	40,000	20	0.115	0.30	250	10	1:4.0
AgNW-T2	0.703	5	1.875	40,000	20	0.115	0.30	250	15	1:4.0

Figure 1. SEM images of AgNWs synthesized by mole ratio of Ag^+ ion:PVP repeat unit, (**a**) 1:1.5; (**b**) 1:3.0; (**c**) 1:4.0; and (**d**) 1:6.0.

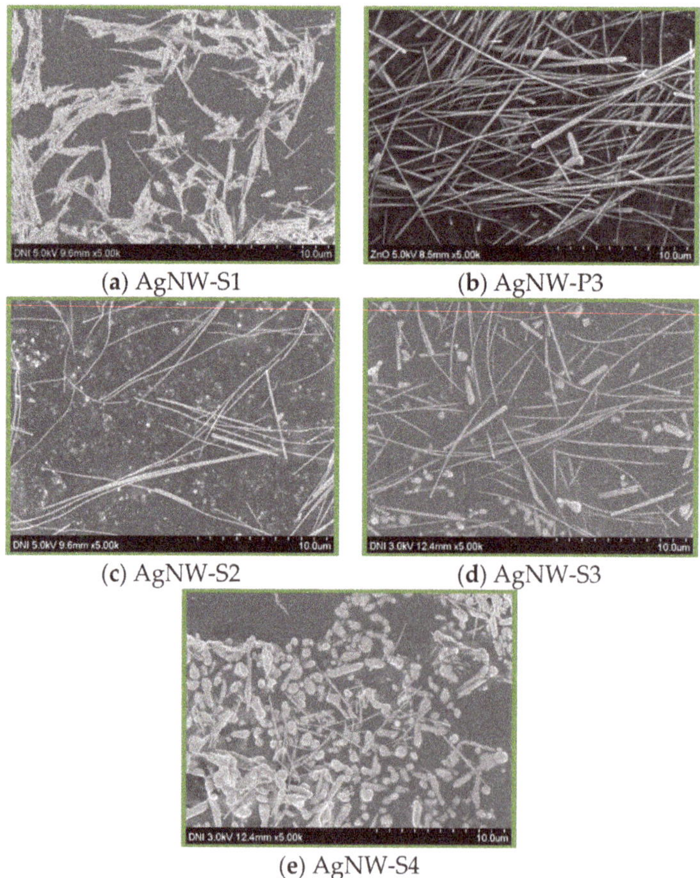

Figure 2. SEM images of AgNWs synthesized by varying the amount of SDS, (**a**) 0.24; (**b**) 0.30; (**c**) 0.36; (**d**) 0.42; and (**e**) 0.48 mM.

Other interesting features of AgNWs synthesis in closed reactor (under the reaction condition designated in Table 1) were the reaction time and the stirring speed of magnetic bar. As shown in Figure 3 the morphology of AgNWs was strongly dependent on the speed of magnetic stirrer. At 80 rpm a large amount of amorphous Ag nanoparticles was observed due to the uneven supply of Ag^+ ions to the crystal growth site. At a high stirring speed of 500 rpm the silver nanoparticles instead of AgNWs were mainly observed due to breakdown of the synthesized AgNWs to Ag nanoparticles. This suggests that AgNWs can be destroyed easily by high shearing force exerted by a magnetic stirrer in contact with the bottom of three-neck flask reactors. The effect of reaction time on AgNWs morphology in Figure 3 can also be explained similarly by high shearing forces during an excessive reaction time period.

Figure 3. SEM images of AgNWs synthesized by varying magnetic stirrer speed after addition of $AgNO_3$ in EG solution at (**a**) 80 rpm; (**b**) 250 rpm; and (**c**) 500 rpm and by varying reaction time; (**d**) 5 min; (**e**) 10 min; and (**f**) 15 min.

3.2. Scale-Up and Rapid Synthesis of AgNWs

With the data, obtained in the light of chemical and physical reaction parameters utilizing the small scale three-neck flask synthesis of AgNWs, we tried to scale up the rapid synthesis of AgNWs both in closed and continuous flow type reactors. First the AgNWs were synthesized in a resin kettle type closed reactor with mechanical stirrer instead of magnetic bar stirrer which caused direct contact of AgNWs with the bottom glass of three-neck flask reactor. The reactant materials were increased to eight times as compared to small scale synthesis in three-neck flask and mechanical stirrer with a different impeller structure was used to prepare AgNWs in high concentration at a rapid speed. The composition and ratios of reactant were kept as close as possible to the three-neck flask reactor while adjusting the chemical and physical reaction parameters in the case of scaling up AgNWs synthesis.

The experimental conditions and evaluation of AgNWs synthesized in the batch-type resin kettle reactor with different impeller structure, reaction time, and molecular weight of PVP are shown in Table 2 and Figure 4. As far as the shape of mechanical stirrer is concerned, the denticulated impeller was found to destroy the formed AgNWs faster than the screw type impeller as shown in Figure 4a–f. The denticulated impeller will encounter much more impact with the AgNWs compared to the screw type impeller under high rotating speed. The effect of reaction time in the resin kettle type reactor with mechanical stirrer can also be explained by impact exerted by mechanical stirrer, although the impact to the formed AgNWs will be much lower in the case of screw type impeller. Even with the screw-type impeller the yields of AgNWs were lower than 80% and the ratio of AgNWs to Ag nanoparticles was lower than 80:20. This was considered to be due to the high viscosity of the AgNW reaction mixture with PVP of M_W 40,000. In order to increase the yield and ratio of AgNWs and to decrease the diameter of AgNWs, we reduced the viscosity of reaction medium by decreasing the molecular weight of PVP to 10,000 g/mol. This improved the yield of AgNWs over 90% and the ratio of AgNW to Ag particles to 90:10 ratio. This could be achieved due to the proper transport of Ag^+ ions to the growing AgNWs crystal under reduced reaction medium.

Although the chemical and physical reaction conditions were established through the scale-up of AgNW synthesis in the resin kettle reactor by employing smooth screw type impeller and proper transport of Ag^+ ion to the AgNW crystal growth sites by reducing viscosity of reaction medium, improvement is still needed from the view point of rapid synthesis of AgNWs, minimum damage on the formed AgNWs, and widening of the synthetic process window for the efficient synthesis of desirable AgNW materials.

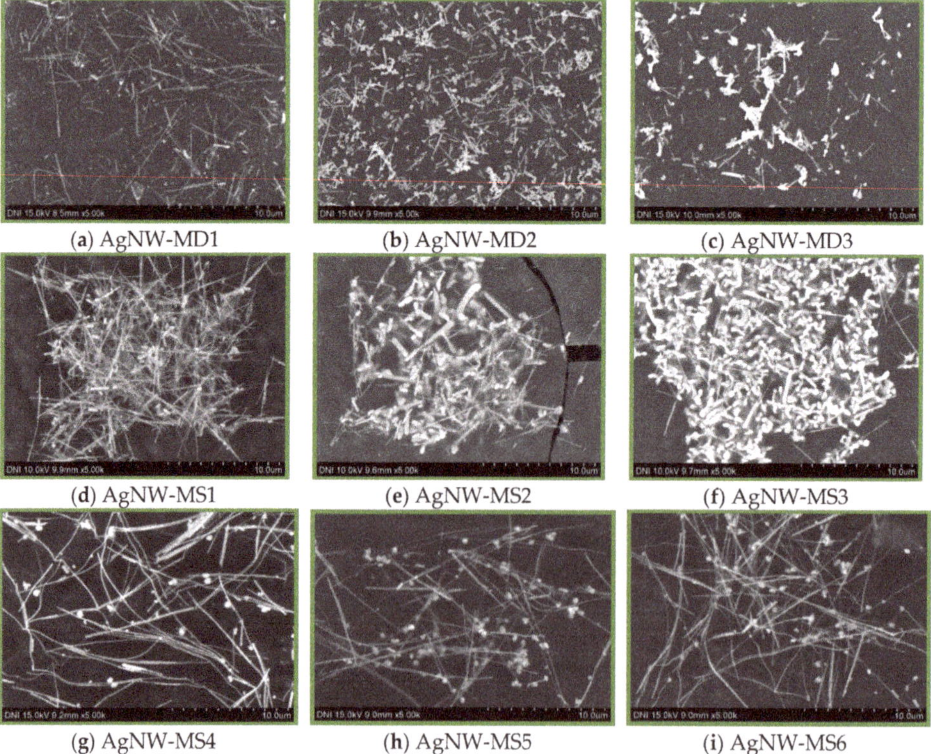

(**a**) AgNW-MD1 (**b**) AgNW-MD2 (**c**) AgNW-MD3
(**d**) AgNW-MS1 (**e**) AgNW-MS2 (**f**) AgNW-MS3
(**g**) AgNW-MS4 (**h**) AgNW-MS5 (**i**) AgNW-MS6

Figure 4. SEM images of AgNWs synthesized by varying impeller structure and reaction times, (**a**–**c**) denticulated-type impeller; (**d**–**f**) screw-type impeller and PVP M_W 40,000; (**g**–**i**) screw-type impeller and PVP M_W 10,000.

Table 2. Synthesis of AgNWs in batch type resin kettle reactors.

Sample No.	$AgNO_3$ Solution		PVP/SDS Solution				Stirrer (rpm)	Reaction Time (min)	Impeller Type	Yield (Wire:Particle)
	$AgNO_3$ (g)	EG (mL)	PVP (g)	PVP (M_W)	EG (mL)	SDS (g)				
AgNW-MD1	5.624	40	15	40,000	160	0.92	150	5	denticulated	NA
AgNW-MD2	5.624	40	15	40,000	160	0.92	150	10	denticulated	NA
AgNW-MD3	5.624	40	15	40,000	160	0.92	150	15	denticulated	NA
AgNW-MS1	5.624	40	15	40,000	160	0.92	150	5	screw	70.0% (75:25)
AgNW-MS2	5.624	40	15	40,000	160	0.92	150	10	screw	79.4% (80:20)
AgNW-MS3	5.624	40	15	40,000	160	0.92	150	20	screw	45.7% (70:30)
AgNW-MS4	5.624	40	15	10,000	160	0.92	150	5	screw	92.0% (90:10)
AgNW-MS5	5.624	40	15	10,000	160	0.92	150	10	screw	94.0% (90:10)
AgNW-MS6	5.624	40	15	10,000	160	0.92	150	15	screw	94.2% (90:10)

Therefore, we designed a tapered tubular type reactor for the continuous flow reaction for the synthesis of AgNWs with minimal damage after formation of AgNWs. The schematic illustration of continuous flow reaction system is shown in Figure 5. In this continuous flow system, the distance ratio of L1 to L2 is kept at 80:20 which will allow synthesized AgNWs to sit at the bottom of the tubular reactor thus preventing a damage from shearing force. The reaction condition were adjusted according to the AgNWs synthetic data in Table 2 except the addition of 0.01 mM of NaBr into PVP/SDS in EG solution for the facile seeding of AgNWs growth. First the tubular reactor was filled with PVP/SDS in EG solution including NaBr seeding agent and then heated to 160 °C followed by dropwise addition of $AgNO_3$ in EG solution into the tapered tubular reactor with a slow rotation of mechanical stirrer with screw impeller. Secondly, after confirming the formation of AgNWs from the initial turbidity of the reaction medium, the reaction mixture in the bottom part of tapered reactor was taken out by opening the valve and then collected in the recovering vessel. The yield and ratio of the AgNWs to Ag nanoparticles were over 90% and 90:10, respectively. The average diameter of the AgNWs was about 88 nm and length of the AgNW was in the range of 30–80 μm as shown in Figure 6.

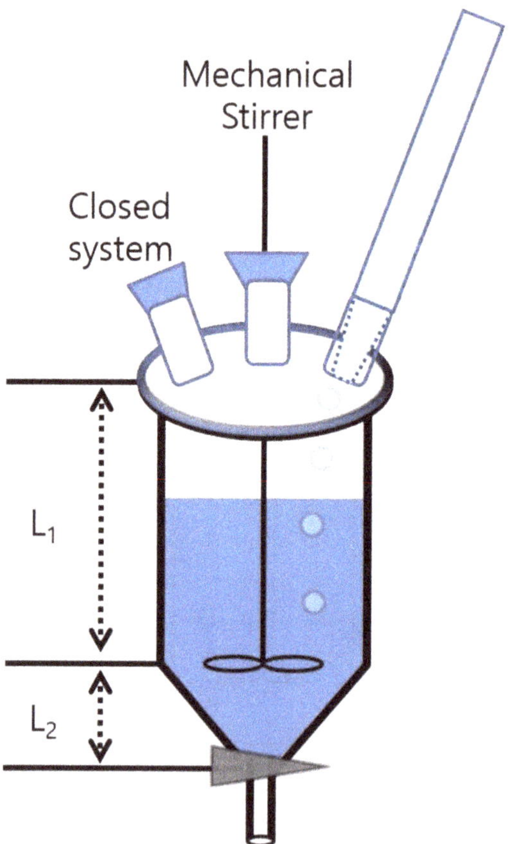

Figure 5. Schematic image of tapered tubular type reactor for continuous flow reaction of AgNWs.

(a)

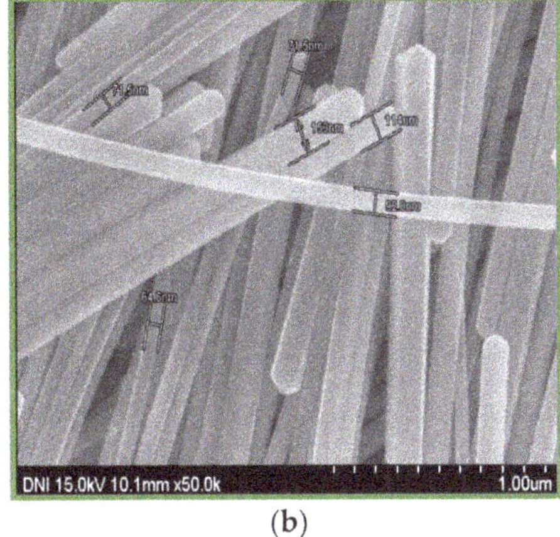

(b)

Figure 6. SEM images of AgNWs synthesized by continuous flow reaction of AgNWs utilizing tapered tubular type reactor: (**a**) low; (**b**) high resolution.

3.3. Recovery and Purification of AgNWs and Properties of AgNW-PET Film

The recovery of the AgNWs has usually been carried out by centrifugation and repeated washing with DI water or ethanol. We first tried to get information on the solvent through the centrifuge experiments with AgNWs reaction mixture obtained as shown in Table 3. The reaction mixture of AgNWs solution (1–3 mL) was diluted to 50 mL with aqueous solution containing NaCl or urea as a hydrogen bonding breaker and then subjected to centrifuge at 3000 rpm for 30–60 min, followed by removal of supernatant by siphon apparatus. The purification of recovered AgNWs was conducted by repeated centrifugation with DI water three times. The recovery data of AgNW in Table 3 shows that 25% NaCl aqueous solution is an effective solvent for recovering AgNWs from the reaction mixture. Although urea is known as an efficient hydrogen bond breaker, the 25 wt. % NaCl aqueous solution performed better in reducing the high viscosity of the EG solution due to the salting out effect [19].

Table 3. Search of recovery solvent for AgNWs from AgNW/EG reaction mixture.

Solvent for Centrifuge		AgNW Solution (mL)	Centrifuge Composition	Rotation Speed 3000 rpm	
				30 min	60 min
H_2O		1	Ag solution 1 g + H_2O 49 g	X	X
NaCl	NaCl 20%	1	Ag solution 1 g + NaCl(aq) 49 g	O	
		2	Ag solution 2 g + NaCl(aq) 48 g	X	O
		3	Ag solution 3 g + NaCl(aq) 47 g	X	X
	NaCl 25%	1	Ag solution 1 g + NaCl(aq) 49 g	O	
		2	Ag solution 2 g + NaCl(aq) 48 g	O	
		3	Ag solution 3 g + NaCl(aq) 47 g	X	O
Urea	20%	1	Ag solution 1 g + Urea(aq) 49 g	X	O
		2	Ag solution 1 g + Urea(aq) 48 g	X	O
		3	Ag solution 3 g + Urea(aq) 47 g	X	X

The efficiency of recovering solvent was examined further with the AgNWs reaction mixture obtained by batch type resin kettle reactor. As shown in Table 4 the 25% NaCl aqueous solution could separate AgNWs in 10 wt. % AGNWs mixture after 50 min centrifugation (80.2% yield) and 4 wt. % AgNWs mixture after 30 min centrifugation (87.7% yield), both at 3000 rpm speed.

Table 4. Recovery of AgNWs from AgNW/EG reaction mixture with centrifuge condition.

Solvent for Centrifuge	AgNW Solution (mL)	Centrifuge Composition	Centrifuge Time (min)	Yield (%)
NaCl 20%	1	AgNW solution 1 g + NaCl(aq) 49 g	30 min	76.2%
	2	AgNW solution 1 g + NaCl(aq) 48 g	30 min	70.2%
	3	AgNW solution 3 g + NaCl(aq) 47 g	50 min	76.9%
NaCl 25%	1	AgNW solution 1 g + NaCl(aq) 49 g	10 min	69.7%
	2	AgNW solution 2 g + NaCl(aq) 48 g	10 min	80.3%
	2	AgNW solution 2 g + NaCl(aq) 48 g	20 min	79.6%
	2	AgNW solution 2 g + NaCl(aq) 48 g	30 min	87.7%
	3	AgNW solution 3 g + NaCl(aq) 47 g	30 min	72.6%
	5	AgNW solution 5 g + NaCl(aq) 45 g	50 min	80.2%

In order to further decrease the recovery cost of AgNWs, precipitation method instead of centrifugation was tried as shown in Table 5. For this purpose the AgNWs reaction mixture obtained from the tapered tubular flow reactor was used, since the precipitation of the AgNWs reaction product can be directly poured into the precipitation tank. This process can reduce both the electrical power and the centrifuge manipulation cost.

After selecting 25% NaCl aqueous solution as precipitation solvent, recovery of AgNWs from the reaction mixture was conducted by natural sedimentation process. The AgNWs reaction mixtures were prepared by the tapered tubular reactor with reactant composition of 40 g PVP (M_W ~ 40,000) in 200 mL EG and 10.14 g $AgNO_3$ in 50 mL EG solutions. A part of the AgNWs reaction product (25 mL) was diluted with 5000 g of 25 wt. % NaCl aqueous solution and stood still for 6 h. After siphoning of supernatant the precipitated AgNWs was purified by washing with DI water three times.

When the composition of AgNWs reactants was 40 g PVP (M_W ~ 40,000) in 300 mL EG and 10.14 g· $AgNO_3$ in 50 mL EG, the precipitation was observed at 4000 g of 25 wt. % aqueous solution as precipitating solvent, while 40 g PVP (M_W ~ 10,000) in 200 ml EG and 10.14 g $AgNO_3$ in 50 mL EG the precipitation occurred at 2000 g of 25 wt. % NaCl aqueous solution. These could be due to the reduced viscosity in the precipitating AgNWs from the reaction mixture. These date indicate that the natural precipitation method is both efficient and cost effective for the recovery of AgNWs.

Table 5. Recovery of AgNWs from AgNW/EG reaction mixture by continuous flow reactor.

Reaction Condition for Synthesis of AgNWs With Tapered Tubular Reactor										Recovery of AgNWs	
$AgNO_3$ Solution		PVP Solution					Temp (°C)	rpm	Reaction Time (min)	25 wt. % NaCl aq. Solution (g)	Precipitation Yield (%) (Wire:Particle)
$AgNO_3$ (g)	EG (mL)	PVP (g)	PVP (M_W)	EG (mL)	SDS (g)	NaBr (mM)					
10.14	50	40.0	40,000	200	1.5	0.05	160	100	10	5000	72.7 (85:15)
10.14	50	40.0	40,000	300	1.5	0.05	160	100	10	4000	77.1 (90:10)
10.14	50	40.0	10,000	200	1.5	0.05	160	100	10	2000	89.9 (90:10)

After purification the AgNWs obtained from the tapered tubular type method was used to make spray coating solution. The AgNW/ethanol solution (0.5 wt. %) was diluted to tenfold (0.05 wt. %) by mixing with ethanol solvent with 0.25 wt. % ethyl cellulose (M_W ~ 80,000) as viscosifying agent. The AgNW/ethanol coating solution was used for the spray coating of AgNWs on top of the optical-grade PET film. The transmittance and sheet resistance of the AgNW-PET film was 86% and 90 Ωsq^{-1}, respectively with 20 times spray coating as shown in Figure 7, which is suitable for the application of touch screen panel (TSP) fabrication.

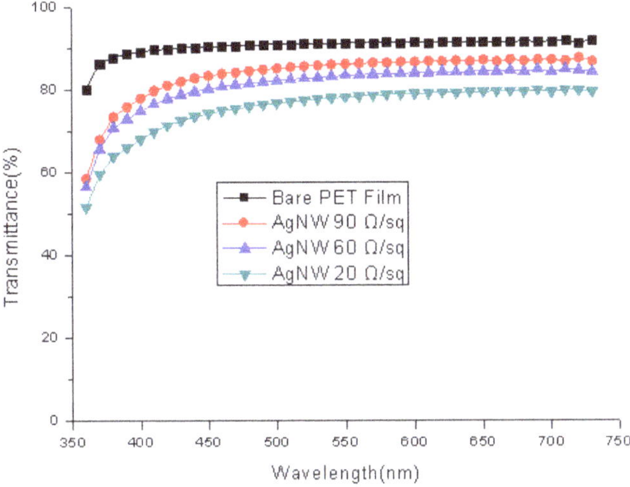

Figure 7. Transmittance and sheet resistance of AgNW-PET film made by spray coating.

The SEM image of the AgNW-PET film and XRD peak of AgNWs are shown in Figure 8 in which the crystal lattice constant of AgNWs was 4.088 Å exhibiting face-centered cubic crystal structure of silver. The bending test of AgNW-PET and ITO-PET film shows that the former is stable up to 10,000 bending stress while the latter shows rapid increase of sheet resistance from 8000 bending motions suggesting that AgNW-PET film can be used in the fabrication of flexible TSP and flexible display substrate.

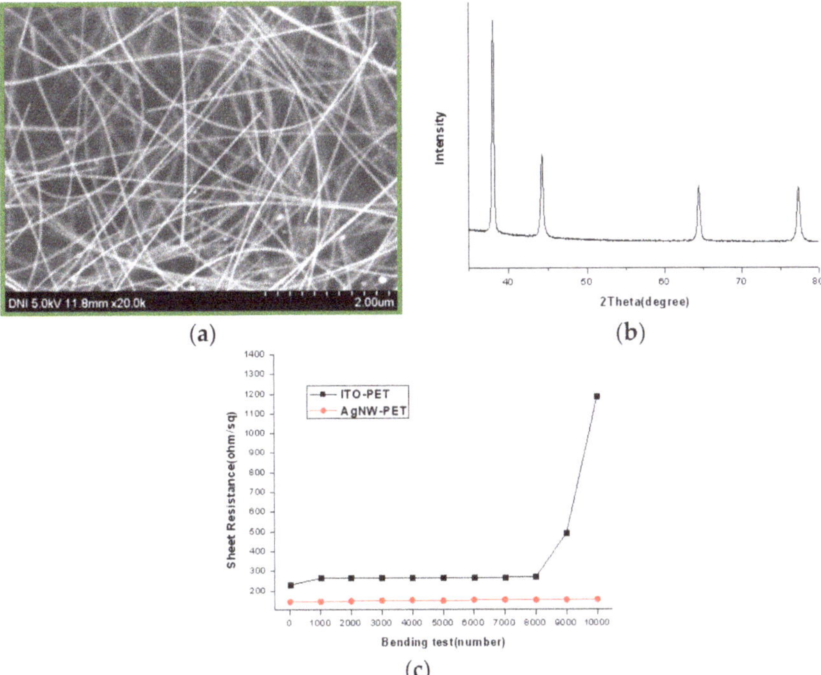

Figure 8. (**a**) SEM image of AgNW-PET film; (**b**) XRD peak of AgNW; and (**c**) bending test of AgNW-PET film.

4. Conclusions

Silver nanowires (AgNWs) were synthesized by the polyol process with high yield and high aspect ratio utilizing a tapered tubular continuous flow reactor. The synthesized AgNWs exhibited less defects in morphology of AgNWs which are commonly caused by excessive shearing for long time in the closed reactor. After reaction the AgNWs were precipitated in the aqueous solution with the aid of hydrogen bond breaker and could be recovered effectively without using high cost centrifugation process. Dispersion of the AgNWs were used to prepare a transparent conducting electrode (TCE) films by spray coating method which showed 86% transmittance and 90 Ωsq^{-1} sheet resistance applicable to touch screen panel fabrication.

Acknowledgments: This work was supported by the 2014 Research Fund(1.140056.01) of UNIST(Ulsan National Institute of Science and Technology).

Author Contributions: H. D. Yun, D. M. Seo and M. Y. Lee cfontributed to the synthesis of AgNWs and S. Y. Kwon to the analysis of AgNWs.

Conflicts of Interest: The authors declare no conflict of interest.

References

1. Li, B.; Ye, S.; Stewart, I.E.; Alvarez, S.; Wiley, B.J. Synthesis and purification of silver nanowiers to make conducting films with a transmittance of 99%. *Nano Lett.* **2015**, *15*, 6722–6726. [CrossRef] [PubMed]
2. Zhang, D.; Ryu, K.; Liu, X.; Polikarpov, E.; Ly, J.; Tompson, M.E.; Zhou, C. Transparent, Conductive, and Flexible Carbon Nanotube Films and Their Application in Organic Light-Emitting Diodes. *Nano Lett.* **2006**, *6*, 1880–1886. [CrossRef] [PubMed]
3. Geng, H.-Z.; Kim, K.K.; So, K.P.; Lee, Y.S.; Chang, Y.; Lee, Y.H. Effect of Acid Treatment on Carbon Nanotube-Based Flexible Transparent Conducting Films. *J. Am. Chem. Soc.* **2007**, *129*, 7758–7759. [CrossRef] [PubMed]
4. Wassei, J.K.; Kaner, R.B. Graphene, a promising transparent conductor. *Materialstoday* **2010**, *13*, 52–59. [CrossRef]
5. Lee, J.H.; Makotchenko, D.W.S.V.G.; Mazarov, A.; Fedorov, B.E.; Kim, Y.H.; Choi, J.-Y.; Kim, J.M.; Yoo, J.-B. One-Step Exfoliation Synthesis of Easily Soluble Graphite and Transparent Conducting Graphene Sheets. *Adv. Mater.* **2009**, *21*, 4383–4387. [CrossRef] [PubMed]
6. Elschner, A.; Lovenich, W. Solution-deposited PEDOT for transparent conductive applications. *MRS Bull.* **2011**, *36*, 794–798. [CrossRef]
7. Na, S.I.; Kim, S.S.; Jo, J.; Kim, D.Y. Efficient and Flexible ITO-Free Organic Solar Cells Using Highly Conductive Polymer Anodes. *Adv. Mater.* **2008**, *20*, 4061–4067. [CrossRef]
8. Hu, L.; Kim, H.S.; Lee, J.; Peumans, P.; Cui, Y. Scalable Coating and Properties of Transparent, Flexible, Silver Nanowire Electrodes. *ACS Nano* **2010**, *4*, 2955–2963. [CrossRef] [PubMed]
9. Lee, J.; Lee, P.; Lee, H.; Lee, D.; Lee, S.S.; Ko, S.H. Very long Ag nanowire synthesis and its application in a highly transparent, conductive and flexible metal electrode touch panel. *Nanoscale* **2012**, *4*, 6408–6414. [CrossRef] [PubMed]
10. Rathmell, A.R.; Bergin, S.M.; Hua, Y.L.; Li, Z.Y.; Wiley, B.J. The Growth Mechanism of Copper Nanowires and Their Properties in Flexible, Transparent Conducting Films. *Adv. Mater.* **2010**, *22*, 3558–3563. [CrossRef] [PubMed]
11. Andrés, L.J.; Menéndez, M.F.; Gómez, D.; Martínez, A.L.; Bristow, N.; Kettle, J.P.; Menéndez, A.; Ruiz, B. Rapid synthesis of ultra-long silver nanowires for tailor-made transparent conductive electrodes: Proof of concept in organic solar cells. *Nanotechnology* **2015**. [CrossRef] [PubMed]
12. Braun, E.; Eichen, Y.; Sivan, U.; Ben-Yoseph, G. DNA-templated assembly and electrode attachment of a conducting silver wire. *Nature* **1997**, *391*, 775–778. [CrossRef] [PubMed]
13. Huang, M.H.; Choudrey, A.; Yang, P. Ag nanowire formation within mesoporous silica. *Chem. Commun.* **2000**, *12*, 1063–1064. [CrossRef]
14. Li, X.; Wang, L.; Yan, G. Review: Recent research progress on preparation of silver nanowires by soft solution method and their applications. *Cryst. Res. Technol.* **2011**, *46*, 427–438. [CrossRef]

15. Xia, Y.; Sun, Y. Large-Scale Synthesis of Uniform Silver Nanowires Through a Soft, Self-Seeding, Polyol Process. *Adv. Mater.* **2002**, *14*, 833–836.
16. Cushing, B.L.; Kolesnichenko, V.L.; O'Connor, C.J. Recent Advances in the Liquid-Phase Syntheses of Inorganic Nanoparticles. *Chem. Rev.* **2004**, *104*, 3893–3946. [CrossRef] [PubMed]
17. Hu, J.Q.; Xie, Z.X.; Han, G.B.; Wang, R.H.; Ren, B.; Zhang, Y.; Yang, Z.L.; Tian, Z.Q. A Simple and Effective Route for the Synthesis of Crystalline Silver Nanorods and Nanowires. *Adv. Mater.* **2004**, *14*, 183–189. [CrossRef]
18. Jana, N.R.; Gearheart, L.; Murphy, C.J. Wet Chemical Synthesis of High Aspect Ratio Cylindrical Gold Nanorods. *J. Phys. Chem. B* **2001**, *105*, 4065–4067. [CrossRef]
19. Mason, S.M.; Cosgrove, D.J. Disruption of hydrogen bonding between plant cell wall polymers by proteins that induce wall extension. *Proc. Natl. Acad. Sci. USA* **1994**, *91*, 6574–6578. [CrossRef]

10

Porous γ-TiAl Structures Fabricated by Electron Beam Melting Process

Ashfaq Mohammad [1,*], Abdulrahman M. Alahmari [1,2], Khaja Moiduddin [2], Muneer Khan Mohammed [1], Abdulrahman Alomar [2] and Ravi Kottan Renganayagalu [3]

Academic Editor: Hugo F. Lopez

1 Princess Fatima Alnijiris's Research Chair for Advanced Manufacturing Technology (FARCAMT), Advanced Manufacturing Institute, King Saud University, Riyadh 11421, Saudi Arabia; alahmari@ksu.edu.sa (A.M.A.); muneer0649@gmail.com (M.K.M.)

2 Industrial Engineering Department, College of Engineering, King Saud University, Riyadh 11421, Saudi Arabia; kmoiduddin@gmail.com (K.M.); alomar.ab@gmail.com (A.A.)

3 Structural Nanomaterials Laboratory, PSG Institute of Advanced Studies, Coimbatore 641004, India; krravi.psgias@gmail.com

* Correspondence: mashfaq@ksu.edu.sa

Abstract: Porous metal structures have many benefits over fully dense structures for use in bio-implants. The designs of porous structures can be made more sophisticated by altering their pore volume and strut orientation. Porous structures made from biocompatible materials such as titanium and its alloys can be produced using electron-beam melting, and recent reports have shown the biocompatibility of titanium aluminide (γ-TiAl). In the present work, we produced porous γ-TiAl structures by electron-beam melting, incorporating varying pore volumes. To achieve this, the individual pore dimensions were kept constant, and only the strut thickness was altered. Thus, for the highest pore volume of ~77%, the struts had to be as thin as half a millimeter. To accomplish such fine struts, we used various beam currents and scan strategies. Microscopy showed that selecting a proper scan strategy was most important in producing these fine struts. Microcomputed tomography revealed no major gaps in the struts, and the fine struts displayed compressive stiffness similar to that of natural bone. The characteristics of these highly-porous structures suggest their promise for use in bio-implants.

Keywords: additive manufacturing; electron beam melting; titanium aluminides; porous structures; bio-implants; compressive stiffness

---

1. Introduction

Porous structures have many advantages over solid ones. For example, implants with porous structures exhibit better bone ingrowth [1]. In addition to this, porous structures are naturally lightweight, and porous implants induce less stress shielding than do fully solid ones [2]. Porous structures are relevant in many other applications as well, including ultra-lightweight structures, thermal equipment, and electrochemical devices [3].

Over the last decade, several researchers have tried to fabricate porous structures by using electron beam melting (EBM), an additive manufacturing process. Most of these reports dealt with Ti-6Al-4V alloy, a popular bio-implant material [4–6], demonstrating that EBM could produce high-quality porous structures. For example, Li *et al.* produced pores as small as 1 mm and struts with a thickness of only 0.5 mm [7]. Other works focused on using EBM to fabricate structures from more materials, such as stainless steels, niobium alloys, and copper [8–10]. Similarly, Cormier *et al.* used EBM to fabricate γ-TiAl, giving some positive results [11]. Processing γ-TiAl by EBM was motivated by the alloy's potential use in aeroturbine manufacturing. Until this EBM study, γ-TiAl was used little because it was difficult to machine, forge, and cast [12,13].

Rivera-Denizard *et al.* argued that γ-TiAl has certain advantages over Ti-6Al-4V as a bio-implant material [14]. For instance, γ-TiAl does not contain vanadium, a toxic element; aluminum oxide, the predominant oxide of γ-TiAl, might prove better than titanium oxide, which erodes easily. These assumptions were tested by *in vitro* experiments that showed γ-TiAl was indeed biocompatible with osteoblast cells. Santiago-Medina *et al.* in a later work suggested that γ-TiAl's biocompatibility could be further improved with suitable surface treatments [15].

Considering the potential uses of γ-TiAl in bio-implants and aircraft turbines, the ability to produce porous structures of this alloy could give designers an edge. However, there are few reports on porous γ-TiAl; in the only report of note, Hernandez *et al.* produced stochastic foams of γ-TiAl by EBM [16]. In the current work, we aim to produce γ-TiAl structures with designed porosity. We attempted to fabricate porous γ-TiAl samples with various pore volume percentages by varying the strut thickness from 1.5 mm to 0.5 mm. From these results, we analyze and discuss how the beam current, strut thickness, and scanning strategy affected the dimensional accuracy, microstructure, and mechanical behavior of EBM-fabricated porous γ-TiAl alloy.

2. Experimental Section

γ-TiAl powder with a nominal composition of Ti-48Al-2Cr-2Nb (at. %) supplied by ARCAM AB was used as feedstock material. The powder particles were 45–150 μm in size with a mean of 110 μm, as measured by laser diffraction particle sizing technique (Mastersizer 2000, Malvern Instruments, Worcestershire, UK) (Figure 1). The particle morphology was essentially spherical, with smaller satellite particles adhering to larger ones (Figure 2). EBM was performed on an ARCAM A2 machine (ARCAM AB, Mölndal, Sweden). Each layer was set to be 90 μm thick. More details on EBM can be found elsewhere [17]. Samples were built on a 10 mm thick stainless steel support plate (area of 100 × 100 mm).

First, the porous structures were designed using CAD software (CATIA, Dassault Systèmes, Vélizy-Villacoublay Cedex, France). These files were saved in STL format and then transferred to Magics (Ver.17) software (Materialise NV, Leuven, Belgium), in which the files were checked for errors, such as missing or inverted triangles. The qualified files were then transferred to Arcam's Build Assembler software (ARCAM AB, Mölndal, Sweden). This software converts the STL geometry into thin slices and writes the output as a machine-specific ABF file. This ABF file contains all the instructions needed for a machine-control program to build the final part layer by layer.

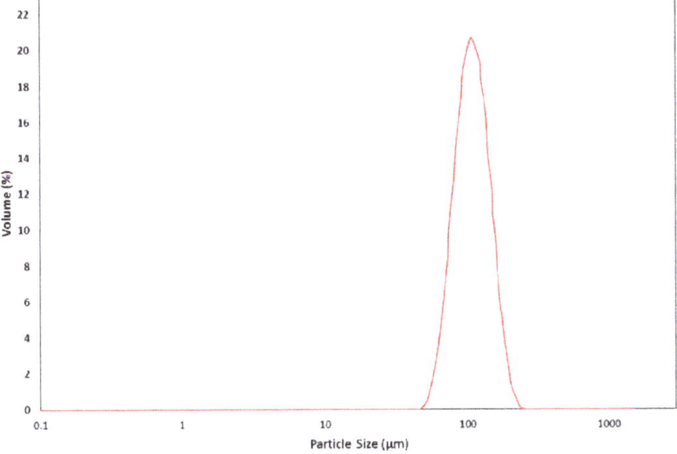

Figure 1. Particle size distribution of the feedstock powder.

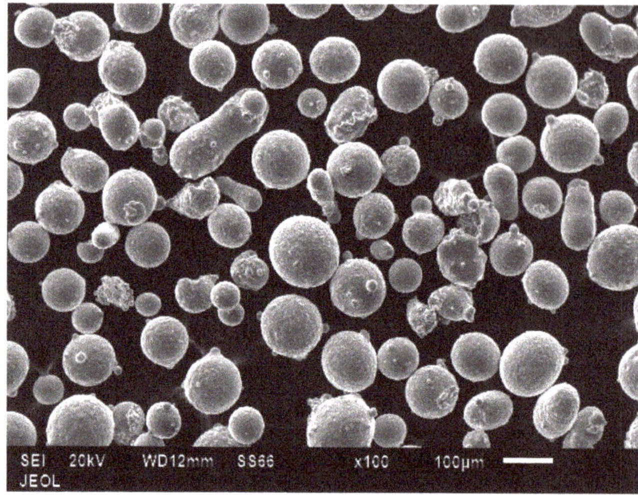

Figure 2. Morphology of γ-TiAl powder particles observed by scanning electron microscope.

Figure 3 shows CAD designs of our porous structure blocks. The blocks had 17 mm sides and a 19 mm height, and the struts had a square cross-section. By fixing the external dimensions of the blocks, the strut thickness was the only design parameter that varied to obtain different pore volume proportions. The theoretical pore volume proportions worked out to be 77%, 68%, and 54% for blocks with struts of 0.5, 1, and 1.5 mm thickness, respectively. These particular pore volumes were chosen because of the fact that at least 50% porosity is necessary in an implant for satisfactory bone ingrowth [18], and at the same time each pore should be in the range 500–2000 μm [6]. In the current study, the pore size of the entire samples was conceived to be 2500 μm in all three dimensions.

Figure 3. Porous designs with various strut thickness: 0.5 mm (**left**), 1 mm (**middle**), and 1.5 mm (**right**).

Fabricating porous structures using EBM is challenging because of such structure's fine features and discrete melt cross-sectional areas, making the scan strategy—how the beam scans when melting a given region—especially important. Typically, in EBM, the beam draws the border (contouring) and then scans the interiors (hatching). However, this "contour-and-hatch" strategy is meant for producing large feature sizes when seen against the small beam diameter (0.2–0.5 mm). Thus, in the modified scan strategy, we completely avoided the contour scan, using only the hatch scan. Table 1 shows the processing parameters that we varied during the experiments.

Table 1. EBM settings for fabricating various porous designs.

Strut Size (mm)	Beam Current (mA)	Scan Strategy
0.5 1 1.5	6, 12, and 18	Standard (contouring + hatching);
0.5 1 1.5	18	Modified (only hatching)

We produced porous γ-TiAl samples using various combinations of processing parameters, and then evaluated these samples' integrity. Discontinuities in strut formation were analyzed using scanning electron microscopy (SEM; JSM-6610LV, JEOL, Tokyo, Japan). After cutting the samples in the middle along vertical (*Z*) and horizontal (*XY*) directions with respect to the build direction, we observed them using optical microscopy (Axio Scope A1, Carl Zeiss, Oberkochen, Germany). This method will reveal asymmetry, if any, between the horizontal and vertical struts. To understand how the beam current and scan strategy affect the grain structure and phase evolution of the samples, we also examined their microstructures with optical microscopy. For this purpose, samples were etched with modified Kroll's agent (10 vol. % HF, 12 vol. % HNO_3, 85 vol. % H_2O).

Microcomputed tomography (μ-CT, General Electric, Wunstorf, Germany) was used to detect and analyze any defects and imperfections formed within the struts. This technique can measure the total open and closed porosity within a sample, and it provides a glimpse of the interior structure (non-destructive evaluation) without physical cutting or polishing. μ-CT studies were performed using a Phoenix VTomeX L240 micro-CT system (General Electric, Wunstorf, Germany) with X-ray settings of 120 kV and 160 μA. Beam filters were used to reduce beam hardening artefacts, and the scan resolution was 20 μm. The compressive strength of the cubes was analyzed using a Zwick Z100 Tester (Zwick Roell AG, Ulm, Germany) attached with a 100kN load cell and at a crosshead travel speed of 1 mm/min.

3. Results and Discussion

Initially, we used the standard scan strategy to produce blocks with struts of various sizes by just varying the current (Figure 4). This strategy produced 1 and 1.5 mm struts, but failed to produce 0.5 mm struts at any beam current. The failed builds had sections of only the horizontal struts; vertical struts that are supposed to connect the horizontal sections were completely absent. The beam diameter in EBM is 0.1–0.4 mm [19–21], which is on a par with 0.5 mm strut dimension. A standard scan strategy in EBM has two stages: contouring followed by hatching. During the contour run, the beam traces the periphery of the cross-section to be melted. Generally, the contour scanning beam has lower intensity. The main purpose of the contour run is to isolate the surrounding powder from the highly intense hatching scan to come. Contour scanning produces a smoother surface because it limits sintering of adjacent powder to the build walls. However, for especially fine cross-sections, such as those of 0.5 mm struts, the contour run would leave no space for hatching, leading to insufficient melting and a failed build.

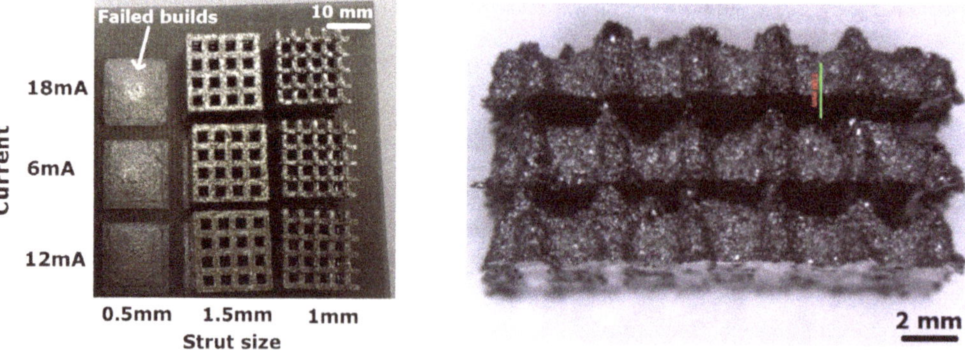

Figure 4. γ-TiAl porous structures built by EBM using the standard scan strategy (**left**); vertically disconnected layers in a failed build of the 0.5 mm strut (**right**).

Figure 5 shows a cross-section of a 1.5 mm strut produced using the standard scan strategy. Figure 6 shows a magnified image of one vertical strut. The vertical strut is ~1.37 mm wide, slightly narrower than the designed size of 1.5 mm. In contrast, the horizontal struts are much thicker (~2.05 mm) than the design. Heinl *et al.* also observed thicker horizontal struts and thinner vertical struts in Ti-6Al-4V porous structures [22]. The beam placement mainly controls the size accuracy of the vertical strut, while the resolution of the stacked layers and the table movement control the size accuracy of the horizontal struts. This process overbuilds or underbuilds the horizontal struts if the designed strut size is not an integer multiple of the layer thickness.

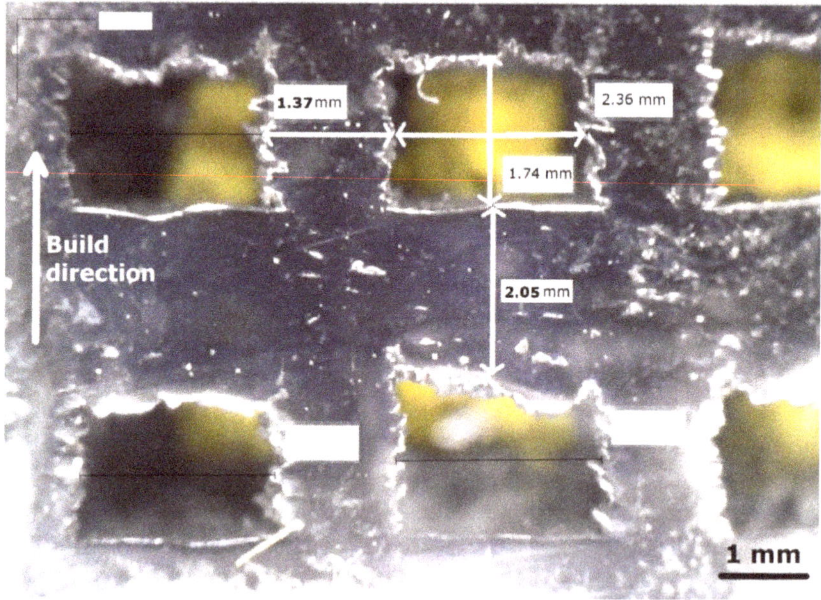

Figure 5. Macro image of 1.5 mm vertical struts produced using the standard scan strategy (cross-section along the Z-direction).

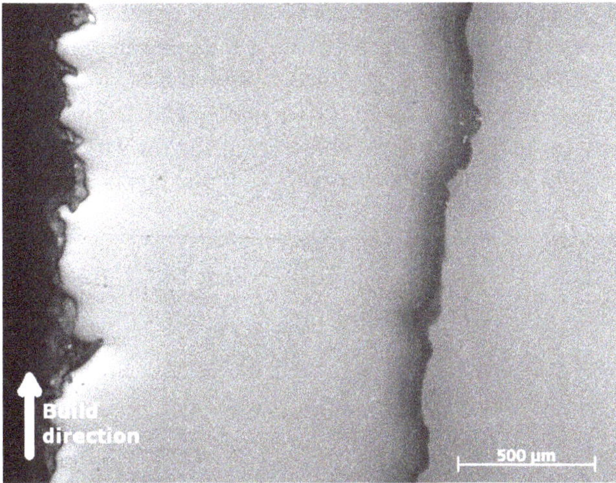

Figure 6. High-magnification optical micrograph showing 1.5 mm vertical strut produced with a standard scan strategy.

Even within a single horizontal strut, differences appeared between the top and bottom surfaces. For example, Figure 5 shows a top surface that is flatter and smoother than the bottom. During a build, the input heat mainly conducts downwards, causing the bottom surface to attract underlying loose powder and sinter it. Since there is no molten layer on the top surface, the colder powder sitting on it does not sinter as much as the bottom surface.

In the *XY* plane, the horizontal struts were uniform in size (Figure 7), and the pores were closer to square in shape (~2.56 × 2.55 mm). In the *Z* plane, though, the pores were more rectangular (~2.36 × 1.74 mm) (Figure 5). This difference means that the pores tend to be anisotropic in EBM structures. If this anisotropy is undesirable, the horizontal and vertical struts should then be designed with different sizes.

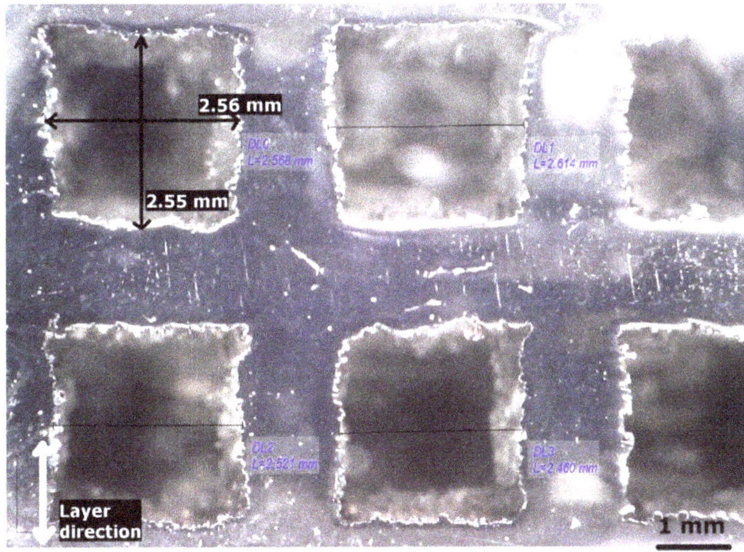

Figure 7. Macro image of 1.5 mm horizontal struts produced using standard scan strategy (cross-section along *XY* direction).

The vertical struts had jagged surfaces (Figure 8). Discrete layers added one over the other lead to the jagged vertical surfaces. This was not the only reason though; complex melt pool dynamics—how much the melt pools wets the underlying solidified layer and how much the melt pool entraps the adjacent powder particles—also determine the surface quality [23]. In general, the surface quality can be improved by reducing individual layer thickness. However, the basic size of the powder particles limits how thin a layer could be spread. Furthermore, very thin layers increase the total build time.

Figure 8. SEM image showing jagged vertical struts produced using standard scan strategy (1.5 mm strut).

3.1. Modified Scan Strategy

To avoid the problem of insufficient space for the hatching scan, the contour scan was removed in the next attempt. EBM experiments were later performed with this modified scan strategy. To ensure sufficient heat input, we selected a maximum beam current of 18 mA. This process successfully produced 0.5 mm struts along with the other strut sizes (Figure 9).

Figure 9. Intact γ-TiAl porous blocks built with 0.5 mm struts by EBM with the modified scan strategy.

Even at such a small length of 0.5 mm, both the horizontal and vertical struts were continuous and displayed no evidence of major defects (Figure 10). The modified scan strategy also improved the quality of the horizontal struts in samples with larger struts (Figure 11). In these samples, only a little powder appeared sintered to the bottom surfaces. The pores were relatively square. These advantages bolster the case for discarding the contour scan altogether while building fine porous structures using EBM.

Figure 10. Intact 0.5 mm struts produced using the modified contour strategy.

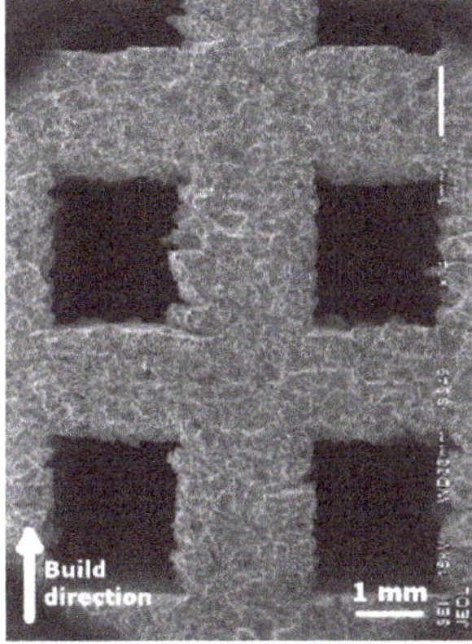

Figure 11. Better-quality surfaces of horizontal struts produced with the modified scan strategy (1.5 mm strut).

Figure 12 shows a vertical strut produced with the modified scan strategy. Here, the average strut size was ~1.8 mm, which is indeed thicker than the intended size of 1.5 mm. Note that a standard scan strategy produced a thinner strut (Figure 6). Since the modified scan strategy has no contour scan, the intense hatching beam scanned the whole cross-section. This process generated a larger melt volume, producing thicker struts.

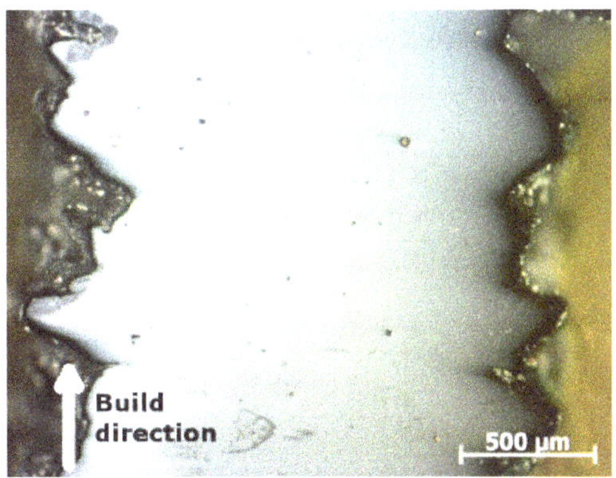

Figure 12. Vertical 1.5 mm strut built with the modified scan strategy.

3.2. *Effect of Beam Current*

Figure 13 shows how the beam current affected the microstructures of the struts. These microstructures came from the middle region of the samples in the build (vertical) direction. Further, the microstructures were almost the same even in the top and bottom regions (not shown here). This was apparently due to the limited height (19 mm) of the samples in our study.

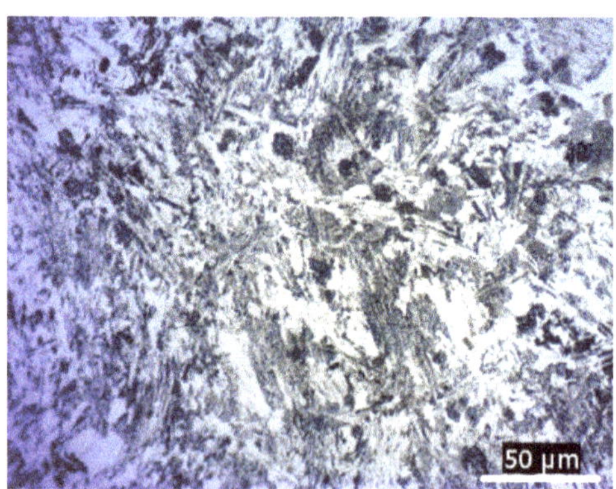

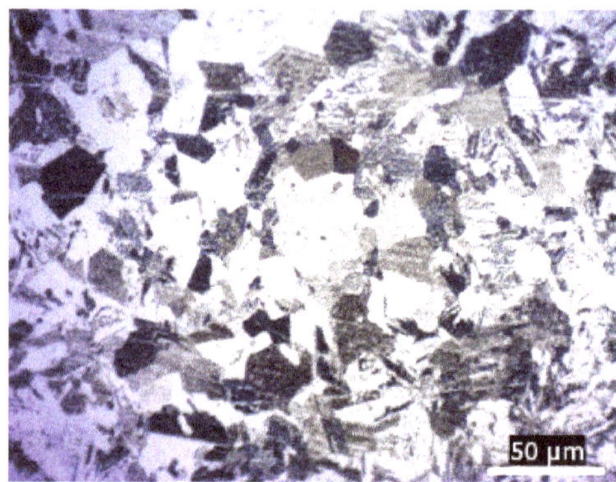

Figure 13. A lamellar microstructure produced at a higher beam current of 18 mA (**left**), compared to an equiaxed structure produced at 6 mA (**right**); both with the standard scan strategy.

Typically, the microstructure for the composition chosen in the current work should consist of two phases: γ(TiAl) and $\alpha 2$(Ti_3Al). Depending upon the dispersion of each of these phases, the microstructure can vary from near gamma to fully lamellar, with a duplex type in between. In the near gamma type, the grains are mainly γ-phase, whereas $\alpha 2$-phase can be found only along the grain boundaries. At the other extreme, these two phases can co-exist side-by-side as lamellae. A mix of these two types of microstructures results in the duplex type, wherein lamellar colonies would be seen amidst the equiaxed γ grains. The microstructure that eventually settles down in the dual phase TiAl compositions ultimately depends on the processing conditions, mainly the temperature. For instance, elemental powders having composition (48% Al-2% Cr-2% Nb) similar to the present work, when consolidated by the powder metallurgy route at a temperature of 1400 °C resulted in a fully lamellar microstructure, while at a lesser temperature of 1200 °C, the same powder resulted in a duplex structure [24]. Likewise, in the current study, a higher beam current produced a microstructure dominated by coarse lamellar structure with few equiaxed grains (Figure 13), but when the current was lowered, more equiaxed regions started to emerge (Figure 13 right). Higher beam currents mean more heating of the current and the underlying layers. The best microstructure from a ductility point of view, however, is the duplex type [25] , and this was seen in the case of a modified scan strategy (Figure 14). It is interesting to note that Biamino *et al.* [25] achieved the duplex microstructure in EBM produced γ-TiAl after heat treating the near equiaxed microstructure. The fact that we observed duplex microstructure in as-produced EBM samples suggests that the electron beam subjected the underlying regions to the same kind of heat treatment. Hence, it cannot be overemphasized how important was the beam current and scan strategy in obtaining an optimum microstructure. In other words, both these parameters influenced the processing temperature.

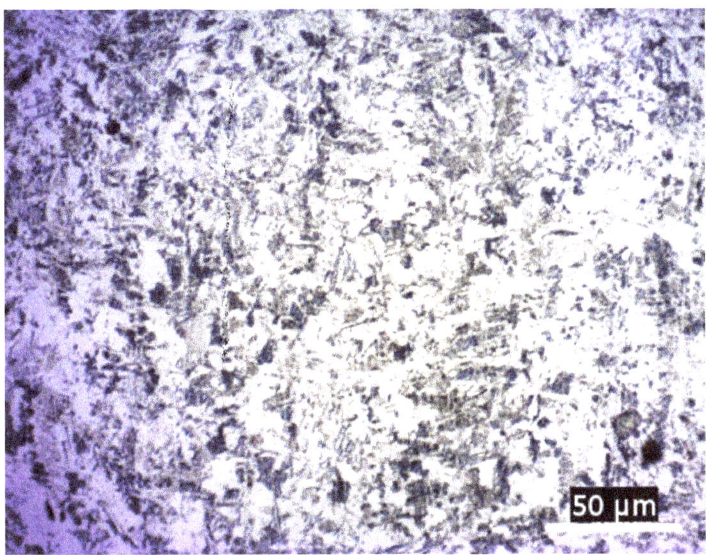

Figure 14. Duplex microstructure produced with the modified scan strategy.

Table 2 shows the µ-CT analysis of the porous γ-TiAl blocks. The 0.5 mm struts had a surface-to-volume ratio almost twice that of the 1 mm struts. Such a high ratio is desirable for material-to-bone adhesion in bio-implants [26].

Table 2. Surface-to-volume ratio estimated by µ-CT.

Strut Thickness (mm)	Scan Strategy	Surface Area/Volume Ratio (mm^{-1})
1	Standard	1.19
1	Modified	4.63
0.5	Modified	8.88

Figure 15 gives a µ-CT image of the 0.5 mm struts. The 2D slice (top left image in Figure 15) shows scattered discontinuities along the span of the struts. This may create an impression that the struts were broken at these positions, which is not true, however. If we take into account that it is a 2D slice, it is quite possible that an adjacent slice could envelop a discontinuity, thus making it a closed pore rather than a total rupture in the strut. Hence, a 3D illustration (bottom right in Figure 15) would prove more valuable in such cases. Indeed, quantitative analysis revealed internal porosity to be minimal (~0.19%) (Table 3). The diameter of the largest internal pore was ~0.43 mm, meaning at least a portion of the material was constantly present along the strut. Since the currently used scan resolution is 20 µm, pores smaller than this dimension could not be detected by the µ-CT test. For detecting such small pores, one has to resort to alternate techniques such as 2-D image analysis.

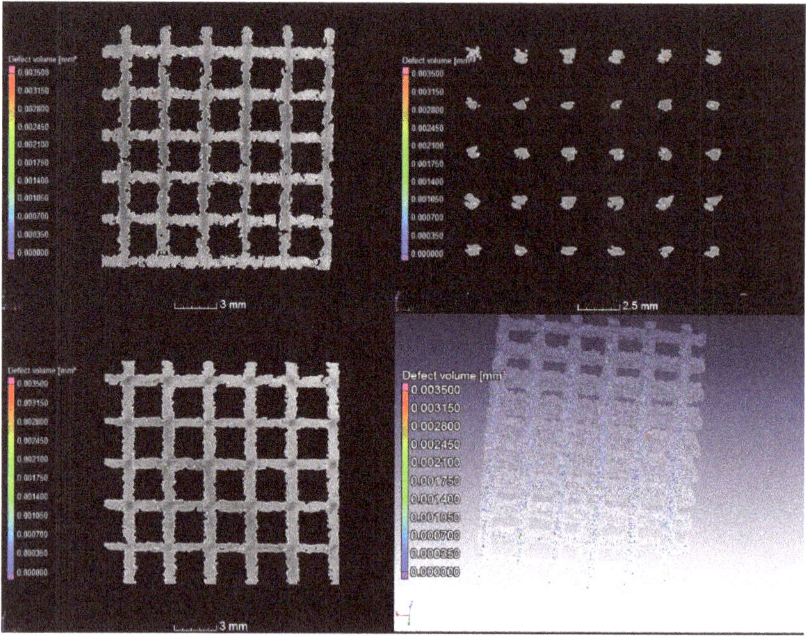

Figure 15. Internal defect volume analyzed by μ-CT (0.5 mm strut).

Table 3. Quantitative assessment of internal porosity by μ-CT.

Strut Thickness (mm)	Scan Strategy	Largest Internal Pore Diameter (mm)	Average Internal Porosity (vol. %)
1	Standard	1.19	0.25
1	Modified	0.52	0.21
0.5	Modified	0.43	0.19

The modified scan strategy proved to be better even for producing thicker struts (Table 3). This strategy reduced both the size and total volume of internal pores. Figure 16 shows the extent of divergence observed in the built specimens from the STL dimensions. The divergence was lower for the samples produced with the modified scan strategy. As discussed earlier, this occurred because the modified scan strategy reduced powder sintering to the bottom of the horizontal struts. This aspect ultimately narrowed down the divergence from the actual values.

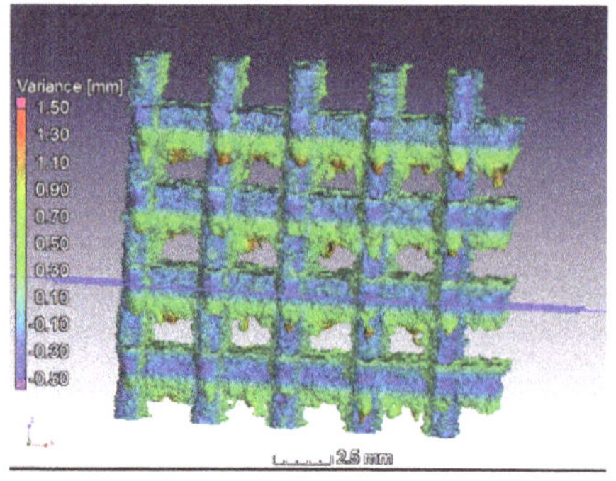

Figure 16. Degree of variance from original STL file (1 mm strut): standard (**left**) and modified (**right**) scan strategies.

3.3. Compression Test

Generally, a compression curve depends on the material and pore structure. For example, Hernández-Nava *et al.* showed that identical structures of Ti-6Al-4V and aluminum had different compression behavior [27]. Table 4 lists the results of the compression test performed on the porous blocks. Figure 17 demonstrates how the strut size affected compressive mechanical behavior of the γ-TiAl samples. The diversity in the shapes of compression curves was caused by the varying pore volume proportion. The strut thickness seems to be the dominant influence in mechanical properties, such as modulus and peak stress. Structures with 1.5 mm struts had better mechanical properties. For a given strut size, the beam current had no significant effect.

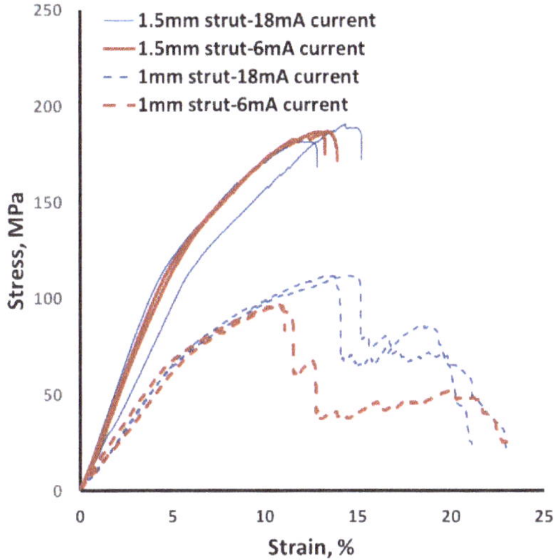

Figure 17. Effect of strut size on compressive properties in γ-TiAl porous structures produced using a standard scan strategy (Note: This scan strategy failed to build 0.5 mm strut samples as indicated in Table 4).

Table 4. Compression test results for different porous designs.

Sample Number	Sample Conditions			Young's Modulus (GPa)	Peak Compressive Stress σ (MPa)
	Strut Size (mm)	Current (mA)	Scan Strategy		
1	1.5	18	Standard	2.39 ± 0.51	186 ± 6.22
2	1.5	12	Standard	Not tested	Not tested
3	1.5	6	Standard	2.59 ± 0.02	186 ± 0.20
4	1	18	Standard	1.46 ± 0.04	111 ± 0.07
5	1	12	Standard	1.71 ± 0.02	111 ± 1.07
6	1	6	Standard	1.35 ± 0.07	96 ± 0.71
7	0.5	18	Standard	Build failed	Build failed
8	0.5	12	Standard	Build failed	Build failed
9	0.5	6	Standard	Build failed	Build failed
10	1.5	18	Modified	2.67 ± 0.16	255 ± 5.59
13	1	18	Modified	1.91 ± 0.07	125 ± 7.76
16	0.5	18	Modified	0.81 ± 0.05	26 ± 0.33

In general, the compressive stress-strain curve of a porous sample has three regions. The first is the linear elastic deformation region, in which the cells are bent or compressed elastically. The second stage is the plateau region, in which the cell edges yield by bending. In the third stage, the cell experiences strain hardening; it is in this stage that the stress maximizes [28].

The compressive stress-strain graphs initially curve up (Figures 17–19). After the upward curving ends, the linear elastic region sets in. The upward curve is generally associated with porous structures because it takes a certain time for the machine platens to come into full contact with the specimen faces [27]. None of the curves had a distinct plateau region. The elastic region was followed by a linear plastic region. Beyond this point, the structures did not deform as a whole. Instead, they experienced local fractures, reflected in the serrations at the end of curves. This behavior of the γ-TiAl porous structures was similar to that of a porous Ti-6Al-4V structure [27,29], as both the materials are brittle and barely tolerate defects.

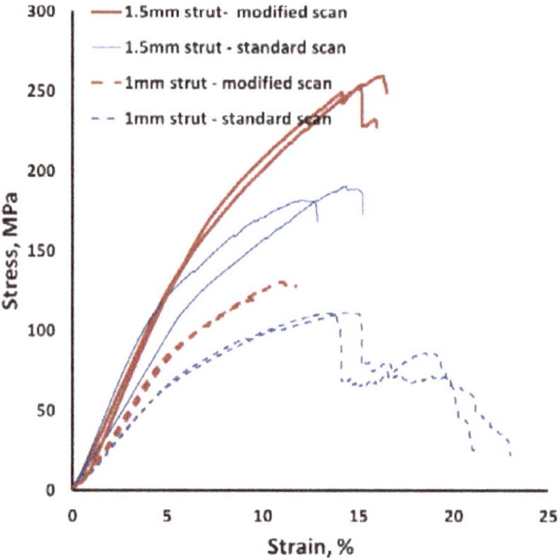

Figure 18. Effect of scan strategy on compressive properties in γ-TiAl porous structures (beam current of 18 mA).

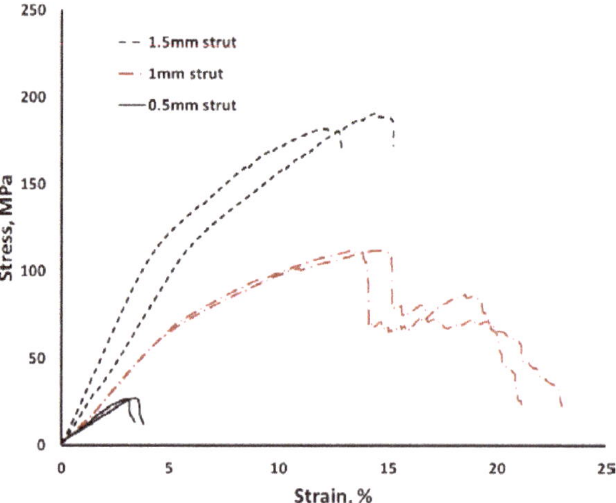

Figure 19. Effect of strut size on compressive properties of γ-TiAl porous structures (beam current of 18 mA).

Figure 18 explains the role of the scan strategy. For a given strut size and beam current, the sample produced with the modified scan strategy showed a larger peak stress than that produced with the standard strategy. As was discussed above (Section 3.2), modified scan strategy resulted in a duplex microstructure, whereas the standard scan strategy produced a lamellar microstructure. A duplex microstructure is better as far as stress and ductility are concerned [24]. Thus, the appropriate microstructure could be said to have contributed to better compression properties. Add to this the better quality struts, and we can safely attribute the higher compression properties to the modified scan strategy.

The compression results agree well with the Ashby-Gibson theoretical model. Increasing the porosity volume decreased the Young's modulus as well as the peak stress (Table 4 and Figure 20). One of the engineering fields where a relationship between the pore volume and stiffness is of interest is powder metallurgy. Of course, in powder metallurgy the pores are assumed to be round and randomly distributed. Nevertheless, predictive equations that relate pore volume and stiffness, although drawn from a diverse field such as powder metallurgy, can serve as a generic aid for the present case. Allison *et al.* compared many such equations available in the literature related to powder metallurgy [30]. All the equations estimated a fall in stiffness with increasing pore volume fraction, similar to what was observed in the current work. The equation that Allison *et al.* concluded to be most accurate evinced a prominent drop in stiffness when the pore fraction went above 20%. This, too, substantiates our observation about the larger drop in modulus for the highly porous 0.5 mm strut samples (Figure 20). The possibility of producing defined porous components by an additive manufacturing route, quite unlike powder metallurgy, encouraged Choren *et al.* [31] to seek similar relationships, if any available. Again, most of the scrutinized equations predicted a non-linear drop in modulus in regimes up to 40% porosity. Beyond this range, however, Choren *et al.* lamented about a lack of satisfactory models.

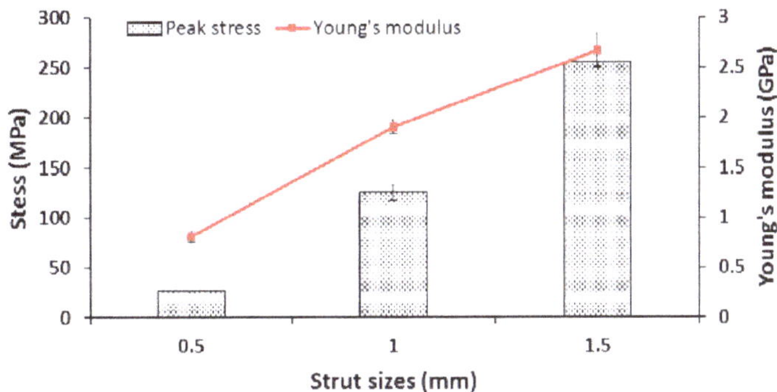

Figure 20. Compressive peak stress based on the total cross-sectional area of the samples (modified scan strategy) (see also Table 4).

The compression values achieved in our study have practical importance in fabricating bio-implants, whose stiffness must be matched to the stiffness of patients' bones to minimize stress shielding. We obtained stiffness of ~0.81 GPa with 0.5 mm struts which is close to that of the trabecular bone (~1.08 GPa) [22]. Jayanthi *et al.* produced Ti-6Al-4V porous structures with a stiffness of 0.57 GPa [6], but their peak stress was a mere 7 MPa. With the same Ti-6Al-4V alloy, Hernandez *et al.* achieved a stiffness of 0.65 GPa and a peak stress of 11.7 MPa [27]. We produced 0.5 mm struts with almost the same stiffness but with a higher peak stress of 26 MPa, allowing these γ-TiAl implants to withstand greater loads while protecting adjoining natural bone from the stress-shielding effect.

Compressive properties were also evaluated taking into account only the strut area that contributes to the load bearing ability of the porous samples (Figure 21). The trend in the peak stress variation was similar to when the total sample cross-section was considered, *i.e.*, the stress decreased with strut size. The drop in the stress from 1 mm to 0.5 struts was more drastic than from 1.5 mm to 1 mm struts. As the struts became thinner and thinner, the surface area increased (Table 2); consequently, the possibility of occurrence of surface defects also increases which ultimately affected the peak stress. The trend in Young's modulus, however, was converse to the peak stress: the Young's modulus was higher in thinner struts. This means that a unit area of material distributed into a greater number of load bearing columns behaved in a stiffer manner than when the same unit area of the material end up in a thicker, but with fewer number of columns. As previously mentioned, the linear elastic region of porous designs is shaped by how the cells bend elastically. The framework in 0.5 mm strut samples was fashioned from more number of struts and pores than other samples. Greater number of such cells, then, must have supported each other from easily bending and, hence, becoming stiffer in the process.

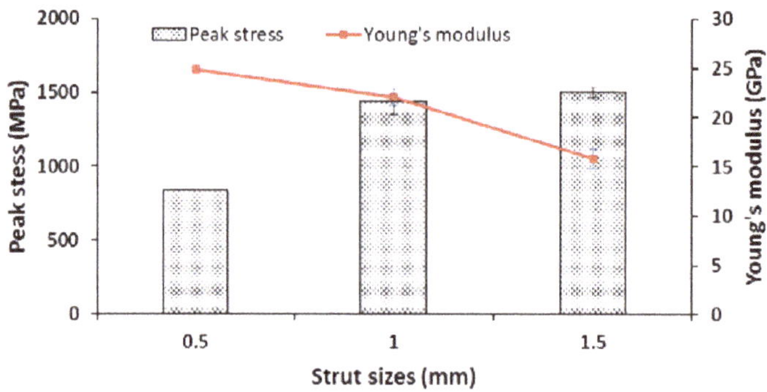

Figure 21. Compressive properties after correcting for the designed cross-sectional area of the struts (modified scan strategy).

4. Conclusions

In this study, we used electron beam melting to produce porous structures from γ-TiAl. Samples with overall pore volumes of 54%, 68%, and 77% were produced. Keeping the pore dimensions constant, we only changed the strut size to determine the final pore volume proportion. Three beam currents and two scan strategies were analyzed for their effect on the strut quality. The main conclusions are as follows:

1. The standard scan strategy (contour and hatch) produced all the struts except the 0.5 mm-thick ones, which failed because the scan strategy could not build the thin vertical struts. Even higher beam current could not create the 0.5 mm struts using this scan strategy.
2. With a standard scan strategy, powder sintered to the underside of the horizontal struts, making them bulkier than designed. The differences in thickness between the vertical and horizontal struts caused anisotropic pore shapes.
3. A modified scan strategy (only hatching) produced 0.5 mm struts. This was possible because, in the modified scan strategy, the intense hatching beam imparted greater energy to the narrow regions required to build thin vertical struts.
4. The modified scan strategy gave better results than the standard strategy, even for thicker struts. The modified strategy reduced the amount of powder sintered to the bottoms of the horizontal struts which, in turn, produced pores closer to the design in all directions.
5. The stiffness of the 0.5 mm strut structures was ~0.8 GPa, which is closer to the stiffness of a trabecular bone. Additionally, these fine structures had a decent peak stress of 26 MPa. Thus, by tuning the thicknesses of the struts and modifying the scan strategy, we could produce porous structures of γ-TiAl with the desired mechanical properties by using EBM. This attribute will play a critical role in manufacturing bio-implants specific to each patient.

In the current study, probably because the samples were small, we could remove any loose powder, even from the interiors, by high-pressure powder blasting. However, this method may not work well enough for larger samples; thus, other powder-removal techniques should be considered in any future works.

Acknowledgments: This project was funded by the National Plan for Science, Technology and Innovation (MAARIFAH), King Abdul-Aziz City for Science and Technology, Kingdom of Saudi Arabia, Award Number (11-ADV1494-02).

Author Contributions: Ashfaq Mohammad conceived and designed the study. Khaja Moiduddin, Muneer Khan Mohammed, and Abdulrahman Alomar joined Ashfaq Mohammad in conducting the EBM experiments. Ravi Kottan Renganayagalu and Ashfaq Mohammad characterized and analyzed the microstructure and mechanical properties. Abdulrahman M. Alahmari and Ashfaq Mohammad drafted the manuscript and critically reviewed it.

Conflicts of Interest: The authors declare no conflict of interest.

References

1. Roy, S.; Panda, D.; Khutia, N.; Chowdhury, A.R.; Roy, S.; Panda, D.; Khutia, N.; Chowdhury, A.R. Pore geometry optimization of titanium (Ti6Al4V) alloy, for its application in the fabrication of customized hip implants. *Int. J. Biomater.* **2014**. [CrossRef]
2. Ikeo, N.; Ishimoto, T.; Fukuda, H.; Nakano, T. Fabrication and characterization of porous implant products with aligned pores by EBM method for biomedical application. *Adv. Mater. Res.* **2011**, *409*, 142–145. [CrossRef]
3. Xiong, J.; Mines, R.; Ghosh, R.; Vaziri, A.; Ma, L.; Ohrndorf, A.; Christ, H.J.; Wu, L. Advanced micro-lattice materials. *Adv. Eng. Mater.* **2015**, *17*, 1253–1264. [CrossRef]
4. Lv, J.; Jia, Z.; Li, J.; Wang, Y.; Yang, J.; Xiu, P.; Zhang, K.; Cai, H.; Liu, Z. Electron beam melting fabrication of porous Ti6Al4V scaffolds: Cytocompatibility and osteogenesis. *Adv. Eng. Mater.* **2015**, *17*, 1391–1398. [CrossRef]
5. Murr, L.; Gaytan, S.; Martinez, E. Fabricating functional Ti-alloy biomedical implants by additive manufacturing using electron beam melting. *J. Biotechnol. Biomater.* **2012**. [CrossRef]
6. Parthasarathy, J.; Starly, B.; Raman, S.; Christensen, A. Mechanical evaluation of porous titanium (Ti6Al4V) structures with electron beam melting (EBM). *J. Mech. Behav. Biomed. Mater.* **2010**, *3*, 249–259. [CrossRef] [PubMed]
7. Li, S.; Zhao, S.; Hou, W.; Teng, C.; Hao, Y.; Li, Y.; Yang, R.; Misra, R.D.K. Functionally graded Ti-6Al-4V meshes with high strength and energy absorption. *Adv. Eng. Mater.* **2016**, *18*, 34–38. [CrossRef]
8. Martinez, E.; Murr, L.E.; Hernandez, J.; Pan, X.; Amato, K.; Frigola, P.; Terrazas, C.; Gaytan, S.; Rodriguez, E.; Medina, F.; *et al.* Microstructures of niobium components fabricated by electron beam melting. *Metallogr. Microstruct. Anal.* **2013**, *2*, 183–189. [CrossRef]
9. Cormier, D.; Harrysson, O.; West, H. Characterization of H13 steel produced via electron beam melting. *Rapid Prototyp. J.* **2004**, *10*, 35–41. [CrossRef]
10. Ramirez, D.A.; Murr, L.E.; Li, S.J.; Tian, Y.X.; Martinez, E.; Martinez, J.L.; Machado, B.I.; Gaytan, S.M.; Medina, F.; Wicker, R.B. Open-cellular copper structures fabricated by additive manufacturing using electron beam melting. *Mater. Sci. Eng. A* **2011**, *528*, 5379–5386. [CrossRef]
11. Cormier, D.; Harrysson, O.; Mahale, T.; West, H. Freeform fabrication of titanium aluminide via electron beam melting using prealloyed and blended powders. *Adv. Mater. Sci. Eng.* **2008**. [CrossRef]
12. Aspinwall, D.K.; Dewes, R.C.; Mantle, A.L. The Machining of γ-TiAl intermetallic alloys. *CIRP Ann. Manuf. Technol.* **2005**, *54*, 99–104. [CrossRef]
13. Tetsui, T.; Shindo, K.; Kaji, S.; Kobayashi, S.; Takeyama, M. Fabrication of TiAl components by means of hot forging and machining. *Intermetallics* **2005**, *13*, 971–978. [CrossRef]
14. Rivera-Denizard, O.; Diffoot-Carlo, N.; Navas, V.; Sundaram, P.A. Biocompatibility studies of human fetal osteoblast cells cultured on gamma titanium aluminide. *J. Mater. Sci. Mater. Med.* **2008**, *19*, 153–158. [CrossRef] [PubMed]
15. Santiago-Medina, P.; Sundaram, P.A.; Diffoot-Carlo, N. The effects of micro arc oxidation of gamma titanium aluminide surfaces on osteoblast adhesion and differentiation. *J. Mater. Sci. Mater. Med.* **2014**, *25*, 1577–1587. [CrossRef] [PubMed]
16. Hernandez, J.; Murr, L.E.; Gaytan, S.M.; Martinez, E.; Medina, F.; Wicker, R.B. Microstructures for two-phase gamma titanium aluminide fabricated by electron beam melting. *Metallogr. Microstruct. Anal.* **2012**, *1*, 14–27. [CrossRef]

17. Murr, L.E.; Esquivel, E.V.; Quinones, S.A.; Gaytan, S.M.; Lopez, M.I.; Martinez, E.Y.; Medina, F.; Hernandez, D.H.; Martinez, E.; Martinez, J.L.; *et al.* Microstructures and mechanical properties of electron beam-rapid manufactured Ti-6Al-4V biomedical prototypes compared to wrought Ti-6Al-4V. *Mater. Charact.* **2009**, *60*, 96–105. [CrossRef]
18. Arabnejad, S.; Johnston, R.B.; Pura, J.A.; Singh, B.; Tanzer, M.; Pasini, D. High-strength porous biomaterials for bone replacement: A strategy to assess the interplay between cell morphology, mechanical properties, bone ingrowth and manufacturing constraints. *Acta Biomater.* **2016**, *30*, 345–356. [CrossRef] [PubMed]
19. Attar, E. *Simulation of Selective Electron Beam Melting Processes*; VDM Publishing: Saarbrücken, Germany, 2011.
20. Shen, N.; Chou, K. Thermal modeling of electron beam additive manufacturing process: Powder sintering effects. In Proceedings of the ASME 2012 International Manufacturing Science and Engineering Conference, Notre Dame, IN, USA, 4–8 June 2012; pp. 287–295.
21. Arce, A.N. *Thermal Modeling and Simulation of Electron Beam Melting For Rapid Prototyping on Ti6Al4V Alloys*; North Carolina State University: Raleigh, NC, USA, 2013.
22. Heinl, P.; Müller, L.; Körner, C.; Singer, R.F.; Müller, F.A. Cellular Ti-6Al-4V structures with interconnected macro porosity for bone implants fabricated by selective electron beam melting. *Acta Biomater.* **2008**, *4*, 1536–1544. [CrossRef] [PubMed]
23. Körner, C.; Attar, E.; Heinl, P. Mesoscopic simulation of selective beam melting processes. *J. Mater. Process. Technol.* **2011**, *211*, 978–987. [CrossRef]
24. Kothari, K.; Radhakrishnan, R.; Sudarshan, T.S.; Wereley, N.M. Characterization of rapidly consolidated γ-TiAl. *Adv. Mater. Res.* **2012**, *1*, 51–74. [CrossRef]
25. Biamino, S.; Penna, A.; Ackelid, U.; Sabbadini, S.; Tassa, O.; Fino, P.; Pavese, M.; Gennaro, P.; Badini, C. Electron beam melting of Ti-48Al-2Cr-2Nb alloy: Microstructure and mechanical properties investigation. *Intermetallics* **2011**, *19*, 776–781. [CrossRef]
26. Feller, L.; Jadwat, Y.; Khammissa, R.A.G.; Meyerov, R.; Schechter, I.; Lemmer, J.; Feller, L.; Jadwat, Y.; Khammissa, R.A.G.; Meyerov, R.; *et al.* Cellular responses evoked by different surface characteristics of intraosseous titanium implants. *BioMed Res. Int.* **2015**. [CrossRef] [PubMed]
27. Hernández-Nava, E.; Smith, C.J.; Derguti, F.; Tammas-Williams, S.; Léonard, F.; Withers, P.J.; Todd, I.; Goodall, R. The effect of density and feature size on mechanical properties of isostructural metallic foams produced by additive manufacturing. *Acta Mater.* **2015**, *85*, 387–395. [CrossRef]
28. Li, X.; Wang, C.; Zhang, W.; Li, Y. Fabrication and characterization of porous Ti6Al4V parts for biomedical applications using electron beam melting process. *Mater. Lett.* **2009**, *63*, 403–405. [CrossRef]
29. Wauthle, R.; Ahmadi, S.M.; Yavari, S.A.; Mulier, M.; Zadpoor, A.A.; Weinans, H.; van Humbeeck, J.; Kruth, J.P.; Schrooten, J. Revival of pure titanium for dynamically loaded porous implants using additive manufacturing. *Mater. Sci. Eng. C Mater. Biol. Appl.* **2015**, *54*, 94–100. [CrossRef] [PubMed]
30. Allison, P.G.; Horstemeyer, M.F.; Brown, H.R. Modulus dependence on large scale porosity of powder metallurgy steel. *J. Mater. Eng. Perform.* **2012**, *21*, 1422–1425. [CrossRef]
31. Choren, J.A.; Heinrich, S.M.; Silver-Thorn, M.B. Young's modulus and volume porosity relationships for additive manufacturing applications. *J. Mater. Sci.* **2013**, *48*, 5103–5112. [CrossRef]

11

Residual Stress Distribution and Microstructure at a Laser Spot of AISI 304 Stainless Steel Subjected to Different Laser Shock Peening Impacts

Wenquan Zhang, Jinzhong Lu * and Kaiyu Luo

Academic Editor: Hugo F. Lopez

School of Mechanical Engineering, Jiangsu University, Zhenjiang 212013, China; nicequan1991@163.com (W.Z.); kyluo@ujs.edu.cn (K.L.)

* Correspondence: jzlu@mail.ujs.edu.cn

Abstract: The effects of different laser shock peening (LSP) impacts on the three-dimensional displayed distributions of surface and in-depth residual stress at a laser spot of AISI 304 stainless steel were investigated by X-ray diffraction technology with the $\sin^2\varphi$ method and MATLAB 2010a software. Microstructural evolution in the top surface subjected to multiple LSP impacts was presented by means of cross-sectional optical microscopy (OM) and transmission electron microscopy (TEM) observations. Experimental results and analysis indicated that residual stress distribution and microstructure at a laser spot depended strongly on the number of multiple LSP impacts, and refined grain and ultra-high strain rate play an important role in the improvement of compressive residual stress of AISI 304 stainless steel. The analysis of treatment of the extended surface was presented to obtain uniform surface properties on the top surface of AISI 304 stainless steel.

Keywords: multiple laser shock peening; residual stress distribution; microstructure; ultra-high strain rate

---

1. Introduction

Laser shock peening (LSP) is a kind of advanced surface enhancing method which employs the mechanical effect of laser shock wave induced by a laser beam to improve the fatigue resistance, damage tolerance, corrosion resistance of metallic materials and alloys. More and more researchers have been paying comprehensive attention to LSP due to its high-pressure (in the scale of GPa), high-energy (more than 1 GW/cm^2), ultrafast (several tens of nanoseconds) and ultra-high strain rate (more than 10^6 s^{-1}) [1–3].

Over the last two decades, substantial research efforts have yielded significant insights into the effects of overlapping LSP treatment and multiple LSP impacts on the mechanical properties and microstructure of metallic materials and alloys. For example, Wang *et al.* investigated the effects of overlapping LSP impacts on stress corrosion behavior of 7075 aluminum alloy laser welded joints, and results showed that the elongation, time of fracture and static toughness were improved by 11.13%, 20% and 100% after overlapping LSP impacts, respectively [4]. Trdan *et al.* confirmed the insignificant influence of LSP without coating treatment direction on micro-hardness distribution, indicating essentially homogeneous conditions in both longitudinal and transverse directions [5]. Jia *et al.* presented that repeated LSP impacts have a very beneficial effect on surface hardening, and the high-cycle fatigue life of LSPed Ti834 alloy increases due to the fact that the introduction of compressive residual stress can delay the initiation and growth of the fatigue crack [6]. Correa *et al.* established a good correlation between simulations, residual stress measurements and fatigue life of

Al2024-T351 specimens treated by LSP [7]. In our previous work, we reported that the improvement of the stress corrosion cracking (SCC) resistance of AISI 304 stainless steel was caused by compressive residual stress and grain refinement after massive LSP impacts [8]. We also indicated that typical microstructure and residual stress in the surface layer of LY2 aluminum alloy [9] and AISI 304 stainless steel have a close relation with the number of multiple LSP impacts. A single LSP impact can refine the original grain in the near-surface layer mainly by mechanical twins (MTs) in a single direction and multiple LSP impacts can refine the original grain mainly by MTs in multi-directions, which are the direct reason why LSP can improve the nano-hardness, elastic modulus, and residual stress of AISI 304 stainless steel [10]. When laser shock wave acts on the surface layer, local plastic deformation occurs in the LSPed region and the affected region. These investigations mainly focused on the interaction of adjacent LSP impacts during overlapping LSP treatment and multiple LSP impacts. In practice, the surfaces of metallic components are often overlappingly treated by multiple LSP treatment. Uneven residual stress distribution on the component surface often deteriorates the surface properties of the treated component [11], and the uniformity of residual stress in the surface layer of metallic component after massive LSP impacts depends strongly on residual stress distribution and microstructure at a laser spot. However, over the years, only few researchers have investigated how the influence mechanism of a single LSP impact and the number of multiple LSP impacts on residual stress distribution at a laser spot affect the uniformity of surface properties on the extended surface.

This paper aimed to investigate the effects of different LSP impacts on surface and in-depth residual stress distributions of AISI 304 stainless steel at a laser spot. The three-dimensional residual stress displayed distribution as a function of the number of multiple LSP impacts was highlighted. In particular, microstructure in the top surface under different LSP impacts is characterized and analyzed. Finally, the underlying influence mechanism of typical substructure induced by laser shock wave on residual stress distribution at a laser spot was determined and we also analyzed how the underlying influence mechanism affects the surface properties' uniformity on the extended surface. These topics can provide some insight into the uniformity of compressive residual stress generated by massive overlapping LSP impacts, which has a lot of practical use in engineering application.

2. Experimental Procedures

2.1. Specimen Preparation

AISI 304 austenitic stainless steel was chosen in this study, and its chemical composition was 0.06 C, 1.54 Mn, 18.47 Cr, 0.30 Mo, 8.3 Ni, 0.37 Cu, 0.48 Si, 0.027 Nb and balance Fe (wt. %). The mechanical properties of AISI 304 stainless steel are shown in Table 1 [12]. All specimens from the same steel sheet were cut into dimensions of 5 mm × 5 mm × 2 mm (width × length × thickness). Before LSP treatment, all specimen surfaces were progressively mechanically polished using SiC paper with grit numbers from 150 to 1800 (including 150#, 360#, 600#, 800#, 1200# and 1800#) to achieve smooth surfaces. Subsequently, these specimens were cleaned with acetone and subjected to ultrasonic vibration to degrease their surfaces.

Table 1. Mechanical properties of AISI 304 austenitic stainless steel.

Type	Value
Tensile strength, (σ_b) (kgf/mm^2)	520
Specific gravity, (d) (g/cm^3)	7.93
Yield strength, (σ) (kgf/mm^2)	205
Vickers-hardness (HV)	200
Elongation, (δ) (%)	40

2.2. LSP Experiment

LSP experiments were carried out by a Q-switched Nd: YAG laser produced by Thales Company in Paris, France with a wavelength of 1064 nm at Laser Technology Institute in Jiangsu University. For this application, the laser spot diameter was 3 mm, and a pulse laser with a duration of 10 ns and a pulse energy of 9.3 J was used as LSP energy source. The output beam of single flat-top laser pulse was attenuated and sent to a charge-coupled-device (CCD) camera produced by SVC ASSET MANAGEMENT INC in California, CA, United States, and its profile can be seen in Figure 1. To ensure the mechanical effect of laser shock wave, professional aluminum tape produced by 3M company in St Paul, MN, USA, with a thickness of 100 μm was used as an ablation medium for plasma initiation, and a water film with a thickness of 1–2 mm was set as a transparent confining layer. To investigate the effects of the number of LSP impacts on microstructure and residual stress distribution at a laser spot, three types of LSPed specimens were prepared in the present study, and these specimens were treated by different LSP impacts at a single laser spot. According to the number of LSP impacts, we defined these specimens subjected to one LSP impact, two LSP impacts, and three LSP impacts as single LSP impact, dual LSP impacts, and triple LSP impacts, respectively.

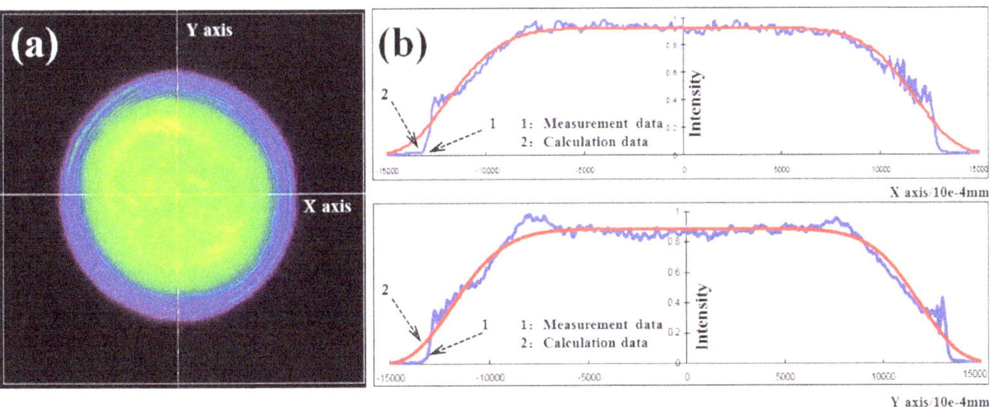

Figure 1. The output beam of single flat-top laser pulse and corresponding spatial profiles along horizontal and perpendicular directions. (**a**) Along horizontal direction; (**b**) along perpendicular direction.

2.3. Residual Stress Measurement and Microstructural Observations

Surface residual stresses of all specimens were measured by X-ray diffraction technology with the $\sin^2\varphi$ method [13]. The X-ray beam diameter was about 2 mm, and the X-ray source was the Cr-Kα ray. The feed angle of the ladder scanning was 0.1° per second, and the scanning starting angle and terminating angle were 145° and 153°, respectively. To measure residual stress along the depth direction, material removal by electropolishing was used, and the specimen surface was removed by a depth of 0.05 mm to the top surface every time. After each removal with a depth of 0.2 mm, residual stress values at a regular interval of 0.5 mm at the center line were recorded. Thus, the surface layer with a thickness of 1 mm will be removed five times. All measurements were repeated thrice for each point, and the average value was obtained and recorded. According to these experimental data, cross-sectional residual stress distributions at the plane perpendicular to the top surface along the center line can be presented by MATLAB 2010a software developed by The Mathworks in Natick, MA, USA.

After LSP treatment, the as-machined specimen and three types of LSPed specimens used for metallographic investigation were cut as the sections perpendicular to the specimen surface, and then subjected to several successive steps of grinding and polishing. After that, the vertical sections of the samples were etched using a professional reagent that consists of 15 cc of HCl, 10 cc of HNO_3, 10 cc of acetic acid, and 2/3 (two thirds) drops glycerine, and then characterized by cross-sectional OM observations. Cross-sectional thin foils for TEM observations were also prepared, and microstructure in the top surface of all specimens was observed by a JEM-2100 transmission electron microscopy (TEM) produced by JEOL, Tonkin, Japan operated at a voltage of 200 kV.

3. Result and Discussions

3.1. Surface Residual Stress Distribution at a Single Laser Spot

Figure 2 shows surface residual stress three-dimensionally displayed distribution at a single laser spot as a function of the number of LSP impacts. For the above three cases, we choose a diameter of every laser spot, as shown in Line 1 in Figure 2a, Line 2 in Figure 2b, and Line 3 in Figure 2c. Figure 3 shows surface residual stress distributions along the diameters of laser spots subjected to single LSP impact, dual LSP impacts, and triple LSP impacts, respectively.

From Figures 2a and 3, it can be seen that there is a maximum of −239 MPa in residual stress at the center of the laser spot subjected to a single LSP impact. The diameter of the LSP-affected region is about 4 mm. After dual LSP impacts, the maximum of −239 MPa is increased to −336 MPa, and the center region with a radius of ~0.8 mm exists in high-level compressive residual stress of more than −300 MPa. The LSP-affected region keeps an increase to a diameter of approximately 4.5 mm, as shown in Figures 2b and 3. After a third LSP impact, the maximum of surface residual stress is increased to −371 MPa, and the center region with a radius of ~1.3 mm exists in high-level compressive residual stress of more than −300 MPa. The LSP-affected region continues to increase to a diameter of approximately 5 mm, as shown in Figures 2c and 3.

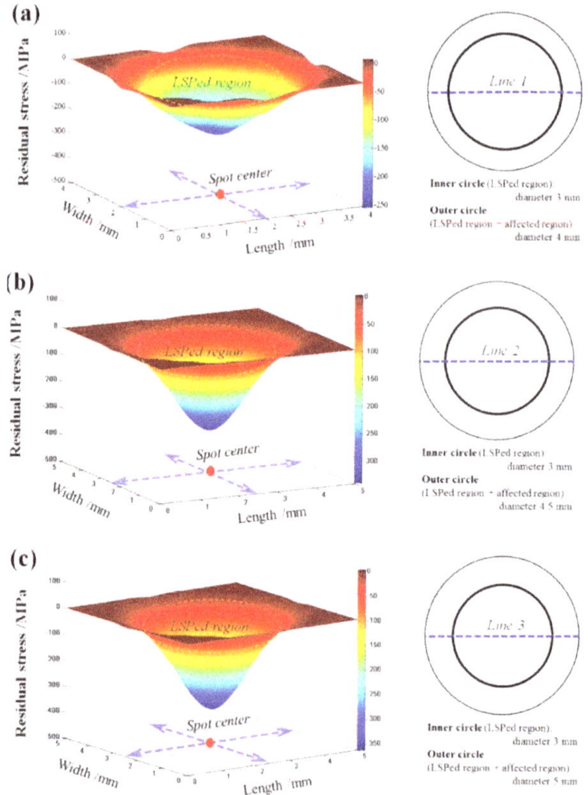

Figure 2. Surface residual stress three-dimensionally displayed distributions at a single laser spot as a function of the number of LSP impacts. (**a**) Single LSP impact; (**b**) dual LSP impacts; and (**c**) triple LSP impacts.

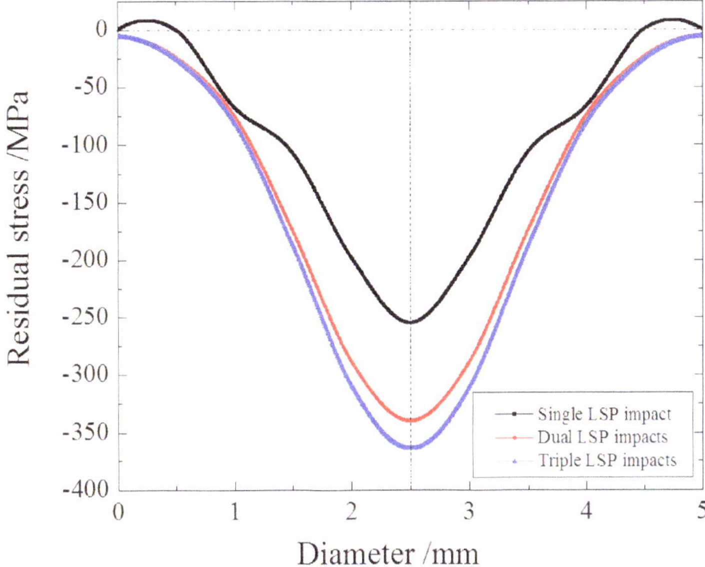

Figure 3. Surface residual stress distributions along the diameters of laser spots subjected to different LSP impacts.

From the above experimental data, the following two facts can be obtained. On the one hand, the maximum of compressive residual stress increases with the increasing number of LSP impacts, but the increasing rate decreases with the increasing number of LSP impacts. On the other hand, while the number of LSP impacts increases, the LSP-affected region increases. In addition, there is only a little change in the LSP-affected region when the number of LSP impacts varies from two to three. We may speculate that the LSP-affected region can reach a peak with the increasing number of LSP impacts. Among the fringe field of the LSP-affected region, the recoiling pressure wave decreases steeply and it cannot reach the yield limit of AISI 304 austenitic stainless steel, which leads to the inadequate plastic deformation of the region far away from the spot center. Hence, surface residual stress along the diameter of the laser spot also decreases gradually with the increasing distance to the spot center. It is worth noting that multiple LSP impacts cause more and more uneven surface residual stress distribution at a laser spot with an increasing number of LSP impacts.

3.2. Residual Stress Distribution in Depth Direction along the Diameter of Laser Spot

Figure 4 shows in-depth residual stress three-dimensionally displayed distributions at Line 1 in Figure 2a, Line 2 in Figure 2b, and Line 3 in Figure 2c. Figure 5 shows in-depth residual stress distributions of the spot center subjected to single LSP impact, dual LSP impacts, and triple LSP impacts, respectively. From Figure 4a–c, in-depth residual stress distributions at the diameter as a function of the number of LSP impacts can be clearly presented. After single LSP impact, the residual stress of the spot center is −239 MPa, and the affected depth of the compressive residual stress is about 0.78 mm, as shown in Figure 5. After dual LSP impacts, the residual stress of the spot center is increased to −336 MPa, and the affected depth of the compressive residual stress reaches 1.05 mm, as shown in Figure 5. After triple LSP impacts, the residual stress value of the spot center and its affected depth are −371 MPa and 1.18 mm, as shown in Figure 5.

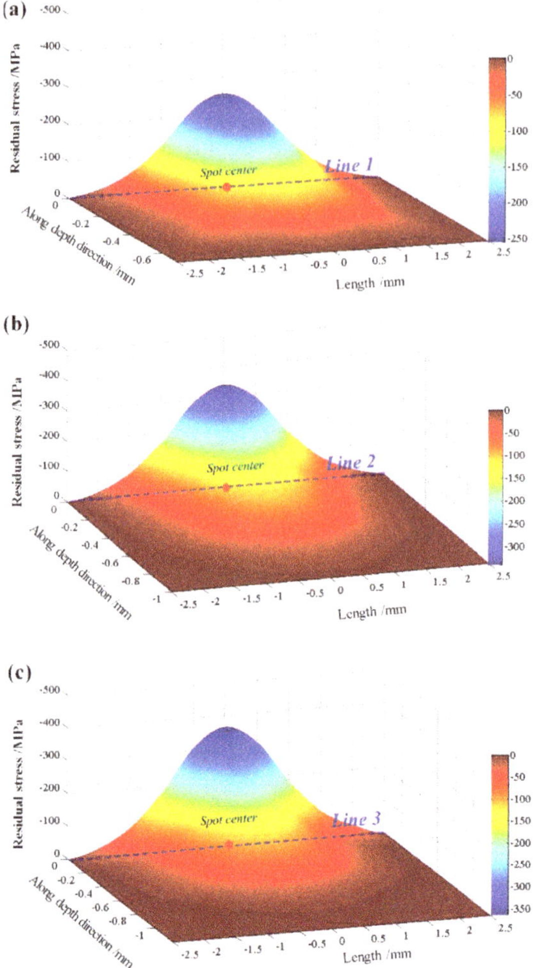

Figure 4. In-depth residual stress three-dimensionally displayed distributions along the diameter of laser spots with different LSP impacts. (**a**) Single LSP impact; (**b**) dual LSP impacts; and (**c**) triple LSP impacts.

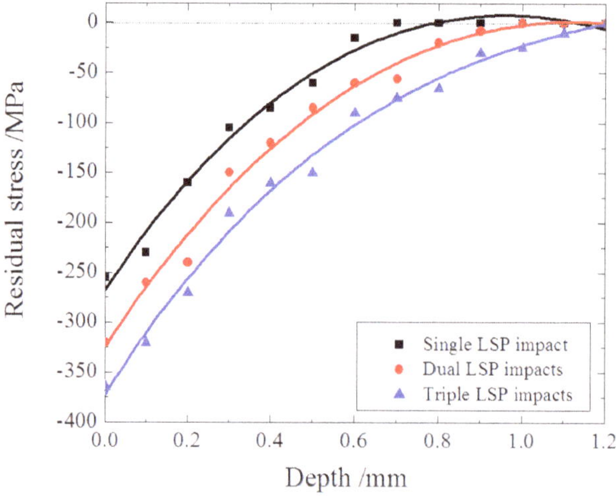

Figure 5. In-depth residual stress distributions of the spot center subjected to different LSP impacts.

From the above experimental data, it can be concluded that after triple LSP impacts, the residual stress value of the spot center and the affected depth reach a peak, and the compressive residual stress keeps a constant and the affected depth also reaches a saturated value.

3.3. Metallographic Observation and Microstructural TEM Characterization

Metallographic observation of the top surface of the as-machined specimen and three types of LSPed specimens can be seen in Figure 6. Figure 7 shows typical TEM images of the top surface of the as-machined specimen and LSPed specimens. The grain size, grain boundary, and grain shape can be recognized after being immersed in the professional etching reagent for 30 s at room temperature.

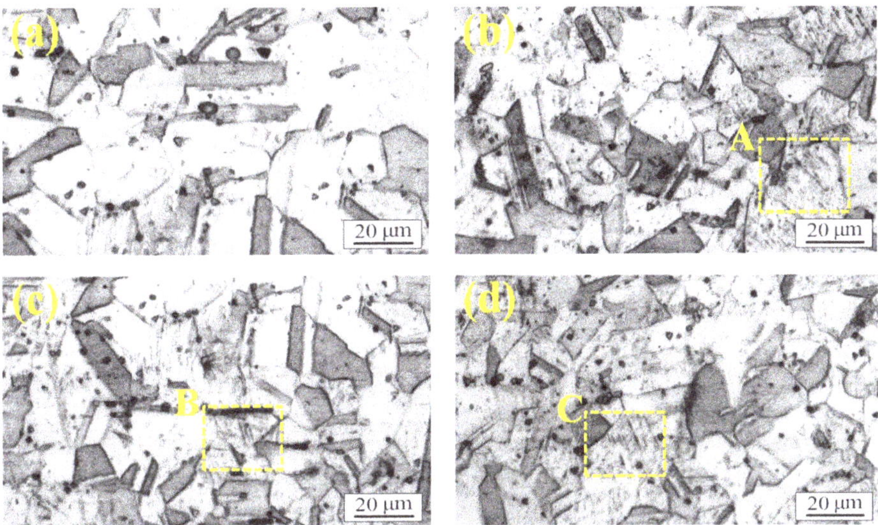

Figure 6. Typical metallographic observations in the top surface of as-machined specimen and three types of LSPed specimens. (**a**) As-machined specimen; (**b**) single LSP impact; (**c**) dual LSP impacts; and (**d**) triple LSP impacts.

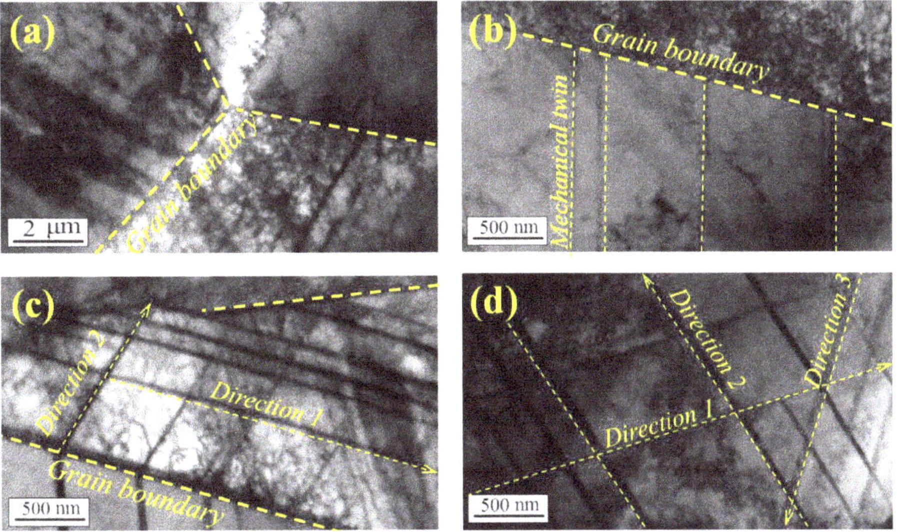

Figure 7. Typical TEM microstructures in the top surface of as-machined specimen and three types of LSPed specimens. (**a**) As-machined specimen; (**b**) single LSP impact; (**c**) dual LSP impacts; and (**d**) triple LSP impacts.

From Figure 6a, it can be seen that the original coarse grain in the top surface of the as-machined specimen varies from 20 to 60 μm. For three types of LSPed specimens, evidence of plastic deformation is obvious in the surface layer, as shown in Figure 6b–d. After a single LSP impact, an overwhelming majority of coarse grains are obviously refined into grains with an average size of 10–20 μm, as shown in Figure 6b. Some coarse grains are not refined, but some mechanical twins (MTs) with one direction can be found within the coarse grain. These MT boundaries are parallel to each other and they subdivide the coarse grain into thin twin-matrix (T-M) lamellae at the top surface, whose width ranges from 40 nm to 700 nm, as can be clearly seen in Rectangle (A) in Figures 6b and 7b. After a second LSP impact, laser shock wave causes the presence of refined grain with an average size of 5–15 μm (as shown in Figure 6c), and some MTs with two directions can be found within these grains. Moreover, the intersections of MT-MT in two directions' result in submicron rhombic blocks and the average dimension of the submicron rhombic block is about 100 nm × 500 nm, as presented in Rectangle (B) in Figures 6c and 7c. After triple LSP impacts, the refined grain turns more uniform and smaller, and some MTs with three directions can be found within these grains. Under the effect of the third impact, submicron rhombic blocks are transformed into triangular blocks with the dimension of about 130 nm × 130 nm, as shown in Rectangle (C) in Figures 6d and 7d. Hence, it can be concluded that when the number of LSP impacts varies from one to three, the coarse grains are refined by T-M lamellae, submicron rhombic blocks and triangular blocks, respectively.

According to the previous results, there are two reasons that are responsible for the observed MTs of Face-centered cubic (FCC) metals. The first one is small grain size. With the decrease of grain size, the energy needed to form MTs and stacking faults gradually becomes closer to the energy needed to nucleate a perfect dislocation, so the deformation mechanism controlled by normal slip transforms to that controlled by partial dislocation activity [14]. The other important reason is the ultra-high strain rate. With the increase of the strain rate, dislocations align themselves into walls at the twin boundary, which transform submicron subdivided blocks into subgrain boundaries gradually. When the strain rate is further increased, subgrain boundaries are turned by dislocations and transform into equiaxed refined grains [15,16]. The dislocation-slip process is suppressed when the strain rate is enough high, assisting the formation of deformation MTs [17,18].

The strain rate of metallic materials and alloys exceeds 10^6 s^{-1} during LSP treatment, and refined grains are generated in the top surface. As a result of the grain refinement and ultra-high strain rate, the LSPed region is strengthened according to classical dislocation theory [19]. Moreover, for 304 stainless steel, the grains in the top surface turn smaller and smaller with the increasing number of LSP impacts. The reaction between the laser shock wave and the specimen will be generated in the top surface, leading to the generation of the dislocation and micro-structural deformation (such as MT), which can be explained by the fact that high-level compressive residual stresses are generated in the top surface, and the magnitude of the compressive residual stress in the top surface gradually increases with increasing number of LSP impacts.

The beneficial effect brought by LSP treatment is derived from the generation of a compressive residual stress field and the refined microstructure in the near-surface region at a laser spot. Although the output beam of a single laser pulse exists in flat-top shape (as shown in Figure 1), the uneven distribution of surface residual stress induced by the inhomogeneous deformation at a laser spot is further aggravated when AISI 304 stainless steel is treated by multiple LSP impacts, as discussed in Section 3.1. There is a similar result in in-depth residual stress distribution at a laser spot as a function of the number of LSP impacts, as discussed in Section 3.2. The above fact decreases the effect of multiple LSP impacts. As can be seen in Figure 7d, novel structures with submicron triangular blocks are found at the top surface subjected to three LSP impacts, which is the direct reason of the improvement of the residual stress, nano-hardness, and elastic modulus of AISI 304 stainless steel. Hence, improving the overlapping rate during laser beam sweeping can increase the LSP impact number at a laser spot. When the laser impact overlapping rate is 50%, most of the spots of the treated surface are impacted by laser pulse at least three times. Large numbers of submicron triangular blocks are generated on the top surface of AISI 304 stainless steel and the microstructures become uniform. The residual stress, nano-hardness, and elastic modulus on the top surface of AISI 304 stainless steel are greatly improved and distributed uniformly on the extended surfaces. In engineering applications, the surfaces of metallic components are often overlappingly treated by multiple LSP treatments. For future investigation, the uniformity of the residual stress distribution induced by massive LSP treatment will be in a very important position.

4. Conclusions

The effects of different LSP impacts on surface and in-depth residual stress distributions and microstructure of AISI 304 stainless steel at a laser spot were investigated, and microstructure in the top surface under different LSP impacts was characterized and analyzed. Some important conclusions can be drawn as follows:

(1) The maximum of the surface compressive residual stress at a laser spot increases with the increasing number of LSP impacts, but the increasing rate decreases with the increasing number of LSP impacts. Meanwhile, the LSP-affected region can reach a peak with the increasing number of LSP impacts. Multiple LSP impacts cause more and more uneven surface residual stress distribution at a laser spot with an increasing number of LSP impacts.

(2) After triple LSP impacts, the residual stress value of the spot center and the affected depth reach a peak, and the compressive residual stress remains constant and the affected depth also reaches a saturated value.

(3) When the number of LSP impacts varies from one to three, the coarse grains in the top surface of AISI 304 stainless steel are refined by three types of substructures: T-M lamellae, submicron rhombic blocks and triangular blocks, respectively.

(4) The underlying influence mechanism of typical substructure induced by laser shock wave on residual stress distribution at a laser spot was revealed. Refined grain and ultra-high strain rate are two important factors to generate high-level compressive residual stresses in the top surface of AISI 304 stainless steel.

(5) Most of the spots of the treated surface are impacted by laser pulse at least three times when the laser impact overlapping rate is 50%. Large numbers of submicron triangular blocks are generated on the top surface of AISI 304 stainless steel and microstructures become uniform. The residual stress, nano-hardness, and elastic modulus are distributed uniformly on the extended surfaces.

Acknowledgments: The authors are grateful for the support provided by National Natural Science Foundation of China (No. 51575242), Natural Science Foundation of Jiangsu Province in China (Nos. BK20140012 and BK20151341), China Postdoctoral Science Special Foundation (No. 2015T80504), and Six Major Talent Peak of Jiangsu Province (2014-ZBZZ-015).

Author Contributions: J.L., W.Z. and K.L. conceived and designed the experiments; W.Z. performed the experiments; J.L. and W.Z. analyzed the data; K.L. contributed reagents/materials/analysis tools; W.Z. and J.L. wrote the paper.

Conflicts of Interest: The authors declare no conflict of interest.

References

1. Ye, C.; Suslov, S.; Fei, X.L.; Cheng, G.J. Bimodal nanocrystallization of NiTi shape memory alloy by laser shock peening and post-deformation annealing. *Acta Mater.* **2011**, *59*, 7219–7227. [CrossRef]
2. Trdan, U.; Grum, J. SEM/EDS characterization of laser shock peening effect on localized corrosion of Al alloy in a near natural chloride environment. *Corros. Sci.* **2014**, *82*, 328–338. [CrossRef]
3. Trdan, U.; Grum, J. Evaluation of corrosion resistance of AA6082-T651 aluminium alloy after laser shock peening by means of cyclic polarisation and EIS methods. *Corros. Sci.* **2012**, *59*, 324–333. [CrossRef]

4. Wang, J.T.; Zhang, Y.K.; Chen, J.F.; Zhou, J.Y.; Ge, M.Z.; Lu, Y.L.; Li, X.L. Effects of laser shock peening on stress corrosion behavior of 7075 aluminum alloy laser welded joints. *Mater. Sci. Eng. A* **2015**, *647*, 7–14. [CrossRef]
5. Trdan, U.; Porro, J.A.; José, L.; Ocaña, J.G. Laser shock peening without absorbent coating (LSPwC) effect on 3D surface topography and mechanical properties of 6082-T651 Al alloy. *Surf. Coat. Technol.* **2012**, *208*, 109–116. [CrossRef]
6. Jia, W.J.; Hong, Q.; Zhao, H.Z.; Li, L.; Han, D. Effect of laser shock peening on the mechanical properties of a near-α titanium alloy. *Mater. Sci. Eng. A* **2014**, *606*, 354–359. [CrossRef]
7. Correa, C.; de Lara, L.R.; Díaz, M.; Porro, J.A.; García-Beltrán, A.; Ocaña, J.L. Influence of pulse sequence and edge material effect on fatigue life of Al2024-T351 specimens treated by laser shock processing. *Int. J. Fatigue* **2015**, *70*, 196–204. [CrossRef]
8. Lu, J.Z.; Luo, K.Y.; Yang, D.K.; Cheng, X.N.; Hu, J.L.; Dai, F.Z.; Qi, H.; Zhang, L.; Zhong, J.S.; Wang, Q.W.; Zhang, Y.K. Effects of laser peening on stress corrosion cracking (SCC) of ANSI 304 austenitic stainless steel. *Corros. Sci.* **2012**, *60*, 145–152. [CrossRef]
9. Lu, J.Z.; Luo, K.Y.; Zhang, Y.K.; Cui, C.Y.; Sun, G.F.; Zhou, J.Z.; Zhang, L.; You, J.; Chen, K.M.; Zhong, J.W. Grain refinement of LY2 aluminum alloy induced by ultra-high plastic strain during multiple laser shock processing impacts. *Acta Mater.* **2010**, *58*, 3984–3994. [CrossRef]
10. Lu, J.Z.; Luo, K.Y.; Zhang, Y.K.; Sun, G.F.; Gu, Y.Y.; Zhou, J.Z.; Ren, X.D.; Zhang, X.C.; Zhang, L.F.; Chen, K.M.; *et al.* Grain refinement mechanism of multiple laser shock processing impacts on ANSI 304 stainless steel. *Acta Mater.* **2010**, *58*, 5354–5362. [CrossRef]
11. Zhan, K.; Jiang, C.H.; Ji, V. Uniformity of residual stress distribution on the surface of S30432 austenitic stainless steel by different shot peening processes. *Mater. Lett.* **2013**, *99*, 61–64. [CrossRef]
12. Nikitin, I.; Altenberger, I. Comparison of the fatigue behavior and residual stress stability of laser-shock peened and deep rolled austenitic stainless steel AISI 304 in the temperature range 25–600 °C. *Mater. Sci. Eng. A* **2007**, *465*, 176–178. [CrossRef]
13. Withers, P.; Bhadeshia, H. Residual stress Part 2—Nature and origins. *Mater. Sci. Technol.* **2001**, *17*, 65–355. [CrossRef]
14. Li, J.; Liao, Y.L.; Suslov, S.; Cheng, G.J. Laser shock-based platform for controllable forming of nanowires. *Nano Lett.* **2012**, *12*, 3224–3230. [CrossRef] [PubMed]
15. Chu, J.P.; Rigsbee, J.M.; Banas, G.; Elsayed-Ali, H.E. Laser shock processing effects on surface microstructure and mechanical properties of low carbon steel. *Mater. Sci. Eng. A* **1999**, *260*, 260–268. [CrossRef]
16. Tao, N.R.; Wang, Z.B.; Tong, W.P.; Sui, M.L.; Lu, J.; Lu, K. An investigation of surface Nano crystallization mechanism in Fe induced by surface mechanical attrition treatment. *Acta Mater.* **2002**, *50*, 4603–4616. [CrossRef]
17. Bohn, R.; Haubopld, T.; Birringer, R.; Gleiter, H. Nanocrystalline intermetallic compounds—An approach to ductility? *Scr. Metall. Mater.* **1991**, *25*, 811–816. [CrossRef]
18. Lu, K.; Wang, J.T.; Wei, W.D. A new method for synthesizing nanocrystalline alloys. *J. Appl. Phys.* **1991**, *69*, 522–531. [CrossRef]
19. Chen, M.W.; Ma, E.; Hemke, K.J.; Sheng, H.W.; Wang, Y.M.; Cheng, X.M. Deformation twinning in nanocrystalline aluminum. *Science* **2003**, *300*, 1275–1277. [CrossRef] [PubMed]

12

Aluminium Foam and Magnesium Compound Casting Produced by High-Pressure Die Casting

Iban Vicario [1,*], Ignacio Crespo [2,†], Luis Maria Plaza [2], Patricia Caballero [1,†] and Ion Kepa Idoiaga [3,‡]

Academic Editor: Hugo F. Lopez

1 Department of Foundry and Steel making, Tecnalia Research & Innovation, c/Geldo, Edif. 700, E-48160 Derio, Spain; patricia.caballero@tecnalia.com
2 Department of Aerospace, Tecnalia Research & Innovation, c/Mikeletegi 2, E-20009 Donostia, Spain; ignacio.crespo@tecnalia.com (I.C.); luismaria.plaza@tecnalia.com (L.M.P.)
3 Industrias Lebario, c/Arbizolea 4, E-48213 Izurza, Spain; jidoiaga@lebario.com
* Correspondence: iban.vicario@tecnalia.com
† These authors contributed equally to this work.
‡ This author supervised this work.

Abstract: Nowadays, fuel consumption and carbon dioxide emissions are two of the main focal points in vehicle design, promoting the reduction in the weight of vehicles by using lighter materials. The aim of the work is to evaluate the influence of different aluminium foams and injection parameters in order to obtain compound castings with a compromise between the obtained properties and weight by high-pressure die cast (HPDC) using aluminium foams as cores into a magnesium cast part. To evaluate the influence of the different aluminium foams and injection parameters on the final casting products quality, the type and density of the aluminium foam, metal temperature, plunger speed, and multiplication pressure have been varied within a range of suitable values. The obtained compound HPDC castings have been studied by performing visual and RX inspections, obtaining sound composite castings with aluminium foam cores. The presence of an external continuous layer on the foam surface and the correct placement of the foam to support injection conditions permit obtaining good quality parts. A HPDC processed magnesium-aluminium foam composite has been developed for a bicycle application obtaining a suitable combination of mechanical properties and, especially, a reduced weight in the demonstration part.

Keywords: high pressure die casting (HPDC); hybrid magnesium aluminium foam cast composite; aluminium foam core; magnesium cast composite

---

1. Introduction

The need of decreasing the weight of the components in the transport industry [1] by the substitution of steel and iron casting components by plastics, carbon fiber, or aluminium and magnesium alloys has become one of the major boosters for transport industries. In the case of the bicycle industry, the substitution of materials such as steel, aluminium, and titanium by carbon fiber for high performance bicycles is a clear tendency.

Magnesium components produced by HPDC are already used for many automotive and bike applications, but the industry continues to look for new parts where the balance of lightness and mechanical properties provided by magnesium-lightened structures may be a solution. HPDC is a high-productivity process that is economically feasible for large production series (more than about 5000–10,000 parts/year) [2], where the high cost of production dies are paid off. In HPDC, the molten metal is poured into a steel shot sleeve. Then the metal is forced to enter into a closed metallic die under high metal velocities (from about 30 to 100 m/s for magnesium alloys) [3] and high dimensional accuracy complex components are produced, presenting a low surface roughness by the subsequent extremely short filling time (from 10 to 150 ms approx.) [4,5]. During solidification the metal contracts and can produce shrinkage porosity in the casting. To solve this physical phenomenon, metal is pressed in the liquid or semi-solid state using high pressures into the die cavity (specific pressures over the injected part of about 60 to 100 MPa). Steel dies can be designed to obtain very complex geometries and the total cycle time of the process is usually lower than 60 s.

However, HPDC is a complex process due to the large amount of parameters that influence the final casting quality. There are multiple parameters independent and interdependent that can have influence over the final part quality [6]. The main parameters that affect the mechanical properties of HPDC components are related to injection parameters (pressures, speeds, starting points) and die temperatures, but there are other essential aspects such as the castability of the alloy, the geometrical complexity of the parts, the cooling rate, and the type of equipment used to produce the components that affect the soundness and properties of injected parts. There are other parameters that affect the flow and solidification behavior of the cast parts as the viscosity and the gate, runner, and mold cavity design [7]. In order to maintain a stable die temperature, cooling and tempering circuits are employed, but to obtain a good quality part the lubrication of the die should also be adjusted to the selected alloy, working temperatures, and cycle time. They are several types of lubricants (water-based, oil-based, or dry lubricants), that should be selected for a determined application [8–10]. The quantity and percentage of lubricant should also be adjusted to the process for avoiding quality defects.

The HPDC process also presents some drawbacks. The main negative aspect is related to the internal porosity of the components, mainly derived from the turbulent flow in which the molten metal is forced to get into the die. This aspect renders HPDC components more difficult to be heat-treated and the mechanical properties attainable are, therefore, lower than those obtained in other casting processes. Contamination of the alloy by deposits of lubricants and oxides, and problems related to high maintenance costs of the dies submitted to cavitation, wear, and soldering phenomena are additional drawbacks hindering the use of HPDC for the production of highly structural components [11,12]. Several strategies have been followed to try to improve the performance of HPDC components in recent years. New HPDC variants have been developed with the aim of reducing the intrinsic porosity of conventional HPDC components.

In order to decrease the weight of injected parts different strategies have been employed. One is the employ of aluminium-magnesium composites [13], but their applications are quite exotic. One of the best known examples of aluminium-magnesium composite is the BWM aluminium-magnesium block [14] where magnesium is cast over the aluminium core. However, no metallurgical bonding is obtained due to the alumina surface of the aluminium core [15] if the aluminium core is not pre-treated with a special surface layer.

Another way is by obtaining hollow parts by HPDC using salt cores. However, it is difficult to eliminate the salt core, and also some holes in the parts are necessary to eliminate the salt from the cast part [16–18].

The development of new alloys for HPDC with improved properties can also reduce the total weight of the injected parts, by using different alloying elements and compositions [19].

One of the main challenges of obtaining composite castings with aluminium foam cores is to prevent the deformation or crash of the cores, due to the high velocity and specific pressures employed in the HPDC and also the necessity of avoiding defects on the casting.

Among these advances, the development of aluminium foams has a special importance [20]. Three main methods of producing aluminium foams are presented in the market. In Figure 1 we can observe some of the most important process used to produce aluminium foams [21–23].

Figure 1. Some of the most employed processes to produce aluminium foams.

Every method has its advantages and disadvantages, which are summarized in Table 1.

Table 1. Pros and cons of different foams production routes.

Production Route	Advantages	Disadvantages
Liquid	• Convert a liquid metal or alloy into foam without an interruption as long as desired, as wide as the vessel. • Less expensive. • Continuously produced, the lowest densities, and the most homogeneous.	• Necessity for cutting the foam, thereby opening the cells. • Limitation to make shaped parts. • Limitation in the metals that can be used. • Low range temperature control.
Solid	• Near-net shaped and complicated parts. • Economical mass production. • Sandwich panels of multi-materials. • Composite can be deformed.	• Subsequent thermal treatment. • More expensive than the liquid route. • Anisotropic properties.
Deposition	• Open pore foams. • Complex forms. • For Ni, Ni-Cr, Cu. • Thin sheets with low thickness.	• Expensive and slow to manufacture. • Limited in raw materials. • Specific granulometry for the raw materials. • Size limitation.

Only a few processes can be employed to obtain foams with complex shapes and an external aluminium skin. One process is to overcast aluminium foam by low-pressure die casting (LPDC). We can obtain aluminium foam sandwich structures with an inner permanent core of aluminium foam and a dense aluminium outer skin with a thickness of several millimetres [24].

A different possibility is to employ the integral foaming molding process [IFM], which is a very interesting process applicable to LPDC and HPDC, and for both aluminium and magnesium alloys [25]. The main advantage is that we can obtain not only aluminium but also magnesium foams. However, the processed HPDC part has a non-skinned area of the total length of the movement of the cores, and there are limitations in the part geometry. Additionally, dies are more complex and their maintenance more expensive.

In the last years, a novel process named advanced pore morphology (APM) has been developed to obtain composites with aluminium, based on the use of aluminium foam spheres as the material to obtain the aluminium foam core. This process is expensive and it is quite complicated to obtain a good interphase between the aluminium foam spheres [26].

The approach used in the present work has also been related to the analysis and control of the HPDC parameters to obtain good quality parts. The final objective was to develop a process that could be used for the production of a bicycle rod currently produced by forging aluminium, titanium, or magnesium, or by using carbon fibers. In Figure 2 we can observe the objective component after having been redesigned to be adapted to HPDC features.

Figure 2. (**a**) 3D rod design; and (**b**) detail of the placement and example of an aluminium foam core.

The developed composite casting has to comply with the established mechanical requirements, but with the reduction of cast part total weight at a competitive production cost as the main objective. In the present study three different close pore aluminium foams types were used to validate the aluminium foam as a core to be overcast with an AM60B alloy. The selected close pore aluminium foams were the Alporas ALPO-PLA-03 [27,28], a Formgrip-based processed foam [29,30] and the 0.4% TiB2 AlSi12 Alulight foam [31–34].

The results here presented are a part of a work aimed to obtain a light new high-performance bicycle rod. The work consists of validating the development of a HPDC magnesium AM60B alloy with an aluminium foam core, presenting a remarkable reduction in weight to be employed in transport and sport.

2. Experimental Section

Castings were produced with an AM60B alloy and three different aluminium foam cores. Table 2 presents the main properties to be obtained and Table 3 presents the composition of the base alloy.

Table 2. Requirements established for magnesium-aluminium foam-based rod.

Properties/Requirements	Result
Weight reduction	35%
External appearance	Without external defects
Porosity	Low

Table 3. Base composition of the AM60B alloy.

Material	Al	Mn	Zn	Si	Fe	Cu	Ni	Be	Mg
AM60B	6.06	0.305	0.012	0.0085	<0.0012	0.0011	0.001	10 ppm	Bal

Three different aluminium foams were used in the project. Alporas-type foams were developed by the liquid route to obtain low density foams by Alcan. ALPO-PLA-03 aluminium foams with densities ranging from 0.25 to 0.4 Kg/dm^3 were selected with a range of 10% of carbon silicide particles. In this foam there is no external aluminium skin, and it is very difficult to obtain complex shapes. The second type of foams were fabricated at Tecnalia using the Formgrip process, by adding carbon silicide and titanium hydride particles of less than 10^{-4} μm to an AlSi7 and AlCSi aluminium molten alloys. The Formgrip process promotes the formation of close porous material but with a more expensive and complicated process in comparison with others. The process is based in adding foaming powders as precursors of gas porosity and stabilizing the created porous foam by heat treatment. The foaming temperatures were established between 680 °C and 720 °C.

The last type of foams employed in the trials was the Alulight 0.4 TiB2 AlSi12 alloy. In this alloy the content of silicon is adjusted to improve castability and mechanical strength values [35].

For making the Alulight type foams, a metallic die was prepared and the foaming process was studied to obtain soundness of aluminium core foams, studying the quantity of the precursor, the die temperature, the furnace temperature, and the residence time, obtaining a density range a from 0.54 to 1.55 Kg/dm^3.

In Figure 3 we can observe the metallic die employed to obtain the Alulight samples.

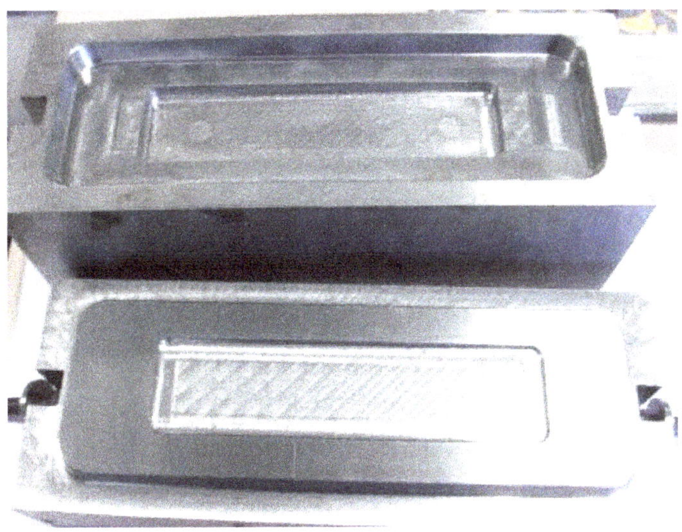

Figure 3. Metallic die to produce Aluminium foams with Alulight.

Different essays at 600, 650, and 670 °C, respectively, were developed to determine the residence time of the Alulight precursor inside the die to obtain the final aluminium foam insert. After the study of the foaming parameters, a manufacturing procedure was established. The die was first heated to 670 °C, after the die was opened, the precursor was placed on the die cavity, the die was closed, and then introduced into the furnace. After 6 min of residence time, the die was extracted from the furnace, opened, and the final foam extracted.

The obtained density varied from 0.25 to 0.4 Kg/dm^3 for Alporas samples, from 0.4 to 0.65 Kg/dm^3 for Formgrip samples, and 0.54 to 1.55 Kg/dm^3 for the Alulight samples.

In order to determine the castability of the composite component, a molten AM60B magnesium alloy was poured over the different types of aluminium foams in a metallic vessel at alloy temperature of 680 °C and 720 °C. In these trials the chemical and temperature resistance of the aluminium foam core when an AM60B alloy was casted over it were analyzed. We can observe in Figure 4 the metallic mold and aluminium foam to obtain the composite component directly by pouring the molten magnesium over the aluminium foam. The foam a steel pin was inserted in order to avoid the floating tendency and to obtain a sample with the aluminium foam in the central section of the cast part.

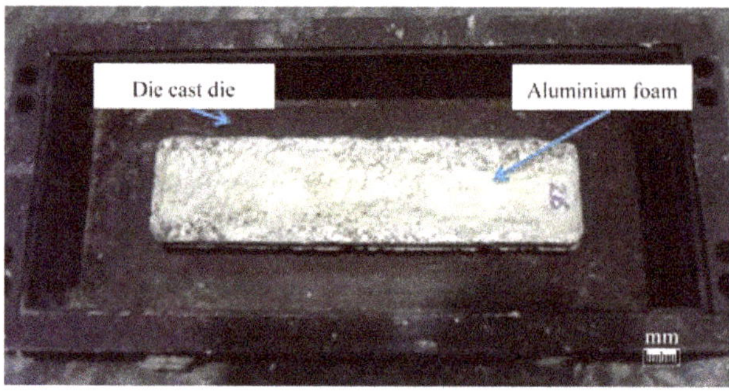

Figure 4. Metallic die to die cast magnesium over the aluminium foam.

A preliminary analysis of the die casting, plastic injection, and HPDC processes were employed to determine the effect of pressure over the selected foams, obtianing the samples' different density ranges. The quality and defects on the quality of the injected parts was determined. This preliminary analysis shows that it was necessary that the aluminium foam has to have a minimum external skin to act as a core without collapsing. Alulight foamed cores supported the magnesium over-injection by HPDC without any damage. Aluminium foams were placed in a Tecnalia's plastic injection test die at room temperature and after preheating the aluminium foam at 60 °C. A transparent-type plastic TPE (THERMOLAST K TF8CGT, KRAIBURG TPE GmbH & Co, Waldkraiburg, Germany) was over-injected in order to observe the effect of injection parameters over the foam, with an injection pressure of 16–40 MPa, slope temperatures of 180 °C, 200 °C, and 220 °C, die temperature of 40 °C, a flow rate of 45 cm^3/s, a cooling time of 40 s, and an injection volume of 46 cm^3.

After having optimized the injection parameters, the maximum pressure and speed injection parameters for a HPDC process were established. We can observe in Figure 5 the plastic injection mold with aluminium foam placed on the die before the injection of the plastic.

Figure 5. The plastic injection mould with an aluminium foam.

To determine the real behavior of aluminium foam cores working with a HPDC process, a series of castings were produced with a 950 ton HPDC machine produced by Pretransa at Tecnalia´s foundry pilot plant. Standard HPDC magnesium injection parameters were used. The pressure ranges used variated from 20 to 80 MPa of specific pressure, 0.25 m/s first phase piston speed, and second phase speed in the range between 20 and 80 m/s. The die used to produce the castings was a multi-cavity die with a bicycle rod and a squeeze pin cavity. We can observe in Figure 6 the process to produce the composite casting with the aluminium foam core and the AM60B magnesium alloy.

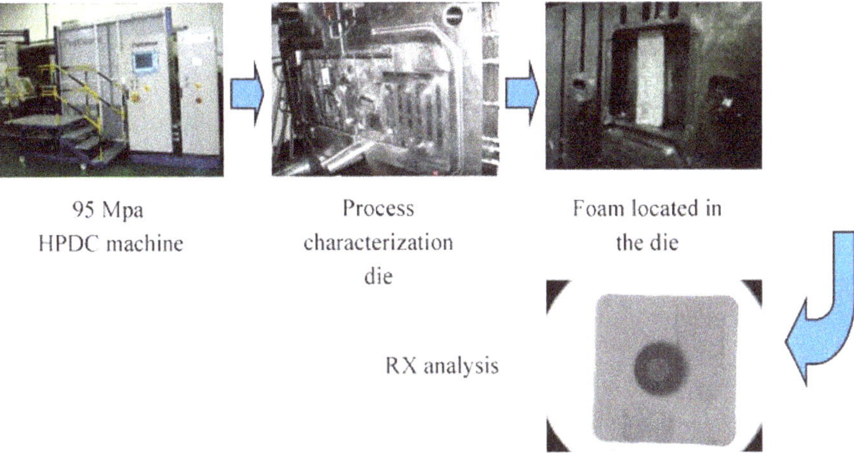

Figure 6. HPDC process in order to obtain the magnesium-aluminium foam core composite.

In order to fix the aluminium foam core to the die in the bicycle rod cavity, the die was modified. Four pins of 1 mm diameter and 1 mm height were located in the fix part of the die, as we can observe in Figure 7. The aluminium foam was inserted on the pins, by pressing the foam in the die cavity.

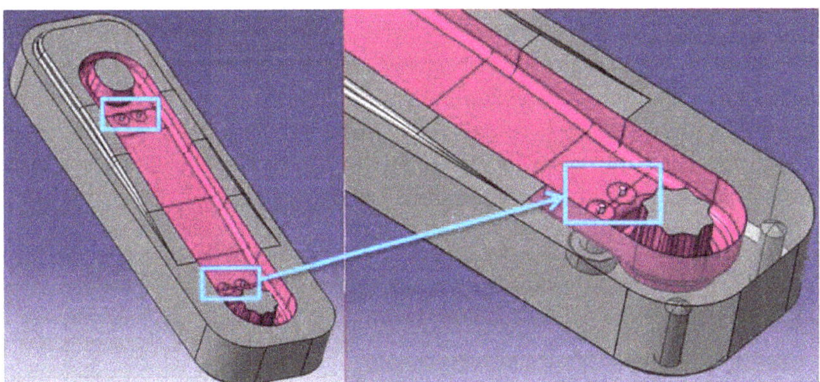

Figure 7. Detail of fixing pins in the fixed die cavity for placing the aluminium foam.

The bicycle rod is composed by a central straight body of 210 mm length, 40 mm width, and from 12 to 16 mm thickness, with two cavities in the extremes that are the sockets that connect the rod with the pedal and the shaft. The rods had been submitted to X-ray analysis employing a General Electric X-cube 44XL (GE Sensing & Inspection Technologies GmbH, Ahrensburg, Germany) at 160 kV to confirm the absence of broken cores that could invalidate the results of the tests, to control the soundness of the components, and to identify potential defects.

Finally, in order to determine the mechanical properties of the composite rods, three tensile test were carried out in accordance with the UNE-EN ISO 6892-1 B:2010 standards at room temperature with a crosshead speed of 5 mm/min using an Instron 3369 electromechanical testing machine. The tensile stress, ultimate tensile strength, and elongation were calculated from obtained stress-strain diagrams.

3. Results

3.1. Determination of Metal Casting Temperature

Different experiments have been developed in order to determine the accuracy of different types of aluminium foams and casting parameters.

The first casting trials were made by casting the AM60B alloy at 680 °C and 720 °C over an aluminium foam core.

At 680 °C short fill/cold shut defects were detected as shown in Figure 8a. In Figure 8b we can observe some gas liberation from the aluminium foam.

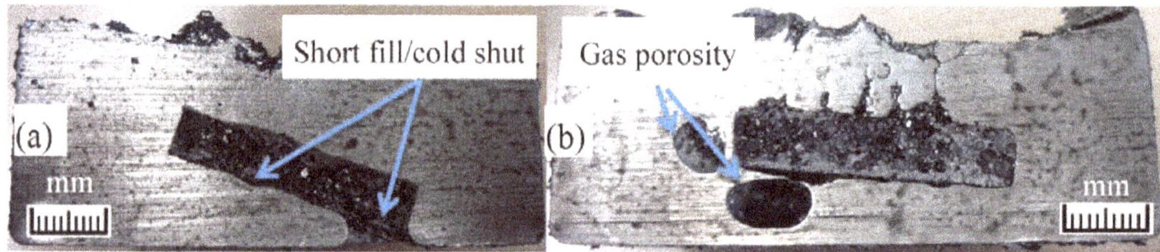

Figure 8. (**a**) Short fill and cold shut defects; and (**b**) gas porosity defects.

The three different types of aluminium foams did not show any casting defects (short fill/poor filling) or any damage on the foam when the magnesium alloy was poured at 720 °C, so a minimum pouring temperature of 720 °C was needed to avoid any poor filling defect. However as expected, there was not an interphase between the aluminium foam and the over-casted AM60B alloy.

As we can observe in Figure 9 the Alulight-type foam is cover with AM60B magnesium without gas porosity, because the external skin of about 0.5 mm thickness of the Alulight type foam avoids the expansion of the gas bubbles inside the magnesium casting.

Figure 9. Central aluminium core covered with AM60B.

3.2. Determination of Injection Pressure

In order to simulate only the pressure action of an injected material over different aluminium foams, a transparent plastic was injected over the three types of foams at pressures between 16–40 MPa, with the objective of determining which kind of aluminium foam, density, and combination of different foams is necessary to support the standard 40 MPa specific pressure for magnesium HPDC. We can observe in Figure 10 different configurations and combinations of aluminium foams.

Figure 10. Different configurations for plastic injection over the aluminium foams.

In Table 4 we can observe the obtained integrity results for the different types and densities of the foams.

Table 4. Integrity of aluminium foams after plastic injection.

Process	Density (Kg/dm^3)	Integrity at 16 MPa	Integrity at 40 MPa
Alporas	0.25–0.4	NO	NO
Formgrip	0.4–0.55	NO	NO
Alulight	0.54–1.55	YES	YES

Despite the different densities, it is shown that only the Alulight foam resists the standard 40 MPa specific pressure for magnesium HPDC.

In order to simulate the industrial application, HPDC trials were carried out. First Alporas and Formgrip foams were over-injected with standard HPDC magnesium injection parameters and with and alloy injecting temperature of 680 °C, in order to confirm the obtained results in the plastic injection machine. In Figure 11a we can observe Alpora's foams and Formgrip foams in Figure 11b. Both types of foams were destroyed over the standard injection conditions, as shown in Figure 12, so they were discarded from being employed in HPDC.

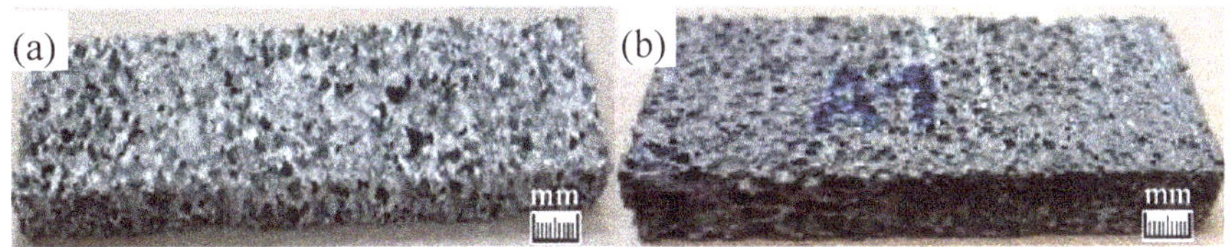

Figure 11. (**a**) Alpora's foam (0.25 to 0.4 Kg/dm^3); and (**b**) Formgrip's foam (0.4 to 0.65 Kg/dm^3).

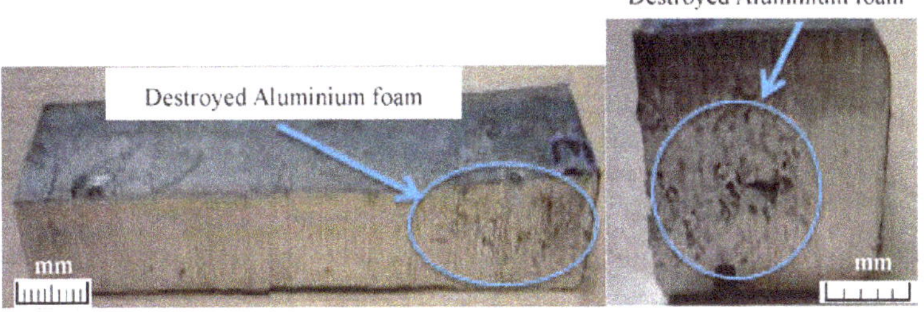

Figure 12. HPDC part with totally destroyed aluminium foam.

In order to determine the possibility of employing foams with non-skinned areas, Alulight aluminium foam samples cut at their center were introduced in the squeeze pin cavity of the HPDC die and over-injected with the magnesium alloy at 680 °C. We can observe in Figure 13a how the sample is introduced in the die and in Figure 13b how the non-skinned area promoted the formation of gas porosity inside the cast part.

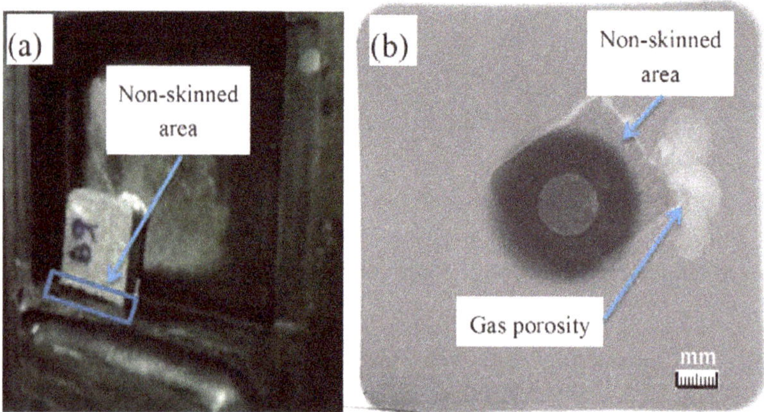

Figure 13. (**a**) Placement of a skinned foam with a non-skin area in the die; and (**b**) release of gas from the foam in the non-skinned area.

As we have observed above, open skin foams can liberate the foaming gas in contact with molten magnesium, promoting gas porosity inside the cast part. Thus, foams which contain areas without an external skin cannot work properly as cores in the HPDC process.

3.3. Determination of Squeeze Pin Suitability for Composite Cast Parts

In order to study the possibility of employing up to 200 MPa of specific pressure in specific areas using an squeeze pin with composite components for reducing shrinkage porosity, several aluminium foam cores with densities from 0.6 to 1.55 Kg/dm^3 were placed into the squeeze pin cavity, and the action of the squeeze pin pressure was evaluated on the foam integrity, with the same casting parameters employed in previous tests. We can observe in Figure 14 how the pressure acts over the foam core.

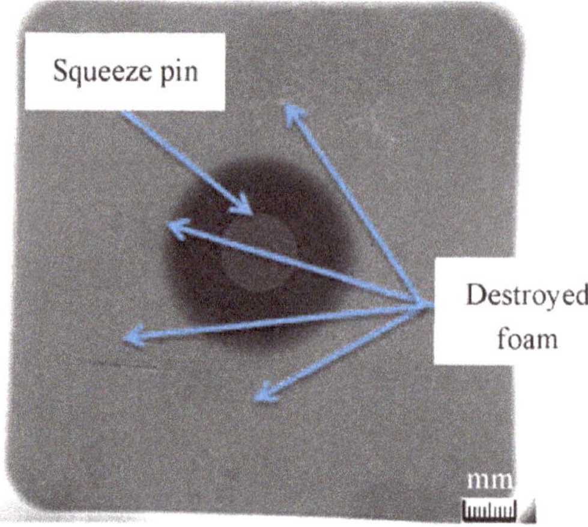

Figure 14. 1.55 Kg/dm^3 Aluminium foam after squeeze pin application.

As we can notice, the foam is totally destroyed in small pieces. Thus, the squeeze pin application in areas near the aluminium foam is not possible, despite increasing foam density.

3.4. Determination of Core Placement

The position of the foam core has been also studied in order to determine if it can be employed in any position or if the high injection speeds and pressures can damage it as a function of the position of the foam in the die with respect to the metal flow into the cavity. Two different configurations were tested. We can observe in Figure 15 the behavior of a core placed horizontally to the flow .

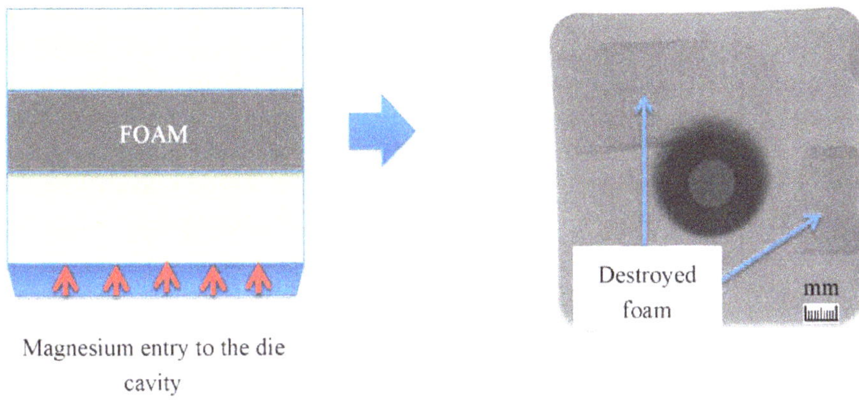

Figure 15. Horizontal placement to the metal flow of the core.

The impact of the metal flow with the foam core causes a shear rupture of the core foam. It is necessary to select the positioning of the core in the die or the metal flows over the core by re-designing the runners in order to decrease the direct impact of injected metal on the core.

We can observe in Figure 16 how a core placed vertically to the flow with the rod demonstrator does not present any damage on the aluminium foam, with only small shrinkage porosity areas.

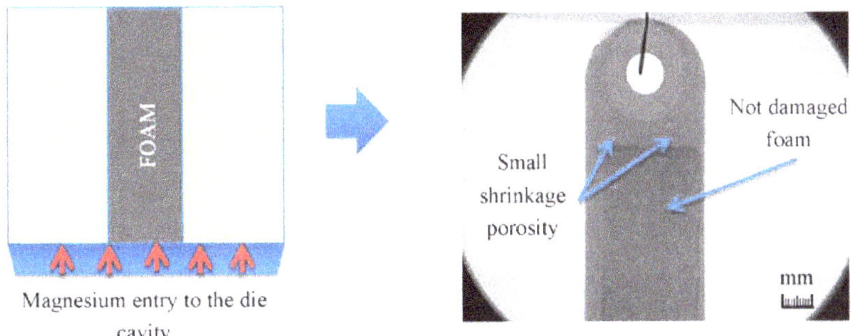

Figure 16. Horizontal core placement to metal flow.

An optimal placement of the foam is critical to avoid damages in the aluminium foam and to obtain sound parts. In Figure 17 we can observe a complete magnesium rod without any external quality defect, with the internal aluminium foam core.

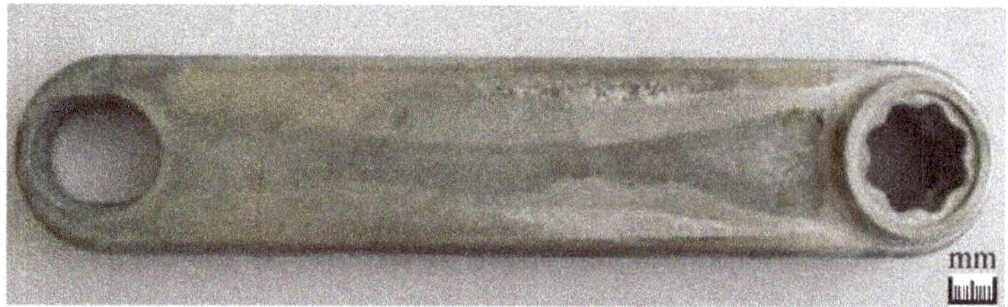

Figure 17. Rod made by magnesium HPDC with the internal core of aluminium foam.

3.5. Determination of Injection Speed

Another critical HPDC parameter is the injection speed. The metal speed in the gates was varied from 20 to 80 m/s and the specific injection pressure over the cast part from 20 to 80 MPa, with the aluminium foam and an alloy temperature of 680 °C. We can observe in Figure 18 how, if we reduce the first injection speed, the part quality is adversely affected.

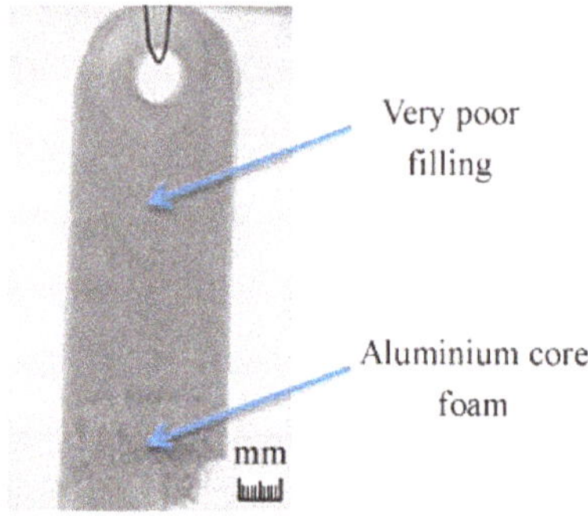

Figure 18. Reduced second phase speed (20 m/s) HPDC cast part.

With magnesium HPDC standard parameters, establishing the second injection speed at 80 m/s and the specific injection pressure at 80 MPa, sound parts were obtained. Only some small solidification porosity was detected. We can observe in Figure 19 an example of the quality obtained on the final composite cast part.

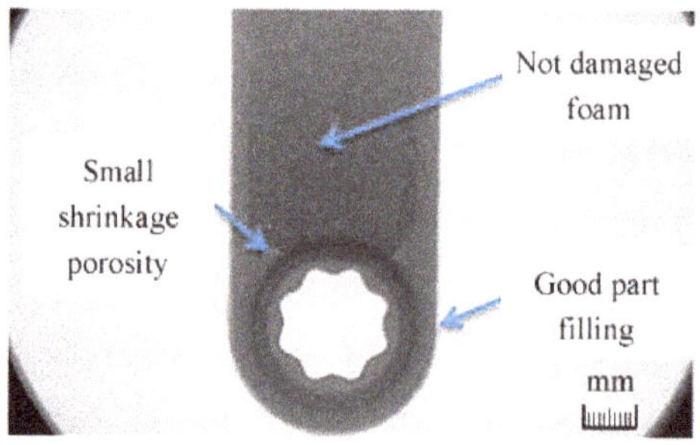

Figure 19. Injected HPDC with core foam at standard parameters.

In the case of employing the Alulight-type aluminium foam of 0.56 Kg/dm^3 density, a total weight reduction is of about a 35% is obtained.

3.6. Determination of Mechanical Properties

In order to determine the mechanical properties, tensile tests were made over three AM60B alloy injected rods, and compared with three aluminium foam-magnesium composite injected rods. The percentage of aluminium foam area in comparison with the total area in the central region of the injected part is about 47%. We can observe the obtained results in Table 5, with an extrapolation of obtained properties of composite rods in relation with the reduction of weight obtained with the use of aluminium foam cores:

Table 5. Obtained properties of composite rods.

Sample	YS (MPa)	UTS (MPa)	Elongation (%)
AM60B	141 ± 4.3	219 ± 3.6	3 ± 0.5
Composite	79 ± 1.75	122 ± 2.16	3 ± 0.5
Extrapolation of composite properties (47% of area reduction)	149 ± 3.3	230 ± 4.1	3 ± 0.5

4. Discussion

The absence of an external minimum and continuous skin on the aluminium foams promotes the presence of internal gas porosity in the composite cast part, due to the gas liberation in the injection process from the surface of the foam. This liberation is promoted by the presence of small cracks in some of the external pores forced by the metal injection speed and pressure.

The absence of an external continuous skin on the foam surface decreases the mechanical resistance to the injection speed and pressure, destroying the foam. The presence of a continuous skin of about 1 mm gives the necessary robustness to the foam to support the injection process. The higher the density, the better the mechanical resistance of the foam, but only when there is a continuous skin because, despite increasing the density of foams, they are collapsed by the injection conditions. However, the increase in tensile test values over Alulight-type foams is very limited [35].

Positioning the aluminium foam opposite to the entry of the metal in the cavity results in a resistance to the free entry of the molten metal into the die cavity, and shear fractures of the aluminium foams due to the combination of high speeds and pressures are produced.

The use of squeeze pins on areas to reduce shrinkage porosity on areas closer to aluminium foams should be avoided because the extremely high pressures (up to 200 MPa) employed by the squeeze pins cannot be supported by the foams. With specific pressures up to 800 MPa in the injection trials the aluminium foam cores supported the applied pressure without any damage, well above the standard magnesium HPDC specific pressures employed in the industry.

There is not a chemical bonding between the aluminium foam core and the over-casted magnesium alloy, because the aluminium foam is totally covered with an alumina surface that prevents the direct contact between the aluminium of the foam and the molten magnesium. This lack of a chemical bonding decreases the final mechanical properties of the composite, reducing the mechanical properties to these related to the cast magnesium. The increase in the obtained mechanical values in the composite cast part in comparison with the correlated magnesium cast part mechanical values could be related to the internal presence of the aluminium foam, with its intrinsic mechanical properties, and by the higher cooling rate on the surface in contact with the molten magnesium, creating an internal surface in the magnesium with an smaller grain size in comparison with a solidification without a core.

The next work will focus on studying the application of surface treatments over the aluminium foam for obtaining a continuous interphase between the aluminium foam and the cast magnesium. This could increase the bond strength between the two materials, increasing the mechanical interlocking. By studying, for example, Zn-based coatings for the aluminium foams it is estimated that the ultimate tensile strength could be slightly improved (approx. 3–5 Mpa).

5. Conclusions

HPDC magnesium aluminium foam composite has been developed for applications requiring a reduction on part weight. Optimum aluminium core foam produced by the Alulight process has been defined through an experimental approach.

The Alulight process permits obtaining close pore foams with an external skin that avoids the presence of internal defects in the composite. Densities may have changes for the core foam from 0.54 to 1.55 Kg/dm^3. The regulation of the foam porosity allows tailoring properties and customizing the part for specific performances, but in applications where the main objective is weight reduction, foams of only 0.54 kg/dm^3 overcame casting conditions. The estimated cost on material and production by using an aluminium foam in the bicycle rod is of about 0.3 Euros/part, which is a reasonable cost.

The developed composite casting has been used for the production of a bicycle rod. Tensile tests have subsequently confirmed the validity of the composite casting for this application and its actual potential to substitute currently used aluminium, titanium, or carbon fiber materials. A reduction of 35% in the total weight of the part has been obtained. Higher reductions are possible depending on the part. Magnesium HPDC can be a solution for mass production of composites with foams, as an alternative to plastics and aluminium parts.

The use of aluminium foams does not seem to give higher mechanical properties to composite cast parts. The tensile test shows very similar values to calculated ones when a hollow area in the middle of the rod is used for estimations. However, the process permits to reduction in the weight of cast parts where the core area has only relative influence over the final mechanical properties. It would be also possible to obtain improvements on vibration/sound control, crash performance, or compression properties in other applications or final parts.

Acknowledgments: This work has been partially funded by the Spanish government through the project CDTI: MAGNO CENIT project CEN-20081028. The authors are also much grateful to the personnel of Grupo Antolín, company that collaborated in aspects related to the development and analysis of the rod and to Industrias Lebario, who collaborated in the development of aluminium foam dies.

Author Contributions: I.V. and I.C. performed the experiments; I.V., L.M.P. and P.C. analyzed the data; I.V. and P.C. wrote the paper; J.K.I. supervised the work.

Conflicts of Interest: The authors declare no conflict of interest.

References

1. Schultz, R. *Aluminium Association Auto and Light Truck Group 2009 Update on North American Light Vehicle Aluminium Content Compared to the other Countries and Regions of the World. Phase II*; Ducker Worlwide LLC: Troy, MI, USA, 2008; pp. 8–19.
2. Bonollo, F.; Urban, J.; Bonatto, B.; Botter, M. Gravity and low pressure die casting of aluminium alloys: A technical and economical benchmark. *Metall. Ital.* **2005**, *97*, 23–32.
3. Luo, A.A. Magnesium casting technology for structural applications. *J. Magnes. Alloys* **2013**, *1*, 2–22. [CrossRef]
4. Gertsberg, G.; Nagar, N.; Lautzker, M.; Bronfin, B. Effect of HPDC parameters on the performance of creep resistant alloys MRI153M and MRI230D. *SAE Tech. Pap.* **2005**, *1*, 1–7.
5. Otarawanna, S.; Laukli, H.I.; Gourlay, C.M.; Dahle, A.K. Feeding mechanisms in high-pressure die castings. *Metall. Mater. Trans. A* **2010**, *41*, 1836–1846. [CrossRef]
6. Bonollo, F.; Gramegna, N.; Timelli, G. High-pressure die-casting: Contradictions and challenges. *JOM* **2015**, *67*, 901–908. [CrossRef]
7. Yim, C.D.; Shin, K.S. Semi-solid processing of magnesium alloys. *Mater. Trans.* **2003**, *44*, 558–561. [CrossRef]
8. Sabau, A.S.; Dinwiddie, R.B. Characterization of spray lubricants for the high pressure die casting processes. *J. Mater. Process. Technol.* **2008**, *195*, 267–274. [CrossRef]
9. Kimura, R.; Yoshida, M.; Sasaki, G.; Pan, J.; Fukunaga, H. Characterization of heat insulating and lubricating ability of powder lubricants for clean and high quality die casting. *J. Mater. Process. Technol.* **2002**, *130–131*, 289–293. [CrossRef]

10. Puschmann, F.; Specht, E. Transient measurement of heat transfer in metal quenching with atomized sprays. *Exp. Therm. Fluid Sci.* **2004**, *28*, 607–615. [CrossRef]
11. Rogers, K.J.; Savage, G. In-cavity pressure sensors-errors, robustness and some process insights. *Die Cast Eng.* **2000**, *44*, 76–80.
12. Kong, L.X.; She, F.H.; Gao, W.M.; Nahavandi, S.; Hodgson, P.D. Integrated optimization system for high pressure die casting processes. *J. Mater. Process. Technol.* **2008**, *201*, 629–634. [CrossRef]
13. Keber, K.; Bormann, D.; Möhwald, K.; Hollander, U.; Bach, W. Compound casting of aluminium and magnesium alloys by High Pressure Die Casting. In Proceedings of the Magnesium 8th International Conference on Magnesium Alloys and Their Applications, Weimar, Germany, 26–29 October 2009; pp. 390–397.
14. Kunst, M.; Fischersworring-bunk, A.; Liebscher, C.; Glatzel, U.; Esperance, G.L.; Plamondon, P.; Baril, E.; Labelle, P. Microstructural characterization of Die Cast Mg-Al-Sr (AJ) Alloy. In Proceedings of the Magnesium: Proceedings of the 7th International Conference on Magnesium Alloys and Their Applications, Dresden, Germany, 6–9 November 2006; pp. 498–505.
15. Vicario, I. Study of high pressure die cast AZ91D magnesium alloy with surface treated aluminium 6063 cores for Al-Mg multi-material. In Proceedings of the 2nd Annual World Congress of Advanced Materials, Suzhou, China, 5–7 June 2013; pp. 7–21.
16. Yaokawa, J.; Miura, D.; Anzai1, K.; Yamada, Y.; Yoshii, H. Strength of salt core composed of alkali carbonate and alkali chloride mixtures made by casting technique. *Mater. Trans.* **2007**, *5*, 1034–1041. [CrossRef]
17. Moschini, R. *Production of Hollow Components in HPDC through the Use of Ceramic Lost Cores*; XXXI Congreso Tecnico di Fonderia: Vincenza, Italy, 2012; pp. 1–33.
18. Yamada, Y.; Yaokawa, J.; Yoshii, H.; Anzai, K.; Noda, Y.; Fujiwara, A.; Suzuki, T.; Fukui, H. Developments and application of expendable salt core materials for high pressure die casting to apply closed-deck type cylinder block. *SAE Int.* **2007**, *32-0084*, 1–5.
19. Vicario, I.; Egizabal, P.; Galarraga, H.; Plaza, L.M.; Crespo, I. Study of an Al-Si-Cu HPDC alloy with high Zn content for the production of components requiring high ductility and tensile properties. *Int. J. Mater. Res.* **2013**, *4*, 392–397. [CrossRef]
20. Benedick, C. Production and application of aluminium foam, past product potential revisited in the new millennium. *Light Met. Age* **2002**, *60*, 24–29.
21. Bausmesiter, J.; Weise, J. *Structural Materials and Processes in Transportation, Metal Foams*; John Wiley & Sons: Hoboken, NJ, USA, 2013; pp. 415–440.
22. Banhar, J. Metal foams: Production and stability. *Adv. Eng. Mater.* **2006**, *9*, 781–794. [CrossRef]
23. Coleto, J.; Goñi, J.; Maudes, J.; Leizaola, I. Applications and manufacture of open and closed cell metal foams by foundry routes. In Proceedings of International Congress Eurofond, La Rochelle, France, 2–4 June 2004; pp. 1–10.
24. Körner, C.; Hirschmann, M.; Wiehler, H. Integral Foam Moulding of Light Metals. *Mater. Trans.* **2006**, *47*, 2188–2194. [CrossRef]
25. Baumeister, J.; Weise, J. Application of aluminium-polymer hybrid foam sandwiches in battery housings for electric vehicles. *Procedia Mater. Sci.* **2014**, *4*, 301–330. [CrossRef]
26. Vesenjak, M.; Borovinšek, M.; Fiedler, T.; Higa, Y.; Ren, Z. Structural characterisation of advanced pore morphology (APM) foam elements. *Mater. Lett.* **2013**, *110*, 201–203. [CrossRef]
27. Miyoshi, T.; Itoh, M.; Akiyama, S.; Kitahara, A. Aluminium foam, "ALPORAS": The production process, properties and application. In Materials Research Society Symposium Proceedings, Boston, MA, USA, 1–3 December 1998; 1998; pp. 133–137.
28. Miyoshi, T.; Itoh, M.; Akiyama, S.; Kitahara, A. Alporas aluminum foam: Production process, properties, and applications. *Adv. Eng. Mater.* **2000**, *2*, 179–183. [CrossRef]
29. Gegerly, V.; Clyne, T.W. The formgrip process: Foaming of reinforced metals by gas release in precursors. *Adv. Eng. Mater.* **2002**, *2*, 175–178.
30. Gergely, V.; Curran, D.C.; Clyne, T.W. Advances in the melt route production of closed cell aluminium foams using gas-generating agents. In Proceedings of Global Symposium of Materials Processing and Manufacturing Processing & Properties of Lightweight Cellular Metals and Structures, Seattle, WA, USA, 17–21 February 2002; pp. 3–8.

31. Baumeister, J. Production technology for aluminium foam/steel sandwiches. In Proceedings of the International Conference on Metal Foams and Porous Metal Structures, Bremen, Germany, 14–16 June 1999; pp. 113–118.
32. Seeliger, H.W. Cellular Metals: Manufacture, properties, applications. In Proceedings of the MetFoam, Berlin, Germany, 23–25 June 2003; pp. 5–12.
33. Kováãik, J.; Simančík, F.; Jerz, J.; Tobolka, P. Reinforced aluminium foams. In Proceedings of the International Conference on Advanced Metallic Materials, Smolenice, Slovakia, 5–7 November 2003; pp. 154–159.
34. Braune, R.; Otto, A. Tailored blanks based on foamable aluminium sandwich material. In Proceedings of the International Conference on Metal Foams and Porous Metal Structures, Bremen, Germany, 14–16 June 1999; pp. 119–124.
35. Gutiérrez-Vázquez, J.A.; Oñoro, J. Fabricación y comportamiento de espumas de aluminio con diferente densidad a partir de un precursor AlSi12. *Rev. Metal.* **2010**, *46*, 274–284. [CrossRef]

13

Numerical Evaluation of Temperature Field and Residual Stresses in an API 5L X80 Steel Welded Joint Using the Finite Element Method

Jailson A. Da Nóbrega [1], Diego D. S. Diniz [2], Antonio A. Silva [1], Theophilo M. Maciel [1], Victor Hugo C. de Albuquerque [3] and João Manuel R. S. Tavares [4,*]

1 Programa de Pós-Graduação em Engenharia Mecânica, Universidade Federal de Campina Grande (UFCG), Campina Grande-PB 58429-140, Brazil; jailson_engmec@hotmail.com (J.A.D.N.); antonio.almeida@ufcg.edu.br (A.A.S.); theophilo.maciel@ufcg.edu.br (T.M.M.)

2 Universidade Federal Rural do Semi-Árido (UFERSA), Campus Caraúbas, Caraúbas-RN 59700-000, Brazil; diego.diniz@ufersa.edu.br

3 Programa de Pós-Graduação em Informática Aplicada, Universidade de Fortaleza, Fortaleza-CE 60811-905, Brazil; victor.albuquerque@unifor.br

4 Instituto de Ciência e Inovação em Engenharia Mecânica e Industrial, Departamento de Engenharia Mecânica, Universidade do Porto, Porto 4200-465, Portugal

* Correspondence: tavares@fe.up.pt

Academic Editor: Hugo F. Lopez

Abstract: Metallic materials undergo many metallurgical changes when subjected to welding thermal cycles, and these changes have a considerable influence on the thermo-mechanical properties of welded structures. One method for evaluating the welding thermal cycle variables, while still in the project phase, would be simulation using computational methods. This paper presents an evaluation of the temperature field and residual stresses in a multipass weld of API 5L X80 steel, which is extensively used in oil and gas industry, using the Finite Element Method (FEM). In the simulation, the following complex phenomena were considered: the variation in physical and mechanical properties of the material as a function of the temperature, welding speed and convection and radiation mechanisms. Additionally, in order to characterize a multipass weld using the Gas Tungsten Arc Welding process for the root pass and the Shielded Metal Arc Welding process for the filling passes, the analytical heat source proposed by Goldak and Chakravarti was used. In addition, we were able to analyze the influence of the mesh refinement in the simulation results. The findings indicated a significant variation of about 50% in the peak temperature values. Furthermore, changes were observed in terms of the level and profile of the welded joint residual stresses when more than one welding pass was considered.

Keywords: multipass welding; temperature field; residual stress; finite element method; computer simulation

1. Introduction

The materials in welded joints undergo many metallurgical changes due to the intense localized heat input, particularly for fusion welding. This heat input takes place in a nonlinear and transient manner, with the main heat input near the heat source and a lesser heat input coming from the center of the welding zone. Thus, non-uniform elastic and plastic deformations are generated promoting high residual stresses in the welded joint [1]. The level and profile of residual stresses exert a considerable influence on the welded joint properties and their control can avoid possible structural failures [2]. Thus, the influence of the residual stresses on the crack growth of welded joints is the focus of many studies concerning engineering applications [3,4], as well as in the optimization of turbine flanges [5] and in the petroleum industries [6], just to name some examples. Therefore, various destructive and non-destructive techniques have been used to evaluate the residual stresses in welded joints. Currently, one of the non-destructive methods is computational analysis based on analytical methods.

The Finite Element Method (FEM) has been used to evaluate the weld thermal cycle, by various authors such as Cho *et al.* [7] and Robertsson and Svedman [8] who used the commercial software ANSYS® and SIMULIA ABAQUS®, respectively. These authors found excellent results in the thermal and structural analyses compared to the experimental ones obtained by thermocouples inserted in the welded joint for temperature measurements and X-ray diffraction to assess the residual stresses.

Methods of numerical analysis can provide benefits to many areas of engineering, from the development of new products to maintenance services. In manufacturing, the welding parameters can be pre-established to minimize distortions and residual stresses caused by plastic deformation and thermal expansion, without the need to carry out costly laboratory experiments. In some cases, control models can be developed from the inversion of computational models in order to perform real-time corrections in the welding process and prevent variations that can produce weak points in components [9–11].

The purpose of this study was to simulate and evaluate the temperature field and residual stresses in an API 5L X80 steel welded joint using the GTAW (Gas Tungsten Arc Welding) process and SMAW (Shielded Metal Arc Welding) process, through the use of the commercial software ABAQUS®, which is based on FEM. This work studies the evolution of the thermal cycles and residual stresses by employing a single welding process or a combination of multiple welding passes, thereby making the computational simulation more realistic.

2. Mathematical Model

2.1. Thermal Modeling

In the electric arc welding process, an electrical source generates a voltage U between the electrode and the base metal, inducing the formation of an electric arc traversed by a current I. In this process, energy losses occur through several factors, among them, the convection and radiation in the electric arc can be mentioned. Consequently, only a portion of this energy is used to melt the material, and therefore it is necessary to add a variable called power efficiency (η). Hence, the effective weld heat input can be given as:

$$Q = \eta UI \tag{1}$$

In this thermal model, the thermal gradient can be assessed by establishing the related energy balance:

$$\rho(T)c(T)\frac{\partial T}{\partial t} = Q + \frac{\partial}{\partial x}\left[Kx(T)\frac{\partial T}{\partial x}\right] + \frac{\partial}{\partial y}\left[Ky(T)\frac{\partial T}{\partial y}\right] + \frac{\partial}{\partial z}\left[Kz(T)\frac{\partial T}{\partial z}\right] \tag{2}$$

where ρ is the material density, c is the specific heat, Q is the heat input (Equation (1)), K_x, K_y, K_z are the thermal conductivity coefficients in each direction, T is the temperature and t the time. The associated heat flow is not linear, and the thermophysical properties of the material are highly dependent on the temperature.

The heat loss by convection q_c and radiation q_r can be calculated using the following equations:

$$q_c = h_f(T - T_\infty l) \tag{3}$$

$$q_r = \varepsilon\sigma(T^4 - T_\infty^4) \tag{4}$$

where h_f is the convective coefficient, T_∞ is the ambient temperature, σ is the Stefan-Boltzmann constant and ε the emissivity of the body surface.

A phase change occurs during the process and this generates a latent heat, which can be expressed as a function of the enthalpy (H) as:

$$H = \int \rho c dT \tag{5}$$

For computational analysis by FEM, an essential issue in the simulation is the modeling of the heat source. Goldak and Chakravarti [12] proposed an analytical solution for modeling the distributed heat source associated to the arc welding which is commonly used in this type of analysis. Therefore, the temperature field can be determined, computationally, based on a 3D finite Gaussian on a double ellipsoid, as shown in Figure 1. This heat source is analytically defined by the equations [13]:

$$q_f(x,y,z) = f_f \frac{\eta U I}{a_f b d \pi \sqrt{\pi}} 6\sqrt{3} e^{\left(\frac{-3x^2}{a_f{}^2}\right)} e^{\left(\frac{-3y^2}{b^2}\right)} e^{\left(\frac{-3z^2}{d^2}\right)} \tag{6}$$

$$q_r(x,y,z) = f_r \frac{\eta U I}{a_r b d \pi \sqrt{\pi}} 6\sqrt{3} e^{\left(\frac{-3x^2}{a_r{}^2}\right)} e^{\left(\frac{-3y^2}{b^2}\right)} e^{\left(\frac{-3z^2}{d^2}\right)} \tag{7}$$

where q_f and q_r are the volumetric energy distributions before and after the torch (W/m^3), and f_f and f_r are the fractional factors of the distributions of the accumulated heat before and after the torch [14] (Figure 1). Additionally, U, I, η are parameters directly linked to the welding procedures, while b and d are the geometric parameters of the heat source and can be determined by metallographic examination. The other parameters a_f, a_r, f_f and f_r are obtained through the parameters b and d; furthermore, the sum of f_f and f_r is equal to 2 [12]. In the absence of better data, the distance (a_f) from the triad reference point to the front of the heat source is defined as being equal to half the width of the weld and the distance (a_r) from the reference point to behind the heat source is defined as being equal to twice the width; this procedure has led to good approximations [13].

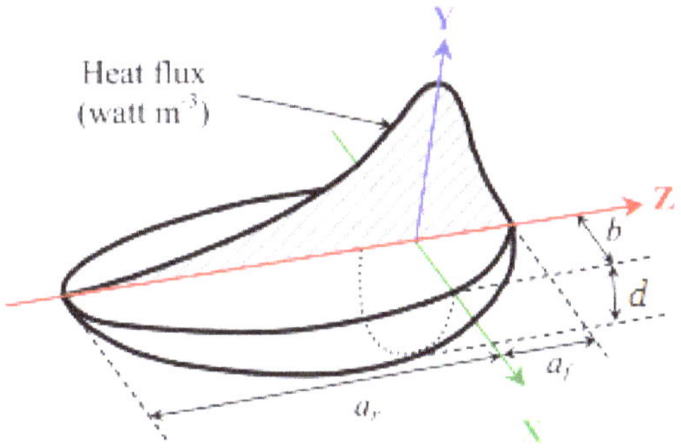

Figure 1. 3D Volumetric Gaussian on a double ellipsoidal of radius a, b and d.

Due to the tighter specifications and control of modern steels, thermal analysis is of great importance to estimate the correct preheating temperature value for reliable welding. For example, Cooper *et al.* [15] compared four methods for calculating the preheating temperature of structural steels, including the API 5L grade X80. The four methods used were: (a) the British Standard Institution method (BS 5135) based on the CE value (IIW); (b) the American Welding Society method (AWS D1.1) that calculates the temperature of preheating through Pcm; (c) the TCE method (Total Carbon Equivalent) which calculates the temperature of preheating as a function of the CE, material thickness, diffusible hydrogen in the weld metal and the heat input, and (d) the CEN method.

In this work, the preheating temperature value was based on the study of Cooper *et al.* [15] and the experimental work conducted by Araújo [16]. The preheating temperature value was defined as being equal to 100 °C corresponding to the BS method, which is the most conservative method of those considered by Cooper *et al.* The interpass temperature value was 150 °C following the N-133J standard [17].

2.2. Residual Stresses Model

The movement of the welding electric arc generates a non-linear thermal gradient and the thermal and mechanical changes arising from the heat source promote thermal stresses and distortions. Therefore, based on the thermal gradients given by Equation (2), the residual stresses based on the deformations caused by the gradients can be determined according to:

$$\sigma_v = \delta_v \lambda \varepsilon_{kk} + 2\mu\varepsilon_v - \delta_v(3\lambda + 2\mu)\alpha T \tag{8}$$

where λ and μ are the Lame constants, which represent the components of the material deformation due to the temperature and are related to the material elasticity modulus E and Poisson's ratio v. The ε_v variable is linked to the deformation and displacement according to:

$$\varepsilon_v = \frac{1}{2}\left(\frac{\partial u_i}{\partial x_j} + \frac{\partial u_j}{\partial x_i}\right) \tag{9}$$

The model used for determining the residual stresses was elastoplastic with isotropic hardening and the correspondent values were obtained from the strains generated during the welding process.

The strains were considered to have elastic, plastic and thermal natures; so, the total strain was determined as:

$$\varepsilon_{ij} = \varepsilon_{ij}^{\mathrm{e}} + \varepsilon_{ij}^{\mathrm{p}} + \varepsilon_{ij}^{\mathrm{th}} \tag{10}$$

where ε_{ij} is the total deformation, $\varepsilon_{ij}^{\mathrm{e}}$ the elastic deformation, $\varepsilon_{ij}^{\mathrm{p}}$ the plastic deformation and $\varepsilon_{ij}^{\mathrm{th}}$ the thermal deformation.

Based on the Hooke's law, the elastic deformation can be written as:

$$\varepsilon_{kl}^{\mathrm{e}} = \sigma_{ij}^{\mathrm{e}} \cdot E(T)^{-1} \tag{11}$$

On the other hand, the deformation due to the thermal effect is given as:

$$\varepsilon_{kl}^{\mathrm{th}} = \alpha_{ij}(T - T_\infty) \tag{12}$$

where α_{ij} is the thermal coefficient of linear expansion and T_∞ is the reference temperature. This thermal deformation depends on the phase of the material.

The theory of plasticity describes the elastic-plastic response of materials through mathematical relationships based on some restrictive assumptions. Among these assumptions, it can be assumed that plastic deformation is the result of a history of tensions that occurred instantaneously, *i.e.*, independently of time [18].

Using the model of the flow rule, which establishes the direction of plasticity for metal materials under conditions of small displacements, the plastic potential function g is equal to the flow area capability f. This relation is known as the associated flow rule, and the direction of plasticity is considered to be normal to the flow surface. The plastic strain ratio is determined as:

$$\mathrm{d}\varepsilon_{kl}^{pl} = \mathrm{d}\lambda \frac{\partial f}{\partial \sigma_{ij}} \tag{13}$$

where λ is a positive constant dependent on the properties of the material. For a perfect elastic-plastic material and for the cases in which a Von Mises surface is used as flow area capability criterion, the parameter λ can be described as:

$$\lambda = \frac{3G s_{ij} s_{kl}}{{\sigma_e}^2} \tag{14}$$

Further details can be seen in the works of Akbari and Sattari-Far [9], Syahroni and Hidayat [10], Queresh [13], Depraudeux [19] and Yao Xin *et al.* [20].

3. Material and Computational Simulation

The computational model was developed in ABAQUS® version 6.12, which is based on FEM. The workpiece was an API 5L X80 steel plate 0.120 × 0.360 × 0.017 m^3, which is the same as the one used in the experimental work of Araújo [16]. The chemical composition of this steel is shown in Table 1.

Table 1. Chemical composition of the API 5L X80 steel.

Percentage (%) by Weight										
C	Mn	Si	P	S	Ni	Mo	Al	Cr	V	Cu
0.084	1.61	0.23	0.01	0.011	0.17	0.17	0.035	0.135	0.015	0.029

The majority of the published works on numerical simulation of welding processes considers that the material properties are highly dependent on temperature. However, it is very difficult to obtain complete information about this, especially for high temperatures. Thus, simplifications are often used in these numerical simulations to overcome this problem. In this study, the thermal and mechanical properties of the API 5L X70 steel as a function of the temperature were defined according to Forouzan *et al.* [6], Figure 2.

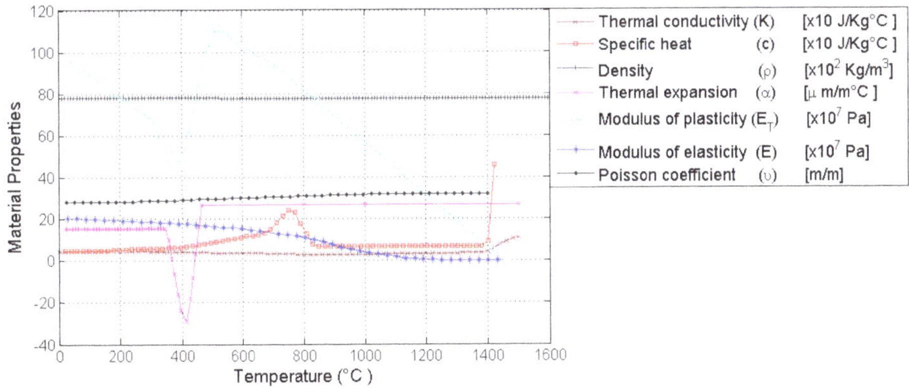

Figure 2. Thermal and mechanical properties of the API 5L X70 steel as a function of the temperature adopted in the simulations performed here.

The plastic deformation of the API 5L X70 steel for the simulations in this work was defined according to [21], which is depicted in Figure 3 as a graph of stress *versus* plastic strain.

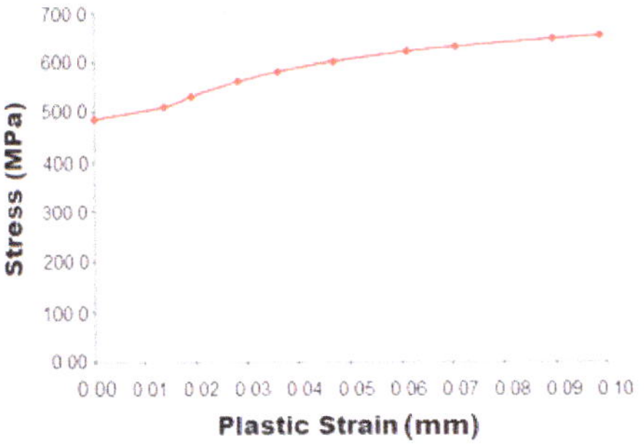

Figure 3. Stress *versus* plastic strain according to the data obtained in an experimental study of API 5L X70 steel [21] that was used in the simulations performed here.

The remaining parameters were considered equal to those of low-carbon steels. Thus, the latent heat for the solidification of the weld pool was considered to be 270 J/g [22], the liquid T_L and solid T_S transformation temperatures were assumed to be 1560 °C and 1440 °C, respectively, and the convective coefficient h_f was considered as a function of the temperature.

In the simulation, the GTAW welding process for the first welding pass and the SMAW process for the second welding pass were considered to have constant speeds, similar to the approach by Araújo [16] who conducted an automated welding. The welds were considered to be done in a horizontal position and the effects of gravity were disregarded. The values of the welding arc efficiency η were taken from [23] and the effect of the material insertion was disregarded.

The simulations were performed according to the experimental conditions adopted by Araújo in [16], who executed a multipass weld with preheating and interpass temperatures of 100 °C and 150 °C, respectively, according to the N-133J standard [17]. The welding variables and geometric parameters of the welding passes used in the simulations are shown in Table 2.

Table 2. Welding parameters used in the simulations.

(1° Pass) GTAW				(2° Pass) SMAW			
ε (%)	***V* (m/s)**	***I* (A)**	***U* (V)**	**ε (%)**	***V* (m/s)**	***I* (A)**	***U* (V)**
65	0.0012	152	12	80	0.0015	69	33
a_f (m)	**a_r (m)**	***b* (m)**	***d* (m)**	**a_f (m)**	**a_r (m)**	***b* (m)**	***d* (m)**
0.00245	0.0098	0.00245	0.00403	0.0043	0.0172	0.0043	0.00376

To represent the required welding conditions, the model of Goldak and Chakravarti [12] was applied and developed in a DFLUX subroutine in FORTRAN and integrated in ABAQUS®.

Figure 4 shows the detailed steps of the computational model used to perform the thermal and thermo-mechanical simulations.

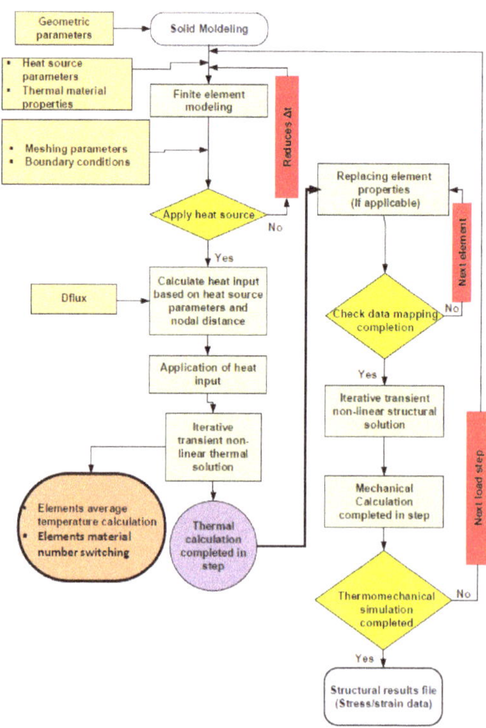

Figure 4. Detailed sequence of preprocessing, processing and post-processing developed in the thermal and thermo-mechanical simulations.

The choice of the mesh used in computer simulations is of fundamental importance as a refined mesh usually leads to results of greater robustness and reliability due to an increase in the number of degrees of freedom. However, the mesh refinement must be carried out carefully, as this also will result in an increased computational cost. In this work, convergence tests were performed on the meshing process. The maximum temperature reached on the weld pool was considered as the reference point for the mesh refinement as was considered by Queresh in [13]; however, the mesh refinement in terms of the thermo-mechanical effects was not considered, due to the associated high computational cost. The response of the refinement process used is shown in Figure 5.

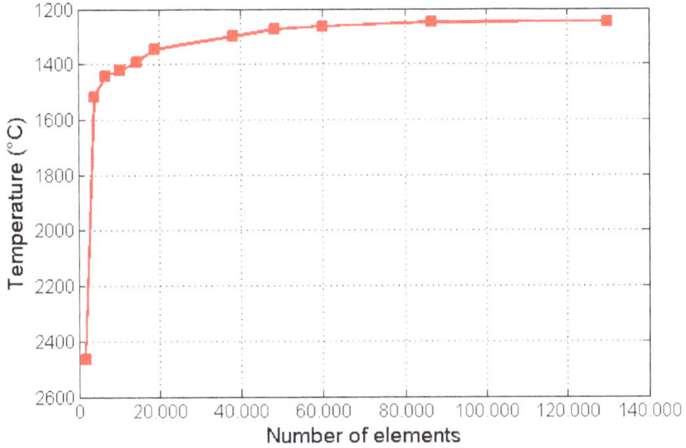

Figure 5. Mesh refinement based on the maximum temperature in the weld pool.

After the convergence tests, the optimized configuration of the mesh refinement led to three main regions, which are depicted in Figures 6 and 7. Each region has a specific number of elements, as shown in Table 3. As aforementioned, for the thermo-mechanical simulation, the mesh refinement was disregarded, since it would increase the computational cost and the simulation time considerably. The elements DC3D8 and C3D8T were used in the thermal and thermo-mechanical simulations, respectively [24].

Table 3. Number of elements of the FEM mesh of each region studied (indicated in Figures 6 and 7).

Model	Number of Elements			
	Region 1	Region 2	Region 3	Total
Thermal	24,480	41,616	-	66,096
Thermo-mechanical	3640	4480	5600	13,720

In the thermal simulation, only half of the plate was considered, since the phenomenon and the conditions used allowed the application of the symmetry theory [25]. In addition, to create a region of higher mesh refinement, the partition (Region 3) at 0.02 m distance from where the heat source passes was defined, due to a higher probability of occurrence of thermal transformations at this location. Such procedure was also adopted in [13,18,25].

The thermal boundary conditions for the heat exchanged by radiation and convection were defined in the model for five sides of the plate; the sixth side supported by the table was excluded. This is in accordance with experimental laboratory procedures, and therefore, the hypothesis of zero heat exchange with the table was adopted. The values of these boundary conditions were defined as: ambient temperature of 25 °C, Stefan-Boltzmann constant equal to 5.67×10^{-8} W$\cdot$ m$^{-2}\cdot$ K^{-4} and emissivity of 0.77, which were obtained from [26], while the h_f value was defined according to [13].

The movement of the heat source is shown in Figure 6. The first welding pass was carried out along the XY plane, with the movement of the heat source along the Y axis. The second welding pass was carried out following the same procedure; however, it was displaced vertically 0.004 m along the Z-axis. The computational thermal evaluation was conducted on the node 0.002 m from the root pass fusion line. The 0.05 m extremities of the plate ends were disregarded, which is in-line with the conditions adopted in a laboratory due to the fixing clips [16].

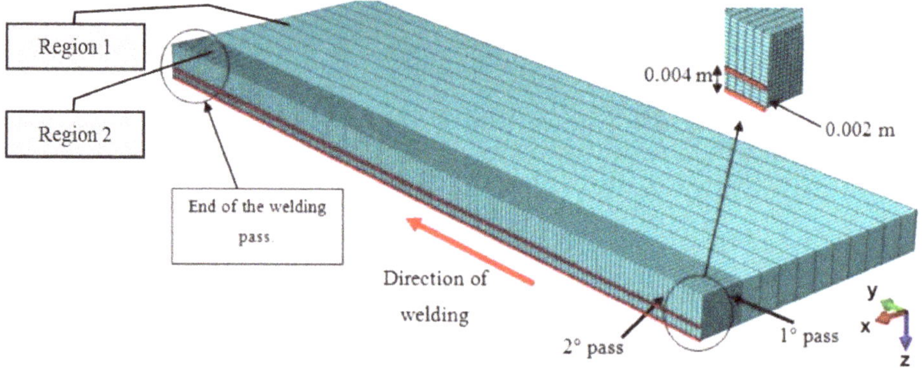

Figure 6. Movement of the welding heat source for the two welding passes. The second pass was initiated when the temperature at the end of the first pass region dropped to 150 °C.

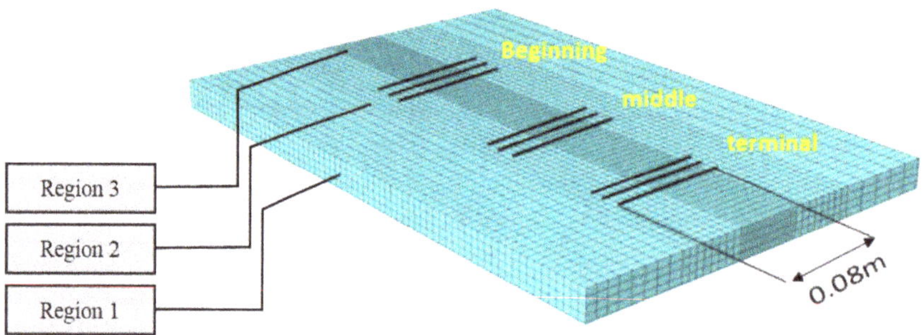

Figure 7. Three positions were analyzed in terms of the transverse residual stresses: beginning, middle and terminal; at each position, the average values based on three reference lines 80 mm long and 0.045 mm equidistant from each other were used.

After the two welding passes, the transverse residual stresses were analyzed in three different positions (beginning, middle and terminal), using the average value for each region, Figure 7. The positions to evaluate these stresses are similar to those followed by Araújo in [16].

4. Results and Discussion

In order to validate the computational model used in the simulations, the welding of an API 5LX70 steel plate with dimensions of 0.1 × 0.1 × 0.019 m^3, and the same experimental parameters and conditions used by Laursen *et al.* [27] were employed here. The authors used a current of 140 A, a voltage of 23 V and an automated speed equal to 0.001 m/s to performing the root pass at room temperature (25 °C). After convergence tests, it was found that the model with 30,000 elements converges to the maximum temperature results in the monitored region. The same conditions of thermal contours described in [27] were employed here, except for the geometric parameters a_f, a_r, b and d, energy parameters f_f and f_r, and the weld pool that were obtained from the literature [13,28]. The virtual thermal cycling was evaluated here in the same region considered in [27], and is shown in Figure 8.

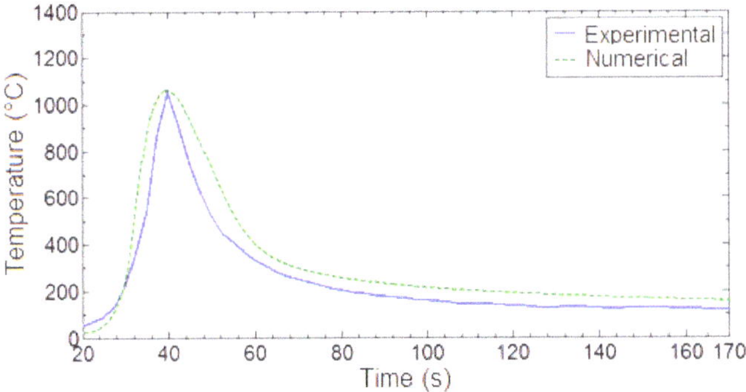

Figure 8. Comparative thermal cycle analysis of the experimental and the numerical method.

The comparative thermal cycle showed a good similarity in terms of the shape of the curves between the experimental and numerical processes (Figure 8); however, the numerical curve had a bandwidth greater than the experimental process. Some assumptions may explain the differences found. The lack of material data and the simplifications adopted may hinder a better approximation between the real and the computational results. According to Heinze *et al.* [29], the diameter of the thermocouple and its position may influence the comparison between the experimental and computer modeling since the computing node has an accuracy greater than the thermocouple positioned in the welded joint. The computational values of the peak temperature and $\Delta t_{8/5}$ were 1062.30 °C and 7.60 s, respectively, while the experimental ones were 1055.55 °C and 6.81 s, causing an error inferior to 1% as to the peak temperature and of 11.60% for the $\Delta t_{8/5}$ value.

The results of the thermal analysis obtained from the simulation were consistent and close to other results obtained computationally, and the errors were in agreement with those obtained by Heinze *et al.* [29], who found errors up to 14%. This close approximation of the peak temperature was due to the Goldak heat source model used [14], which closely portrays the heat source in the experiment and because it also applied accurate values of this heat source; therefore, when the simulated heat source is at the location of where the workpiece is under analysis an error of less than 1% was obtained.

Figures 9 and 10 show the temperature gradients obtained in the simulated welding when the heat source is at the center of the plate. Figure 11 shows the thermal cycles at the beginning, middle and terminal of the welding process for the two welding passes.

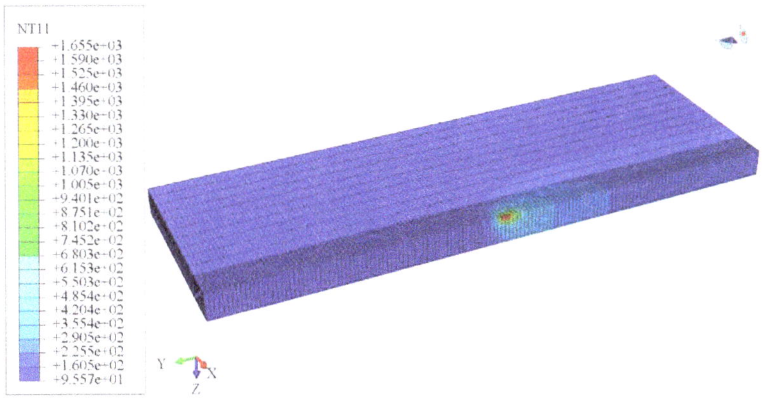

Figure 9. Temperature gradient obtained when the virtual source of heat is in the middle of the first welding pass.

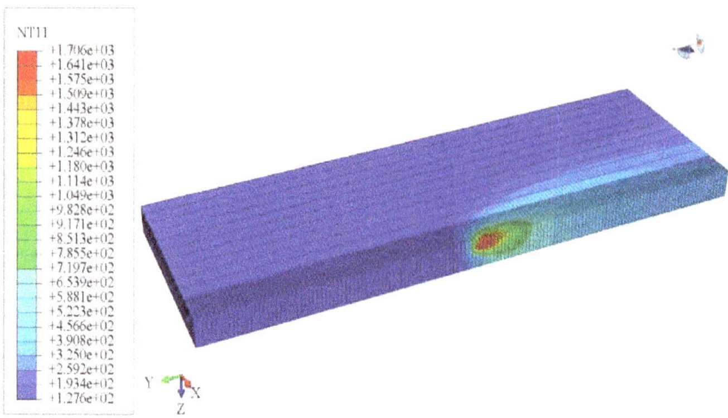

Figure 10. Temperature gradient obtained when the virtual source of heat is in the middle of the second welding pass.

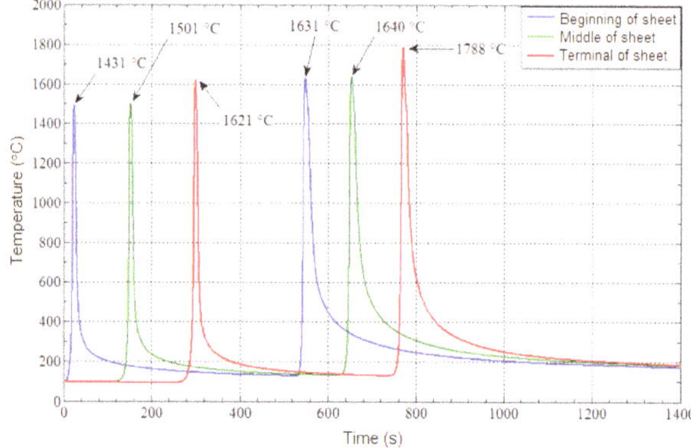

Figure 11. Temperature cycles evaluated at the beginning, middle and end of the welding process at a point located 0.002 m from the heat source.

Figures 9 and 10 show that the maximum temperatures are located close to and just behind the heat source, and that several isotherms with lower temperatures arise from the end of the plate, developing a temperature gradient along the plate, as reported by Kenneth [30]. There was a variation in the maximum temperature between the welding passes: 1655 °C and 1706 °C for the 1st and 2nd passes, respectively. This variation may have occurred due to variations in the welding speed although the welding energy used was constant. This fact was also observed in previous studies, such as the ones by Deng and Kiyoshima [31], Samardžić *et al.* [32], Zhu and Chao [33] and Nandan *et al.* [34].

Table 4 indicates the cooling time between the 800 °C and 500 °C thermal cycles of the curves of Figure 11.

Table 4. Cooling time between 800 and 500 °C ($\Delta t_{8/5}$) values at 0.002 m from the fusion line.

	$\Delta t_{8/5}$ Beginning (s)	$\Delta t_{8/5}$ Middle (s)	$\Delta t_{8/5}$ Terminal (s)	$\Delta t_{8/5}$ Average (s)	Variation (%)
1° pass	3.71	4.02	3.30	3.68	22.00
2° pass	22.50	26.89	27.30	25.56	21.00

Table 4 shows that there were variations of up to 22% in $\Delta t_{8/5}$ when it was assessed at different points in the region under study (beginning, middle and end of the weld joint). The largest cooling time of 27.30 s is located at the end of the second simulated welding pass.

Figure 12 shows a comparison between the transverse residual stresses computationally obtained at the beginning, middle and end of the weld joint after the simulated welding with one pass and two passes until cooling to the ambient temperature.

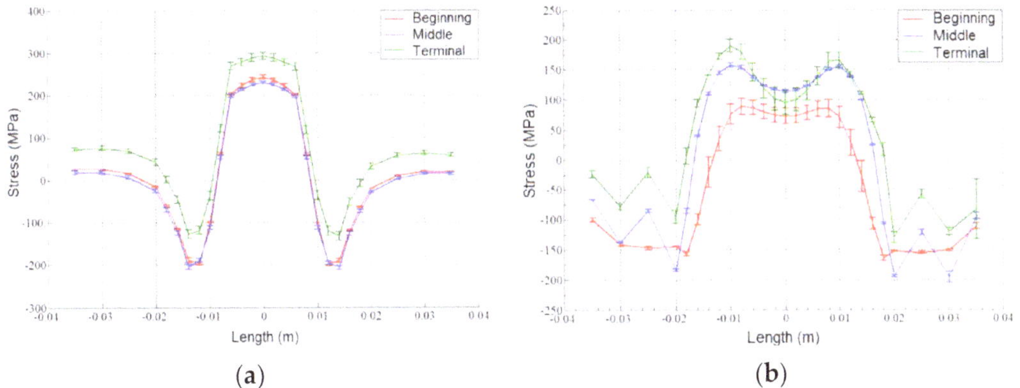

Figure 12. Transverse residual stresses located at the beginning, middle and end, after the first simulated welding pass until cooling to room temperature (**a**); and after two simulated welding passes until cooling to room temperature (**b**).

Figure 12a shows a predominance of traction residual stresses in the center region of the welded joint that corresponds to the weld metal, with stresses varying from 240 to 300 MPa, and compressive stress from −120 to −200 MPa in the regions near the fusion line. This behavior was also observed in the work by Stamenković and Vasovic [35], who performed a computer simulation with a virtual welding pass. On the other hand, Figure 12b shows that there was a change in the level of the residual transverse stresses when the second welding pass was applied on the first one. The traction residual stresses tend to be reduced. Therefore on considering a greater number of passes in the simulated welding, there is a tendency for decreased tensions, and the profile of the residual stresses after the fusion line region becomes more alike to the ones observed experimentally by Araújo [16].

Figure 13 shows the transverse residual stress gradients after the simulated welding with two passes.

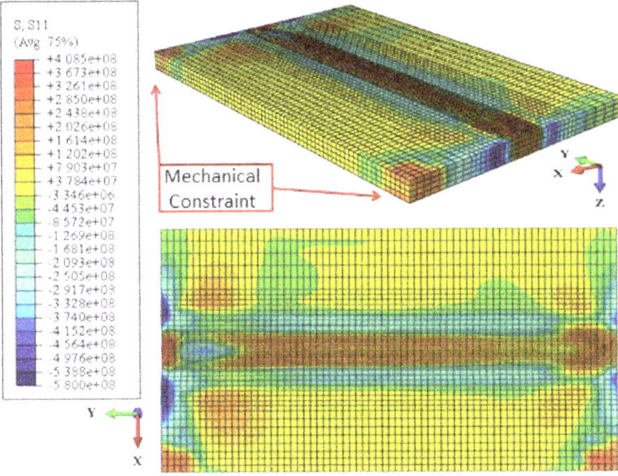

Figure 13. General transverse residual stresses obtained in the simulation with two passes after cooling to room temperature.

A qualitative evaluation shows that the profiles of the transverse residual stresses have a homogeneous behavior, prevailing traction stresses in the center of the plate and compressive stresses in the neighboring regions (supposedly HAZ), and also compressive stresses but to a lesser degree throughout the remainder of the material. After weld solidification, the material shrinkage is prevented by the cooler regions (distant from the weld region), giving rise to tensile stresses along the weld and compression in the more distant regions. In addition, there were higher stresses, close to 400 MPa, in the mechanical constraint regions as well as near the region where the virtual welding was performed. The transverse residual stresses in this region reached values between 100 and 200 MPa.

5. Conclusions

The results of the simulated welding showed that the thermal gradients and thermal cycles vary when evaluated in different regions. This variation is quite possibly caused by the different welding speeds used.

The thermal model applied in the experimental work of Laursen *et al.* [27] showed to be consistent. Therefore, the results of thermal analysis obtained in this simulation were satisfactory and close to the results obtained experimentally with errors ranging from less than 1% to 12%, which are similar in range to these errors found in the literature, such as the works of Heinze *et al.* [29].

Although only two welding passes were performed in the simulation, the residual stresses were satisfactory, since the numerical curve tended to follow the transversal residual stresses similar to the ones found experimentally.

The gradient stresses obtained in all regions of the plate were quite homogenous, and the use of the adopted numerical model allowed the detection of emerging residual stresses in a consistent manner, with the highest levels of traction stresses in the central regions of the welded joint and compressive in more distant regions.

Possible future works could be related to the mesh refinement based on the residual stress which can lead to important insights for more competent welding processes although this type of mesh refinement can lead to high computational costs. In addition, interesting for such a goal would be investigations into the influence of the properties of the temperature-dependent and -independent materials on the simulation results.

Acknowledgments: The authors thank the Academic Unit of Mechanical Engineering and Postgraduate Program in Mechanical Engineering from the Federal University of Campina Grande and the Coordination for the Improvement of Higher Education Personnel (CAPES) in Brazil for the financial support. Victor Hugo C. de Albuquerque acknowledges the sponsorship from the Brazilian National Council for Research and Development (CNPq) through the grants with references 470501/2013-8 and 301928/2014-2.

Author Contributions: Study design: A. Silva and T. Maciel; Experimental work: J. Nóbrega and D. Diniz; Results analysis & Manuscript preparation: All authors; Manuscript proof and Submission: J. Tavares.

Conflicts of Interest: The authors declare no conflict of interest.

References

1. Soul, F.; Hamdy, N. Numerical Simulation of Residual Stress and Strain Behavior after Temperature Modification. In *Welding Processes*; Kovacevic, R., Ed.; InTech: Rijeka, Croatia, 2012; Chapter 10. [CrossRef]
2. Araújo, B.A.; Maciel, T.M.; Carrasco, J.P.; Vilar, E.O.; Silva, A.A. Evaluation of the diffusivity and susceptibility to hydrogen embrittlement of API 5L X80 steel welded joints. *Intl. J. Multiphys.* **2013**. [CrossRef]
3. Carlone, P.; Citarella, R.; Lepore, M.; Palazzo, G.S. A FEM-DBEM investigation of the influence of process parameters on crack growth in aluminum friction stir welded butt joints. *Int. J. Mater. Form.* **2015**, *8*, 591–599. [CrossRef]
4. Citarella, R.; Carlone, P.; Lepore, M.; Palazzo, G.S. Numerical-experimental crack growth analysis in AA2024-T3 FSWed butt joints. *Adv. Eng. Softw.* **2015**, *80*, 47–57. [CrossRef]
5. Jiang, W.; Fan, Q.; Gong, J. Optimization of welding joint between tower and bottom flange based on residual stress considerations in a wind turbine. *Energy* **2010**, *35*, 461–467. [CrossRef]

6. Forouzan, M.R.; Heidari, A.; Golestaneh, S.J.F. FE simulation of submerged arc welding of API 5L-X70 straight seam oil and gas pipes. *J. Comput. Methods Eng.* **2009**, *28*, 93–110.
7. Cho, J.R.; Lee, B.Y.; Moon, Y.H.; van Tyne, C.J. Investigation of residual stress and post weld heat treatment of multi-pass welds by finite element method and experiments. *J. Mater. Proc. Technol.* **2004**, *155–159*, 1690–1695. [CrossRef]
8. Robertsson, A.; Svedman, J. *Welding Simulation of a Gear Wheel Using FEM*; Chalmers University of Technology (Department of Applied Mechanics): Göteborg, Sweden, 2013.
9. Akbari, D.; Sattari-Far, I. Effect of the welding heat input on residual stresses in butt-welds of dissimilar pipe joints. *Int. J. Press. Vessels Pip.* **2009**, *86*, 769–776. [CrossRef]
10. Syahroni, N.; Hidayat, M.I.P. 3D Finite Element Simulation of T-Joint Fillet Weld: Effect of Various Welding Sequences on the Residual Stresses and Distortions. In *Numerical Simulation—From Theory to Industry*; Andriychuk, M., Ed.; InTech: Rijeka, Croatia, 2012; Chapter 24. [CrossRef]
11. Nóbrega, J.A.; Diniz, D.D.S.; Melo, R.H.F.; Araújo, B.A.; Maciel, T.M.; Silva, A.A.; Santos, N.C. Numerical evaluation of multipass welding temperature field in API 5L X80 steel welded joints. *Int. J. Multiphys.* **2014**. [CrossRef]
12. Goldak, J.; Chakravarti, A. A new finite element model for welding heat sources. *Metall. Trans.* **1984**, *15*, 299–305. [CrossRef]
13. Queresh, E.M. Analysis of Residual Stresses and Distortions in Circumferentially Welded Thin-Walled Cylinders. Ph.D. Thesis, National University of sciences and Technology, Karachi, Pakistan, December 2008.
14. Attarha, M.J.; Sattari-Far, I. Study on welding temperature distribution in thin welded plates through experimental measurements and finite element simulation. *J. Mater. Proc. Technol.* **2011**, *211*, 688–694. [CrossRef]
15. Cooper, R.; Silva, J.H.F.; Trevisan, R.E. Influence of preheating on API 5L-X80-pipeline joint welding with self-shielded flux-cored wire. *Weld. Int.* **2005**, *19*, 882–887. [CrossRef]
16. Araújo, B.A. Avaliação do Nível de Tensão Residual e Susceptibilidade à Fragilização Por Hidrogênio em Juntas Soldadas do Aço API 5L X80 Utilizados Para o Setor de Petróleo e Gás. Ph.D. Thesis, Universidade Federal de Campina Grande, Campina Grande, Brazil, September 2013.
17. PETROBRAS. N-133 M: Welding (in Portuguese). Available online: http://sites.petrobras.com.br/CanalFornecedor/portugues/requisitocontratacao/requisito_normastecnicas.asp (accessed on 25 January 2016).
18. Guimarães, P.B.; Pedrosa, P.M.A.; Yadava, Y.P.; Barbosa, J.M.A.; Filho, A.V.S.; Ferreira, R.A.S. Determination of residual stresses numerically obtained in ASTM AH36 steel welded by TIG process. *Mater. Sci. Appl.* **2013**, *4*, 268–274. [CrossRef]
19. Depradeux, L. Simulation Numérique du Soudage—Acier 316L 2004. Validation Sur Cas Tests de Complexité Croissante. Ph.D. Thesis, INSA Lion, Lyon-Villeurbanne, France, 2004.
20. Yao, X.; Zhu, L.-H.; Lim, V. Finite element analysis of residual stress and distortion in an eccentric ring induced by quenching. *Trans. Mater. Heat Treat.* **2004**, *25*, 746–751.
21. Veiga, J.L.B.C. Analysis of Acceptance Criteria of Wrinkles in Pipeline Cold Bends. M.Sc. Thesis, Pontifícia Universidade Católica do Rio de Janeiro, Rio de Janeiro, Brazil, May 2009.
22. Deng, D. FEM prediction of welding residual stress and distortion in carbon steel considering phase transformation effects. *Mater. Des.* **2009**, *30*, 359–366. [CrossRef]
23. Stenbacka, N. On arc efficiency in Gas Tungsten Arc Welding. *Soldagem e Inspeção* **2013**, *18*, 380–390. [CrossRef]
24. Abaqus/CAE User manual, version 6.3. Hibbit, Karlsson & Sorenson Inc. Providence, RI, USA, 2007.
25. Jiang, W.G. The development and applications of the helically symmetric boundary conditions in finite element analysis. *Commun. Numer. Methods Eng.* **1999**, *15*, 435–443. [CrossRef]
26. Incropera, F.P.; DeWitt, D.P.; Bergman, T.L.; Lavine, A.S. *Fundamentals of Heat and Mass Transfer*, 7th ed.; John Wiley & Sons: Hoboken, NJ, USA, 2011.
27. Laursen, A.; Maciel, T.M.; Melo, R.H.F. Influence of weld thermal cycle on residual stress of API 5L X65 and X70 welded joint. In Proceedings of the 2014 Canweld Conference, Vancouver, BC, Canada, 1 October 2014.
28. Gery, D.; Long, H.; Maropoulos, P. Effects of welding speed, energy input and heat source distribution on temperature variations in butt joint welding. *J. Mater. Proc. Technol.* **2005**, *167*, 393–401. [CrossRef]
29. Heinze, C.; Schwenk, C.; Rethmeier, M. Numerical calculation of residual stress development of multi-pass gas metal arc welding. *J. Constr. Steel Res.* **2012**, *72*, 12–19. [CrossRef]

30. Kenneth, E. Fusion welding—Process variables. In *Introduction to the Physical Metallurgy of Welding*, 2nd ed.; Easterling, K., Ed.; Butterworth-Heinemann: Exeter, UK, 1992; Chapter 1; pp. 1–54. [CrossRef]
31. Deng, D.; Kiyoshima, S. FEM prediction of welding residual stresses in a SUS304 girth-welded pipe with emphasis on stress distribution near weld start/end location. *Comput. Mater. Sci.* **2010**, *50*, 612–621. [CrossRef]
32. Samardžić, I.; Čikić, A.; Dunđer, M. Accelerated weldability investigation of TStE 420 steel by weld thermal cycle simulation. *Metalurgija* **2013**, *4*, 461–464.
33. Zhu, X.K.; Chao, Y.J. Effects of temperature-dependent material properties on welding simulation. *Comput. Struct.* **2012**, *80*, 967–976. [CrossRef]
34. Nandan, R.; Roy, G.G.; Lienert, T.J.; DebRoy, T. Numerical modelling of 3D plastic flow and heat transfer during friction stir welding of stainless steel. *Sci. Technol. Weld. Join.* **2006**, *11*, 526–537. [CrossRef]
35. Stamenković, D.; Vasović, I. Finite element analysis of residual stress in butt welding two similar plates. *Sci. Tech. Rev.* **2009**, *59*, 57–60.

14

The Establishment of Surface Roughness as Failure Criterion of Al–Li Alloy Stretch-Forming Process

Jing-Wen Feng [1,2,3], Li-Hua Zhan [1,2,3,*] and Ying-Ge Yang [1,2,3]

Academic Editor: Nong Gao

1 School of Mechanical and Electrical Engineering, Central South University, Changsha 410083, Hunan, China; fengjingwen1@csu.edu.cn (J.-W.F.); 133711022@csu.edu.cn (Y.-G.Y.)
2 State Key Laboratory of High Performance Complex Manufacturing, Central South University, Changsha 410083, Hunan, China
3 2011 Collaborative Innovation Center, Central South University, Changsha 410083, Hunan, China
* Correspondence: yjs-cast@csu.edu.cn

Abstract: Taking Al–Li–S4–T8 Al–Li alloy as the study object, based on the stretching and deforming characteristics of sheet metals, this paper proposes a new approach of critical orange peel state characterizations on the basis of the precise measurement of stretch-forming surface roughness and establishes the critical criterion for the occurrence of orange peel surface defects in the stretch-forming process of Al–Li alloy sheet metals. Stretching experiments of different strain paths are conducted on the specimens with different notches so as to establish the Al–Li–S4–T8 Al–Li alloy, forming limit diagram and forming limit curve equation, with the surface roughness of characteristic critical orange peel structure as the stretch-forming failure criterion.

Keywords: Al–Li–S4 Al–Li alloy; stretch-forming; orange peel; forming limit; surface roughness

---

1. Introduction

Aircraft skin serves as the shape part of an aircraft and constructs the aerodynamic configuration of the aircraft, featuring big size and direct contact with air. Therefore, it requires an accurate shape, smooth streamline, and no surface defects, *etc.* [1]. As a relatively common aircraft skin-forming approach in the field of aeronautics and astronautics manufacturing, the technology of stretch-forming is widely used in the manufacturing of large-scale aircraft skin as part of aircraft aerodynamic configuration [2–4]. In the stretch-forming process, the clamps of the stretch-forming machine clamp both ends of the sheet metal and move along a certain track, or the die goes up to make the sheet metal contact the stretch-forming die, creating uneven plane stretching strain to make the metal sheet conform to the stretch-forming die so as to obtain the required part shape [5–7]. Compared with other approaches to aircraft skin forming, the stretch-forming technology might cause surface defects such as orange peel. Orange peel not only affects the appearance of the aircraft skin, but also damages its surface integrity. Especially in the case of mirror skin, orange peel easily appears due to mirror skin's internal structure and polished surface, seriously affecting its service life, which is usually the main reason for the scrapping of parts [8].

The defect of orange peel is a kind of rough orange peel-like morphology found on the surface of shaped products. In general, coarse and unevenly structured grains on the alloy surface are considered the reason of the orange peel defect appearing on the alloy surface during the stretch-forming process, while the grain size of the alloy has a certain relation to the extent of the deformation. At a certain temperature, when there are relatively small deformations, recrystallization usually does not appear in the alloy and the grains maintain their original state; however, when deformation reaches a certain degree, recrystallized grains will become very coarse. In the manufacturing process of aluminum alloy skin sheet metal, there are usually multiple hot rolling and cold rolling processes as well as several heat treatments including annealing, solution and aging. Because of the imperfection in the control over cold deformation and the choice of heat treatment technology in the manufacturing process of aluminum alloy sheet metal, recrystallized grain structure tends to be coarse and show different sizes, resulting in the piling up of dislocation on large grains and the rapid increases of stress in the follow-up stretch-forming process. Thus, the areas of large grains reach and exceed the elastic limit in advance; the non-synchronous deformation between large grains and small ones gives rise to minor cracks on the surface of the material, manifesting as the orange peel structure at the macro-level [9].

The forming limit of sheet metal is a criterion used to describe whether the sheet metal fails to form or not. To identify the concept of forming limit, one should first determine the failure criterion of sheet metal's forming process. Al–Li alloy as a new type of aluminum alloy has shown a wide application prospect in the fields of aeronautics and astronautics due to its good qualities of low density, high specific strength, and high specific stiffness, and it has become a hot subject in the research field of aluminum alloy materials, being regarded as one of the important candidate materials of modern aeronautic and astronautic structures [10,11]. However, due to the high cost, poor cold plasticity at ambient temperature, evident anisotropy and easy cracking in cold working compared with other regular aluminum alloys, Al–Li alloy at present can only be processed into relatively simple parts and faces great difficulty in the processing and manufacturing of more complex structural parts. Therefore, some of its own attributes also limit Al–Li alloy's application in structural components [12–15]. Relevant researchers have conducted corresponding research on the formability of Al–Li alloy, which, however, mainly focus on the study of the Al–Li alloy's structure property and the aspect of hot forming. Literature [16] explored the 2397-T87 Al–Li alloy with a thickness of 130 mm for the microstructure, stretching property and fracture toughness of layers with different thicknesses and at different orientations. By the uniaxial tension test under different hot deformation conditions, the forming limit test with cracking as a failure criterion, and the stretch-forming tests of 5A90 Al–Li alloy sheet metal, literature [17] built the Al–Li alloy forming limit model and determined the technological parameters for Al–Li alloy to acquire good deformation performance.

Nevertheless, for aviation aircraft skin materials, cracking is not often taken as the criterion for judging whether the skin fails to form or not, especially for Al–Li alloy, the reason of which usually lies in that the forming failure is caused by the appearance of orange peel structure in the stretch-forming process. Currently, there are relatively few studies on the forming limit caused by the defect of orange peel in the stretch-forming process, and a forming failure criterion targeting the orange peel has not yet taken shape. As the occurrence of the orange peel phenomenon is a slowly changing and accumulated process, and all of the previous research on such a phenomenon was conducted by eye measurement, the results have certain randomness.

In this paper, experimental research on the problem of orange peel in the stretch-forming process of Al–Li alloy is conducted, and a novel approach of critical orange peel characterization is proposed on the basis of the precise measurement of stretch-forming surface roughness. Furthermore, the judgment criterion of the orange peel defect is analyzed and established, the stretching tests on specimens of different strain paths are conducted combining the technology of optical deformation measurement, and, ultimately, the stretch-forming limit diagram and its forming limit curve equation are obtained with the surface roughness of orange peel structure with critical characteristics such as the stretch-forming failure criterion.

2. Experiments

2.1. Instrument and Methods

In order to conduct synchronization tests on stress-strain and orange peel surface defect in the stretch-forming process of Al–Li alloy, a stretch-forming test and testing system are established to obtain the critical strain state of orange peel of the product, in which the optical deformation measurement instrument and the universal testing machine operate in collaboration, as shown in Figure 1.

The optical deformation measurement instrument used in the experiment adopts the deformation measurement system based on computer vision technology developed by the German company GOM for three-dimensional deformation analysis. By controlling the synchronized operation of the optical deformation measurement instrument and the universal testing machine, the complete monitoring of the total stretch-forming deformation process of the specimens can be achieved and the computer image processing system can be further utilized to obtain the true strain change rules at different positions on the specimen surface throughout the stretch-forming process.

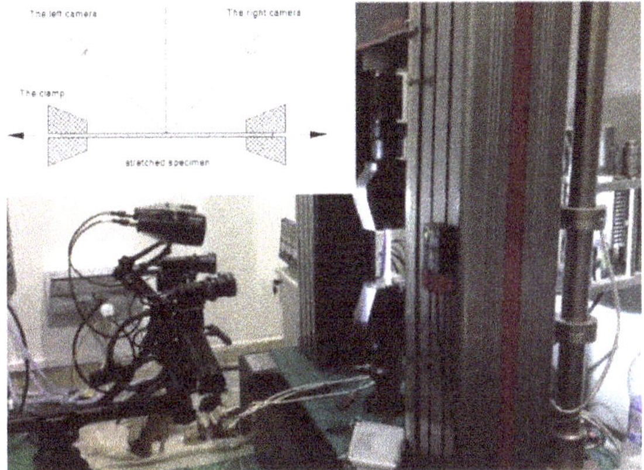

Figure 1. The stretching and testing system.

Compared with the stretch-forming mechanism, this experimental mechanism is to some extent simplified, in particular with the top die removed. However, in the actual stretch-forming process of the sheet metal, with the stretch-forming die going up, the sheet metal gradually bends to attach to the die as shown in Figure 2. Due to the existence of friction, the material flow of the attaching segment AA′ tends to be limited to a relatively small deformation, whereas the free segments without contact with the die AB and A′B′ tend to have relatively greater deformation without the restriction of friction. Therefore, the orange peel structure usually appears first on the free segments between the chucks and the die corners, and then slowly spreads toward the forming surface that attaches to the die. Thus, it can be seen that the deformation of the free segments on the sheet metal is the principal factor affecting the appearance of orange peel structure. As the deformation of the free segments on the sheet metal is similar to its stretch-forming, the conventional stretching test of sheet metal can be used in the research on the influence of the orange peel phenomenon on the stretch-forming of the materials.

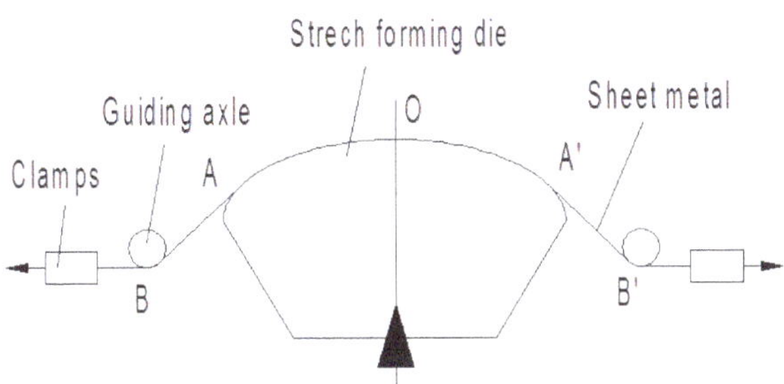

Figure 2. The schematic diagram of stretch-forming die.

2.2. Experimental Design

2.2.1. Experiment Design of Critical State Criterion of Orange Peel Defect

The experimental material was the Al–Li–S4–T8 Al–Li alloy which was used for aircraft skin with a thickness of 2 mm. See Figure 3 for the structure and size of the stretched specimens. The experimental specimens were polished to make the surfaces bright and without obvious scratches.

Meanwhile, in order to establish the same initial surface state for all specimens, the roughness of polished specimens was measured on the optical surface profilometer to ensure similar polishing effects for all the specimens.

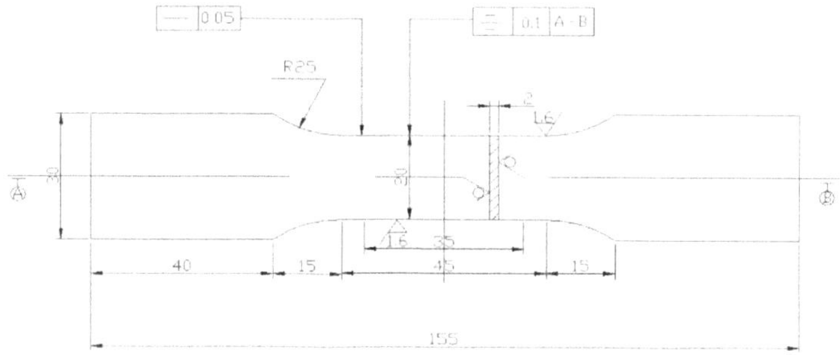

Figure 3. The shape and size of stretched specimen.

Stretching tests of different strains were conducted on polished specimens to investigate the evolution situation of the orange peel phenomenon on material surfaces under the condition of different deformations. See stretching strains in Table 1. Additionally, tests were also conducted at four different strain speeds on T8 alloy to investigate the impacts of different deforming speeds on the orange peel phenomenon on specimen surfaces in the experiment. By contrasting the surface morphology and true strain of the specimen at different speeds, when tiny orange peel phenomena appear on the surface at different strain speeds, the true strain of the Al–Li–S4–T8 Al–Li alloy metal sheet can be measured with the results shown in Table 2. It is observed from the table that when the strain speed is 0.0005/s, the strain at the sampling point is the greatest when tiny orange peel structure on the specimen surface appears. Thus, in actual stretch-forming treatment of aircraft skin, the said speed can be employed to alleviate the appearance of the orange peel phenomenon.

Table 1. Stretching strains of samples.

Experiment Batch No.	1	2	3	4	5
Stretching Strain	0%	3%	6%	9%	12%

Table 2. Critical strain and roughness of samples.

Specimen No.	Strain Speed/s^{-1}	Critical Strain/%	Roughness R_a/nm
1	0.0001	1.78	841
2	0.0005	1.87	827
3	0.001	1.14	973
4	0.0015	1.19	1002

After the stretching tests, both sides of the specimens were cleaned again, and the side used for surface morphology observation was placed under the optical profilometer for surface analysis, obtaining the morphology, nephogram and roughness of the specimen surfaces. Combined with the criterion of the orange peel defect obtained by the experiment, the areas in which critical orange peel structure occurred were found on the specimens, and the strain state of the areas can be also found at the same positions on the other side of the specimens, which actually is the critical strain state of the appearing orange peel defect.

2.2.2. Experiment Design of Stretch Forming Limit Diagram

During the stretch-forming process of aircraft skin, the strain state of sheet metals mainly fell between uniaxial stretching and plane strain with approximately linear strain paths [18]. Thus, stretch-forming specimens that could implement different strain paths were designed with the typical specimen structure and size given in Figure 4, among which the straight-edge stretched specimen was close to the uniaxial stretching state of strain, while the $R = 10$ notched specimen was close to the plane state of strain [19].

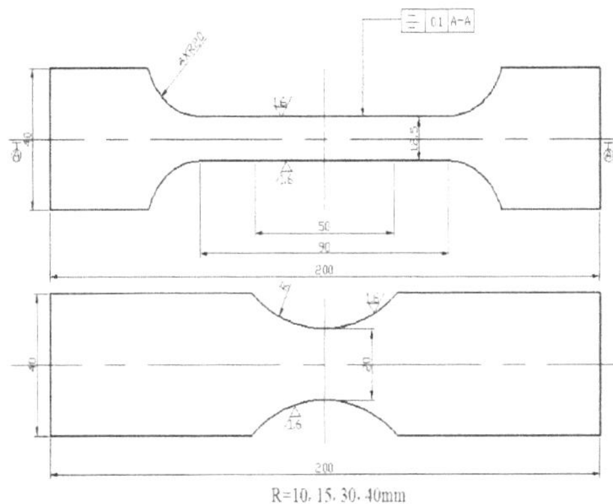

Figure 4. Size of samples with different strain paths.

In order to conduct observation and analysis of the surface morphology and strain state of the specimens, both sides of the specimens were polished until without an obvious scratch, and a profilometer was used to measure the surface roughness to achieve similar polishing effects for each specimen. The specimens with ethanol were cleaned after polishing. As shown in Figure 5, the black and white speckle patterns were sprayed on half of one side of the specimens, so as to analyze the true strain on specimen surfaces during the stretching process; another half was used to observe the evolvement of the orange peel phenomenon on specimen surfaces during the stretching process. Meanwhile, another side, only polished, was used to measure surface morphology and roughness with the profilometer.

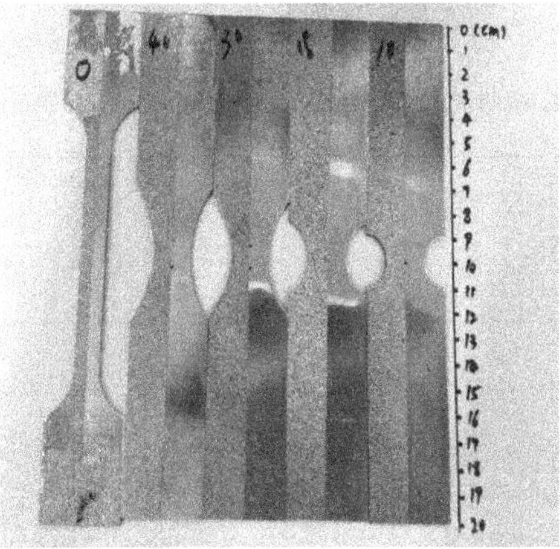

Figure 5. Sample appearances along different strain paths after surface treatment.

3. Results and Analysis

3.1. The Equilibrium Diagram of Tensile Specimen before Deformation

The grain size grade of the specimen can be measured in accordance with the Metal Methods for Estimating the Average Grain Size, as shown in Table 3 below. From the measured grain size grades, as shown in Figures 6 and 7 it is observed that the average grain size of the specimen is 60–80 μm, suggesting that the specimen has already fallen into the category of open grain structure. Meanwhile, judging from the equilibrium diagram, the grain structure of the specimen is extremely uneven with a large difference in size between the large grains and small grains. It is also discovered from the measurement that the size of the large grains is over 100 μm, but that of the small ones is merely around 10 μm. Thus, the equilibrium diagram can explain the appearance of orange peel in the process.

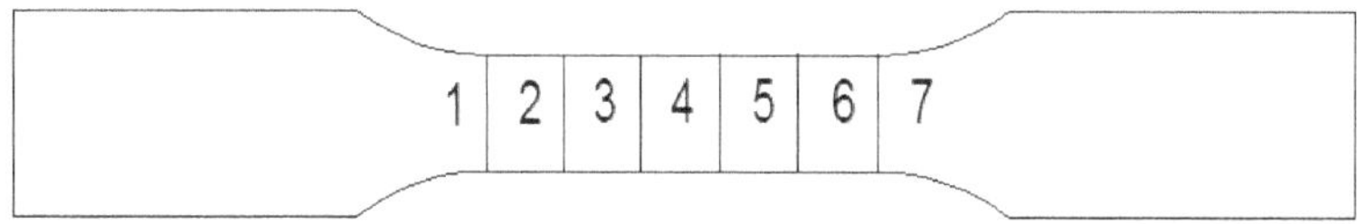

Figure 6. The exact location on the sample.

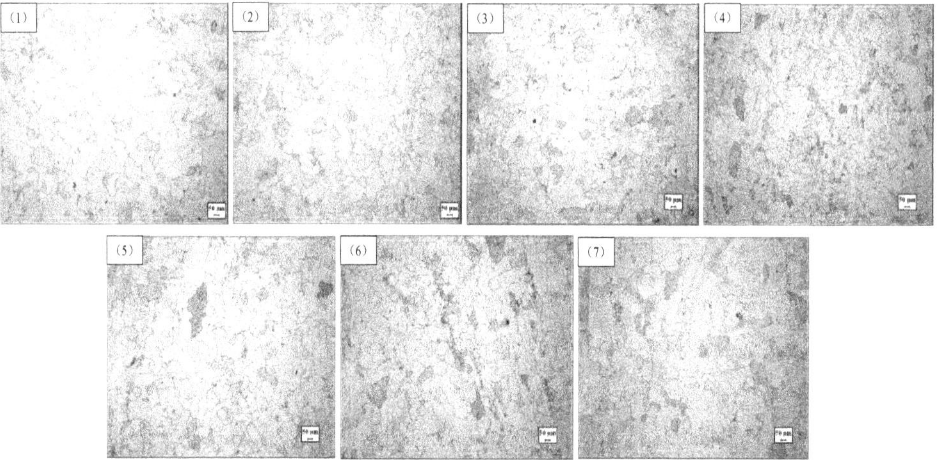

Figure 7. The equilibrium diagram of tensile specimen before deformation. **(1)** is correspondence with 1 in Figure 6; **(2)** is correspondence with 2 in Figure 6; **(3)** is correspondence with 3 in Figure 6; **(4)** is correspondence with 4 in Figure 6; **(5)** is correspondence with 5 in Figure 6; **(6)** is correspondence with 6 in Figure 6; **(7)** is correspondence with 7 in Figure 6.

Table 3. Grain size analysis of tensile specimen.

Heat Treatment Condition	Type of Analysis	Sampling Point							Average
		1	2	3	4	5	6	7	
T8	Grain Size Grade	5.02	5.13	4.51	4.38	4.98	4.62	4.56	4.7
	Average Diameter/μm	62	59	76	77	70	75	75	71

3.2. The Establishment of the Criterion for Critical State of Orange Peel Defect

The specimens were cleaned after stretch processing with ethanol, and surface observation and analysis on the optical profilometer were conducted. The photographed morphology can be seen in Figure 8, which shows that when the strain is 3%, the specimen surface starts to grow rough and form a kind of evenly frosted surface morphology, but without evident minor cracks or significant orange peel phenomenon; when the strain reaches 6%, light black strip areas appear on the surface, which are shallow grooves, and an embryonic form of orange peel structure can be roughly observed. When the strain reaches 9%, there are clear cracks appearing on the specimen surface, which is rather severe despite the fact that they are relatively decentralized and independent; at the macro-level, a very significant orange peel phenomenon is manifested. At this stage, the orange peel structure on the material surface not only affects the appearance of the stretch-forming parts, but also exerts great influence on the performance and service life of the specimen, whereas the minor cracks on the surface can easily give rise to stress concentration and gradually evolve into greater cracks after stress. When the strain amounts to 12%, the surface morphology of the specimen further deteriorates with more and deeper cracks, having already formed gully-like shapes.

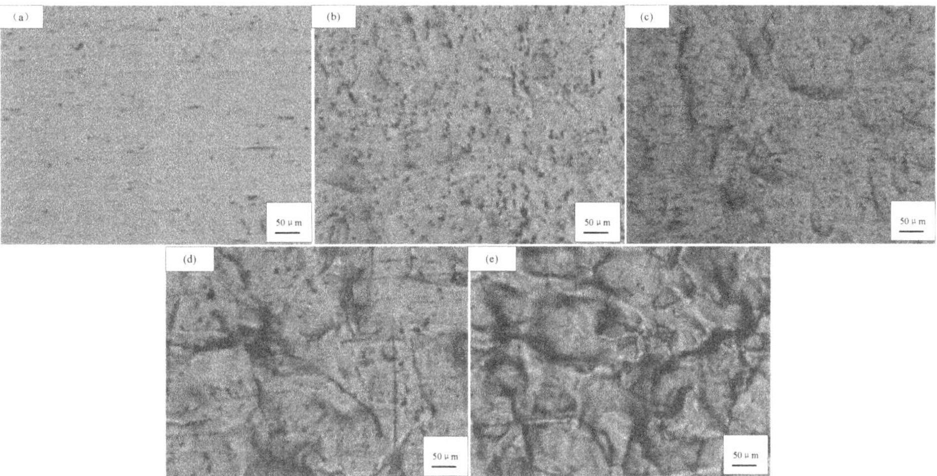

Figure 8. Surface morphology of stretched specimens with different strain variables: (**a**) 0%; (**b**) 3%; (**c**) 6%; (**d**) 9%; (**e**) 12%.

It can be seen from the analysis results that when the strain reaches around 6%, the specimen surface starts to form the orange peel defect in a real sense, but as the gap between the strain variables is relatively large, it is impossible to determine that the specimen is in the critical orange peel state when the strain is 6%. Thus, on the basis of the previous research, strain variables of 5% and 7% were added to a new test to observe their surface morphology after stretching, with the results given in Figure 9. When the strain was 5%, the specimen surface had not yet formed clear grooves, but through comparing the stretched specimens of 6% and 7%, it could be found that, though with roughly the same surface morphology, the scanning image of the stretched specimen with 7% strain clearly showed darker grooves with relatively deep cracks starting to develop.

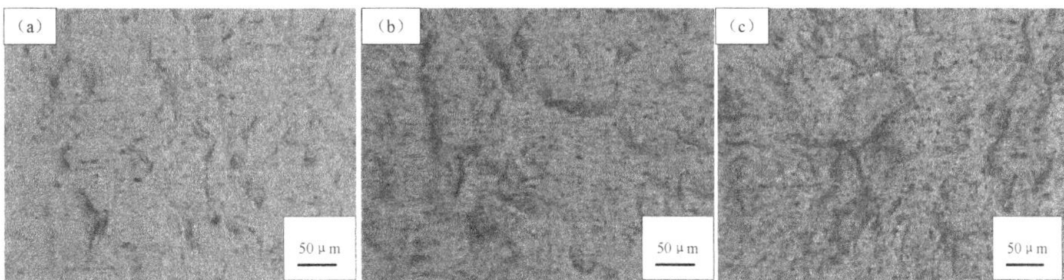

Figure 9. Surface morphology of stretched specimens with different strain variables: (**a**) 5%; (**b**) 6%; (**c**) 7%.

On this basis, the roughness of specimen surfaces with different strain variables is measured, the results of which are as shown in Figure 10, where R_a is the arithmetic mean of the absolute value of the distance between the dot on the profile and the baseline; R_q is the root mean square error of the profile, *i.e.*, the root mean square value of the profile's offset distance within the sampling range; R_z is the micro-irregularity on the material surface, *i.e.*, the sum of the average of the five greatest profile peaks and the average of the five smallest profile valleys within the sampling length. Combined with the surface morphology measured by the profilometer, it can been seen that, at the stage of 0%~5% strain, the main change was that the specimen surface began to wrinkle, with dispersed convexes and concaves appearing as well as a rapid change of surface roughness; at the stage of 5%~6% strain, there was little change in the surface roughness, but dispersed convexes and concaves started to gather to form lumps while grooves appeared and started to develop into cracks, which is also the major stage of qualitative change appearing on the material surface; at the stage of 6%~12% strain, cracks rapidly developed, gradually grew deeper, and joined with each other to form the gully-like morphology, resulting in the rapid increase in R_z of the specimens.

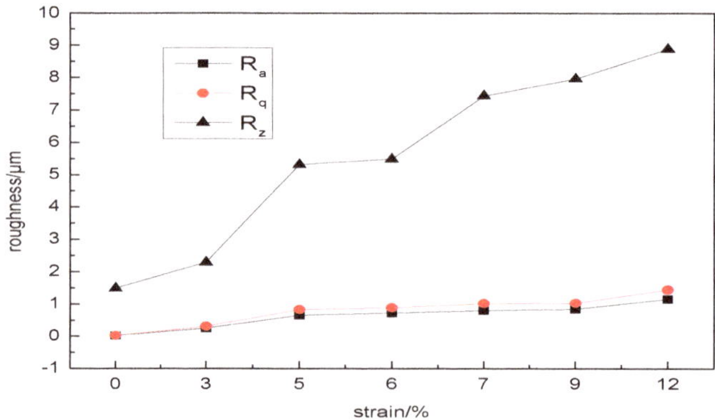

Figure 10. Changing trend of stretched sample surface roughness.

Based on the above experimental analyses, it can be assumed that when a morphology similar to that of the stretched specimen with 6% strain appears on the specimen surface, the orange peel phenomenon comes to a critical state. Thus, a roughness of $R_a = 700 \pm 50$ nm and $R_z = 5.5 \pm 0.5$ μm can be taken as a reference standard for judging whether the orange peel structure on a material surface reaches a critical state.

3.3. The Establishment of Stretch-Forming Limit Diagram

On the basis of obtaining the corresponding strain state and surface roughness of the orange peel structure on the standard specimens, the research on the stretch-forming limit under different strain paths is further conducted, through which strain data of different strain paths is measured with the surface roughness of the characteristic critical orange peel structure as the forming failure criterion, as shown in Table 4.

Table 4. Strain state of critical orange peel structure of specimens under different strain paths.

Sample	Acquisition Point 1		Acquisition Point 2		Acquisition Point 3		Acquisition Point 4		Acquisition Point 5	
	ε_1/%	ε_2/%	ε_1/%	ε_2/%	ε_1/%	ε_2/%	ε_1/%	ε_2/%	ε_1/%	ε_2/%
Straight edge	6.29	−1.94	6.35	−1.98	6.31	−2.08	6.25	−1.84	6.33	−1.88
$R = 10$	3.17	−0.21	3.17	−0.18	2.65	−0.15	2.76	−0.13	3.17	−0.15
$R = 15$	2.95	−0.19	3.04	−0.13	3.27	−0.12	3.31	−0.23	3.09	−0.08
$R = 30$	3.31	−0.55	3.76	−0.58	3.64	−0.77	3.72	−0.57	3.30	−0.55
$R = 40$	4.49	−1.20	4.82	−1.59	4.66	−1.27	3.92	−1.04	4.20	−1.23

Quadratic-multinomial fitting is conducted on measured data so as to establish the forming limit diagram of the orange peel phenomenon on the aluminum alloy surface with the results shown in Figure 11 and the quadratic-multinomial fitting results given in Table 5. It can be seen from the figure that the orange peel structure on the aluminum alloy surface is correlated with the strain state of the materials; meanwhile, the materials at the plane state of strain more easily develop orange peel structure than those at the uniaxial stretching strain state.

Table 5. Binomial fitting results.

Expression	*A*	*B*	*C*
$y = A + Bx + Cx^2$	2.99	−0.389	0.652

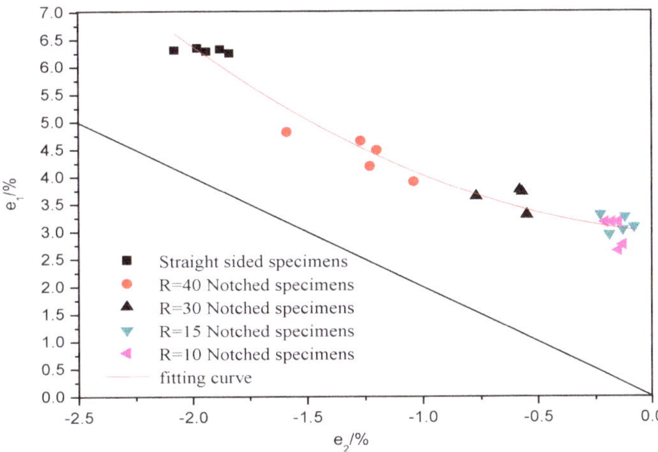

Figure 11. Stretch-forming limit diagram of Al–Li alloy at the condition of Al–Li–S4–T8.

4. Conclusions

(1) A stretch-forming experiment and testing system with the optical deformation measurement instrument and the universal testing machine operating in collaboration are constructed, the surface morphology change rule of stretched specimens with different strain variables is analyzed, and the corresponding relation between critical orange peel defect and the surface roughness of specimens is obtained. It is discovered that when critical orange peel defect appears on Al–Li alloy sheet metal at the condition of Al–Li–S4–T8, the surface roughness is $R_a = 700 \pm 50$ nm and $R_z = 5.5 \pm 0.5$ µm.

(2) By processing different notched specimens, the stretch-forming limit tests with different strain paths are conducted to obtain the forming limit diagram and forming limit curve equation $\varepsilon_1 = 2.99 - 0.389\varepsilon_2 + 0.652{\varepsilon_2}^2$ for Al–Li–S4–T8 Al–Li alloy, with the surface roughness of characteristic critical orange peel structure as the forming failure criterion.

Acknowledgments: This research was financially supported by the National Basic Research Program of China (2014CB046602) and the National Natural Science Foundation of China (51235010).

Author Contributions: Jing-Wen Feng and Li-Hua Zhan conceived and designed the experiments; Jing-Wen Feng and Ying-Ge Yang performed the experiments; Jing-Wen Feng analyzed the data; Jing-Wen Feng wrote the paper.

Conflicts of Interest: The authors declare no conflict of interest.

References

1. Wang, K.; Wan, M.; Hua, C.; Shao, X.F. Determination and application of coarse-grain critical pre-strain curve to aluminum alloy stretch forming. *J. Beijing Univ. Aeronaut. Astronaut.* **2013**, *39*, 508–511.
2. Han, Z.R.; Dai, L.J.; Zhang, L.Y. Current status of large aircraft skin and panel manufacturing technologies. *Aeronaut. Manuf. Technol.* **2009**, *25*, 64–66.
3. Araghi, B.T.; Manco, G.L.; Bambach, M.; Hirt, G. Investigation into a new hybrid forming process: Incremental sheet forming combined with stretch forming. *CIRP Ann. Manuf. Technol.* **2009**, *58*, 225–228. [CrossRef]
4. Kurukuri, S.; Miroux, A.; Wisselink, H.; Boogaard, T.V.D. Simulation of stretch forming with intermediate heat treatments of aircraft skins: A physically based modeling approach. *Int. J. Mater. Form.* **2011**, *4*, 129–140. [CrossRef]
5. General Editorial Board of "The Manual of Aeronautical Manufacturing Engineering". In *The Aviation Manufacturing Engineering Handbook—Aircraft Sheet Metal Process*; Aviation Industry Press: Beijing, China, 1992.
6. Hu, S.G.; Chen, H.Z. *Manufacturing Technology of Aircraft Sheet Metal Parts*; Beijing University of Aeronautics and Astronautics Press: Beijing, China, 2004.
7. Chang, R.F. *Sheet Metal Parts Manufacturing Technology*; National Defence Industry Press: Beijing, China, 1992.
8. Wan, M.; Zhou, X.B.; Li, X.X.; Wu, H. Process parameters in stretch forming of mirror skins. *Acta Aeronaut. Astronaut. Sin.* **1999**, *20*, 326–330.
9. You, Z.H.; Huang, Y.S.; Wu, R.H. A study of orange-like roughness on front side on Airplane. *Aviat. Maint. Eng.* **2001**, *6*, 46–47.
10. Zhang, R.X.; Zeng, Y.S. Development, technological characteristics and application status abroad of aluminum-lithium alloys (In Chinese). *Aeronaut. Manuf. Technol.* **2007**. [CrossRef]
11. Yin, D.F.; Zheng, Z.Q. History and current status of aluminum-lithium alloys research and development. *Mater. Rev.* **2003**, *17*, 18–20.
12. Huo, H.Q.; Hao, W.X.; Geng, G.H.; Da, D.A. Development of the new aero-craft material—Aluminum–lithium alloy. *Vac. Low Temp.* **2005**, *11*, 63–69.
13. Lyttle, M.T.; Wert, J.A. The plastic anisotropy of an Al–Li–Cu–Zr alloy extrusion in unidirectional deformation. *Metall. Mater. Trans. A* **1996**, *27*, 3503–3512. [CrossRef]
14. Li, H.; Tang, Y.; Zeng, Z.; Zheng, Z.; Zheng, F. Effect of ageing time on strength and microstructures of an Al–Cu–Li–Zn–Mg–Mn–Zr alloy. *Mater. Sci. Eng. A* **2008**, *498*, 314–320. [CrossRef]
15. Huang, J.C.; Ardell, A.J. Addition rules and the contribution of δ' precipitates to strengthening of aged Al–Li–Cu alloys. *Acta Metall.* **1988**, *36*, 2995–3006. [CrossRef]
16. Fan, C.P.; Zheng, Z.Q.; Jia, M.; Zhong, J.F.; Cheng, B.; Li, H.P.; Wu, Q.P. Microstructure, tensile property and fracture toughness of 2397 Al–Li alloy. *Rare Met. Eng.* **2015**, *44*, 91–96.
17. Ma, G.S. *Hot Forming Technology of Complex Aluminum Lithium Alloy Parts*; Chemical Industry Press: Beijing, China, 2011.
18. Jin, H.X. *Basic Experimental Research and Simulation on Aluminum Alloy Mirror Skin Tensile Forming*; Beijing University of Aeronautics and Astronautics: Beijing, China, 2009.
19. Wan, M.; Han, J.Q.; Jin, H.X.; Wu, H. Determination of strain criterion of portevin–Le chatelier effect for aluminum alloy sheets. *Trans. Nonferrous Met. Soc. China* **2006**, *16*, 1499–1503.

15

Picosecond Laser Shock Peening of Nimonic 263 at 1064 nm and 532 nm Wavelength

Sanja Petronic [1,*,†], Tatjana Sibalija [2,†], Meri Burzic [1], Suzana Polic [3], Katarina Colic [1] and Dubravka Milovanovic [4,†]

1 Innovation Center, Faculty of Mechanical Engineering, University of Belgrade, Kraljice Marije 16, 11000 Belgrade, Serbia; mburzic@mas.bg.ac.rs (M.B.); kbojic@mas.bg.ac.rs (K.C.)

2 Faculty of Information Technology, Faculty of Management, Metropolitan University, Tadeusa Koscuska 63, 11000 Belgrade, Serbia; tsibalija@gmail.com

3 Central Institute for Conservation in Belgrade, Terazije 26, 11000 Belgrade, Serbia; suzanapolic64@gmail.com

4 Vinca Institute of Nuclear Sciences, University of Belgrade, PO Box 522, 11001 Belgrade, Serbia, dubravka.milovanovic@vinca.rs

* Correspondence: sanjapetronic@yahoo.com

† These authors contributed equally to this work.

Academic Editor: Patrice Peyre

Abstract: The paper presents a study on the surface modifications of nickel based superalloy Nimonic 263 induced by laser shock peening (LSP) process. The process was performed by Nd^{3+}:Yttrium Aluminium Garnet (YAG) picosecond laser using the following parameters: pulse duration 170 ps; repetition rate 10 Hz; pulse numbers of 50, 100 and 200; and wavelength of 1064 nm (with pulse energy of 2 mJ, 10 mJ and 15 mJ) and 532 nm (with pulse energy of 25 mJ, 30 mJ and 35 mJ). The following response characteristics were analyzed: modified surface areas obtained by the laser/material interaction were observed by scanning electron microscopy; elemental composition of the modified surface was evaluated by energy-dispersive spectroscopy (EDS); and Vickers microhardness tests were performed. LSP processing at both 1064 nm and 532 nm wavelengths improved the surface structure and microhardness of a material. Surface morphology changes of the irradiated samples were determined and surface roughness was calculated. These investigations are intended to contribute to the study on the level of microstructure and mechanical properties improvements due to LSP process that operate in a picosecond regime. In particular, the effects of laser wavelength on the microstructural and mechanical changes of a material are studied in detail.

Keywords: laser modification; Nd:YAG; shock peening; Nimonic; superalloy

---

1. Introduction

Nickel based superalloys are materials with remarkable combination of mechanical and physical properties that contribute to their wide application, from gas turbine, aero and rocket engines to chemical processing plants, as presented by Pollock and Tin [1]. Nimonic alloys are the group of Ni-based superalloys, which are widely used for high performance applications, such as disks and blades of either aircraft engines or land-based gas turbines, at temperatures ranging from 650 °C to 1050 °C and in aggressive atmospheres [2]. Nimonic alloy 263 is an important precipitation hardening nickel base superalloy, with high creep strength and oxidation resistance, designed for stationary components like combustion chambers, casings, liners, exhaust ducting, bearing housings and many others [3]. Since the components of this alloy are primarily degraded by creep damage, as presented by Park *et al.* [4], when it is exposed to the influence of harsh environment, it is necessary to retard intergranular cracking and propagation for better performance at elevated temperatures.

Lasers have been used for high precision material processing in micro- and nano-manufacturing operations due to a specific nature of the laser light, high intensity and possibility of controlled surface modification. In the last few decades, surface modifications of different metals and their alloys performed by various types of lasers has gained significant attention. Milanovic *et al.* [5] presented the surface modification of titanium based implant by picosecond laser at both 1064 nm and 532 nm, and found periodic surface structures. Treatment of nickel based superalloys' surfaces by laser irradiation can induce specific changes in the microstructure, which result in improved mechanical properties of the material, as discussed by Petronic *et al.* [6].

The principle of laser shock peening (LSP) implies usage of high intensity laser beam and suitable overlays to generate high pressure shock waves on the surface of a workpiece. Laser shock processing of steel was studied by Yilbas *et al.* [7]. They showed that dislocations are governing mechanism in hardening of the alloy, and the hardness of peened alloy increased 1.7 times compared to the base material. A review on laser shock processing was carried out by Peyre and Fabbro [8]. They presented physical principles of laser shock and induced mechanical effects, and found that higher pressures can be achieved with confinement as compared to direct ablation.

Laser shock processing of metallic surfaces and its applications were presented by Devaux *et al.* [9], theoretically and experimentally. The transient shock waves induce microstructure changes near the surface and alter the stress level, which improve the mechanical properties of material, such as hardness and fatigue strength. Peyre *et al.* [10] presented analytical models for the confined ablation, plastically affected depths and residual stresses induced at the surface. Hong and Chengye [11] showed the improvement in fatigue strength due to the surface residual stresses, while Clauer and Holbrook [12] showed that improvement in hardness is the result of dislocations and other phases' formation arisen by laser shock peening. During LSP process strain rates could reach as high as $10^6\ s^{-1}$, which is very high compared to the conventional strain rates. Amarchinta *et al.* [13] used a finite element technique to predict the residual stresses induced by LSP process, which was followed by comparison between experimental results and simulation. Peyre *et al.* [14] demonstrated that a finite element modeling procedure could be successfully applied to the prediction of residual stresses induced by laser shock processing. Sibalija *et al.* [15] used a hybrid designed method, based on the artificial intelligence techniques, to optimize LSP parameters in order to improve several characteristics of the processed Nimonic 263 sheets: surface characteristics, microstructure, roughness, microhardness and microstructural transformations.

The beneficial effects of LSP include improvement of microstructure, surface quality, *etc.*, which delays the fatigue crack initiation. The surface morphology of metals has a great effect on fatigue behavior, and many laser peened materials showed improvements in fatigue life with LSP, as shown by Ding and Ye [16]. Surface condition has considerable influence on fatigue strength, and the following factors have the most significant effects: surface roughness, residual stresses in the surface layer, work hardening or softening in the surface layer, and change or transformation of the microstructure due to plastic deformation, as presented by Schijve [17].

Although in the last few decades many investigations were devoted to the laser shock peening, most of them were performed in nanosecond laser beam regime. Since the invention of laser, there has been a constant development in terms of shorter pulse times. Some investigation were carried out in femtosecond regime [18,19], including the effect of wavelength on the microstructure characteristics [20]. There is an emerging interest in employing picosecond laser for LSP application, but picosecond laser regime has been rarely discussed in the literature. Therefore, this paper aims to contribute to the investigation of the microstructure features introduced by picosecond LSP. The investigation includes analysis of surface condition after LSP of material, discussion of microstructural changes arisen by the laser treatment, and microhardness testing after LSP. In particular, the impact of 1064 and 532 nm laser irradiation of the superalloy is discussed in detail. The rest of the paper is organized as follows. After the introduction, the experimental set up is presented in the second section. Results of the experiment are presented, analyzed and discussed in the third section, which is divided into three parts: (i) discussion of the microstructural and surface changes arisen by 1064 nm wavelength laser beam; (ii) discussion of the microstructural and surface changes arisen by 532 nm wavelength laser beam; and (iii) comparison of influence of these two wavelengths on the microstructure. The last section brings the main and most important conclusions from the above analysis.

2. Experimental Section

Samples of superalloy Nimonic 263, prepared in the form of sheets with dimensions 10 mm × 50 mm × 1 mm, were polished and cleaned with ethanol prior to application of the protective layer. The chemical composition of Nimonic 263 is listed in Table 1.

Table 1. Chemical composition of alloy Nimonic 263.

Element	C	Si	Mn	Al	Co	Cr	Cu	Fe	Mo	Ti	Ni
%	0.06	0.3	0.5	0.5	20	20	0.1	0.5	5.9	2.2	49.94

The sample sheets were coated with black paint (protective overlay) and immersed in distilled water, which served as the transparent layer. The laser beam was guided by the mirror system and focused below the surface of the sample, at the incidence angle of 90°. The experimental setup for LSP processing of the Nimonic 263 samples is presented in Figure 1.

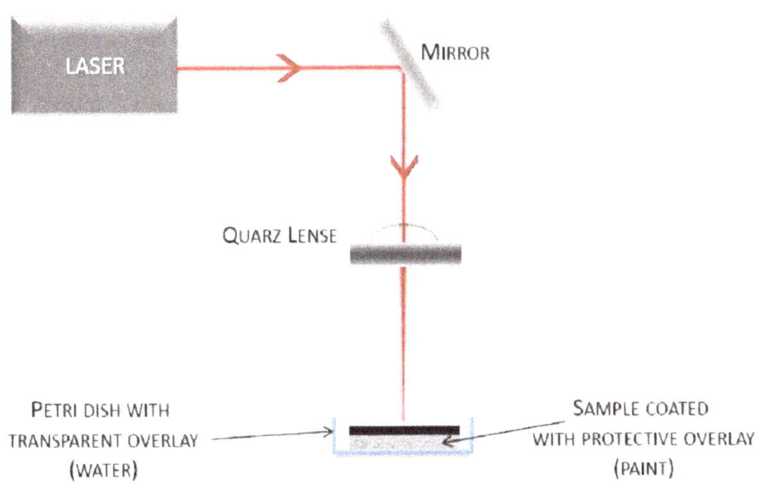

Figure 1. The experimental setup for laser shock peening.

The samples were exposed to a laser light at 1064 nm and 532 nm wavelength, with number of accumulated pulses: 50, 100 and 200. The operating laser was Nd:YAG EKSPLA, model SL212P, (Ekspla, Vilnius, Lithuania). At 1064 nm wavelength, pulse energies were varied from 2 mJ to 15 mJ and corresponding fluencies ranged from 0.1 to 0.3 $J \cdot cm^{-2}$. At laser wavelength of 532 nm, pulse energies were varied from 20 to 35 mJ; corresponding fluencies ranged from 0.28 to 0.4 $J \cdot cm^{-2}$. The spatial distribution of laser energy for both 1064 nm and 532 nm wavelengths is non-homogeneous. Specifically, the highest value of laser energy is at the center of the beam, and energy is decreasing at periphery of the beam. Characteristics of the picosecond laser used in this experiment are listed in Table 2.

Table 2. The characteristics of EKSPLA picosecond laser.

Laser	Nd:YAG
Wavelenght	532 nm, 1064 nm
Pulse duration	150 ps
Mode	about TEM_{00}
Repetition rate	10 Hz

The resulting surface changes were determined by scanning electron microscopy (SEM) model JEOL JSM-5800 (JEOL Ltd., Tokyo, Japan), and compared with non-treated surface. Elemental analysis of the surface was done by energo-dispersive spectroscopy (EDS). Microhardness measurements were performed by Vickers, using the apparatus-model semiautomatic Hauser 249A (Muller Machines SA, Bruegg, Switzerland), and under the load of 0.5 N. In addition, surface morphology changes of the irradiated areas were analyzed using Zygo NewView 7100 optical profiler (Zygo Corporation, Middlefield, CT, USA), and characteristic surface parameters were calculated using MetroPro software (Zygo Corporation, Middlefield, CT, USA). The reflectance spectra of Nimonic 263 with and without protective layer were recorded by Perkin Elmer Lambda 35 UV-Vis spectrophotometer (Perkin Elmer, Waltham, MA, USA). The spectroscopy measurements were done in scanning mode, using tungsten-halogen visible lamp and a deuterium ultraviolet lamp as the light sources. Spectroscopy measurements were non-destructive since deuterium and tungsten-halogen lamp radiation was below damage threshold for Nimonic 263. The grain size mean values (*F*m) were calculated using AutoCAD (2009, Autodesk, Mill Valley, CA, USA) according to the circle method presented by Shuman [21]. Elastic limit of impacted material of 480 MPa was determined in previous experiments [22].

3. Results and Discussion

The surface of Nimonic 263 before coating with the overlay paint is shown in Figure 2. The microstructure is characterized by the grains, sized *F*m = 232.45 μm^2 (Figure 2), as calculated by Petronic *et al.* [23]. The measure of surface deviation from flatness is peak to valley ratio (PV ratio), and the base Nimonic 263 is characterized by PV ratio of 48.5 μm, determined by the above mentioned profilometer.

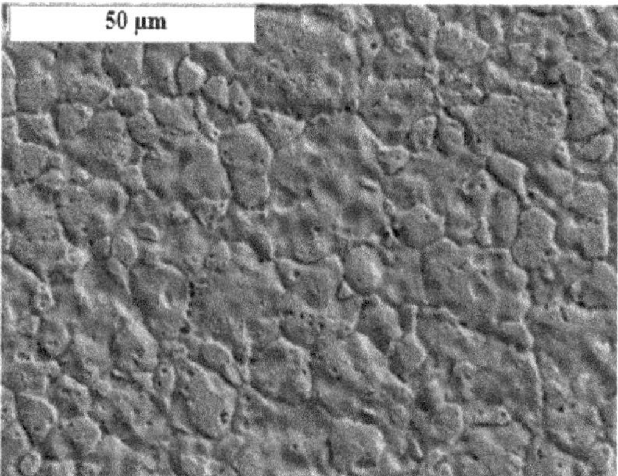

Figure 2. SEM of Nimonic 263 sample prior to application of the paint overlay.

The interaction of picosecond laser radiation with metals is a complex process, as presented by Bauerle [24], because it involves several physical processes. This interaction can be summarized as the absorption of laser energy by free electrons which leads to the initial electron-electron collision. This is followed by energy transfer to the lattice phonons which can lead to deformation and cracking of lattice bonds and finally melting or vaporization, if sufficient amount of energy is applied. The resulting surface effects strongly depend on the specific parameters of the operating laser (wavelength, pulse duration, energy, *etc.*) and surface conditions of the sample (roughness, absorbance coefficient, *etc.*). Treatment of a surface by the pulsed laser light with picosecond pulse duration can be accompanied by the formation of liquid phase which can decrease the precision of processing, as discussed by Bauerle [24]. The protective overlay in laser surface treatment of alloys is used for two reasons: (1) to absorb the incident laser energy, expand abruptly and transfer the shock wave to the metal target; and (2) to protect the metal target from the excessive heat influence of the incident light. The implementation of the transparent layer increases the plasma pressure by a trapping-like effect on the plasma expansion.

The transient shock waves induce microstructure changes near the surface and cause high density of dislocations to be formed. The combined effect of the microstructure changes and dislocation entanglement contributes to an increase in the mechanical properties near the surface. LSP improves fatigue, corrosion and wearing resistance of metals through mechanical effect produced by shock waves [25].

It is important to determine how protective layer influences the value of absorptivity/reflectivity of the sample surface for light wavelengths that were used in this experiment (532 nm and 1064 nm). Comparative reflectance spectra and the results for Nimonic 263 surface with and without protective paint are presented in Figure 3 and Table 3.

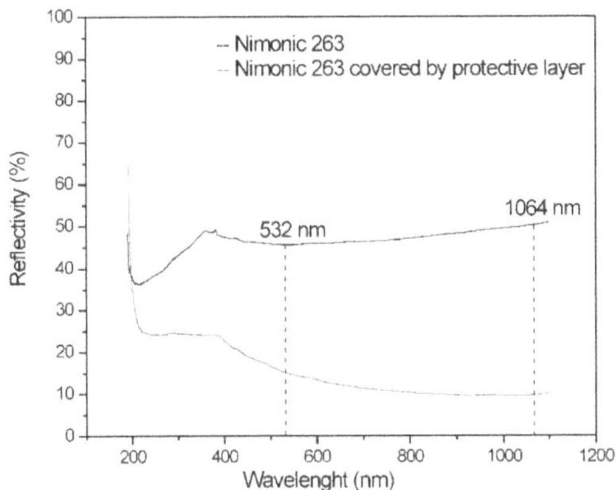

Figure 3. Comparative reflectance spectra of Nimonic 263 surface with and without protective paint layer.

The reflectivity of base material at 532 nm wavelength is lower than at 1064 nm, as shown in Figure 3. After applying the protective paint layer, surface of the sample covered by protective layer exhibits significantly lower values of reflectivity for incoming light than the sample base material, at both observed wavelengths. For 532 nm wavelength, the reflectance decreases by two-thirds, and for 1064 nm wavelength the value of reflectance reduces by four-fifths, as presented in Table 3.

Table 3. Reflectivity of Nimonic 263 with and without protective paint layer.

Wavelength (nm)	Reflectivity (%)	
	Nimonic 263	Nimonic 263 with protective layer
532	45.6	15.0
1064	50.2	9.8

The laser shock peening experiment was carried out with variation of pulse energy: 2 mJ, 10 mJ and 15 mJ at 1064 nm wavelength (respectively, power density values 0.7, 3.3 and 5 GW/cm^2), and 20 mJ, 25 mJ and 30 mJ at 532 nm wavelength (respectively, power density values 1.9, 2.4 and

2.8 GW/cm^2. Beside the pulse energy, the number of pulses was varied as well: 50, 100 and 200, for both groups of experiment.

3.1. Laser Shock Peening of Nimonic 263 at 1064 nm Wavelength

Figure 4 presents the microstructure after picosecond LSP process by the laser light at 1064 nm wavelength, pulse energy of 10 mJ and number of pulses 50, 100 and 200, respectively. When the number of accumulated pulses is increased, the microstructure becomes finer, but after 200 pulses the partial melting of the material occurs (Figure 4c), as the paint was partially removed. For this reason, LSP becomes a thermomechanical process (so called "direct ablation"), as presented in [26].

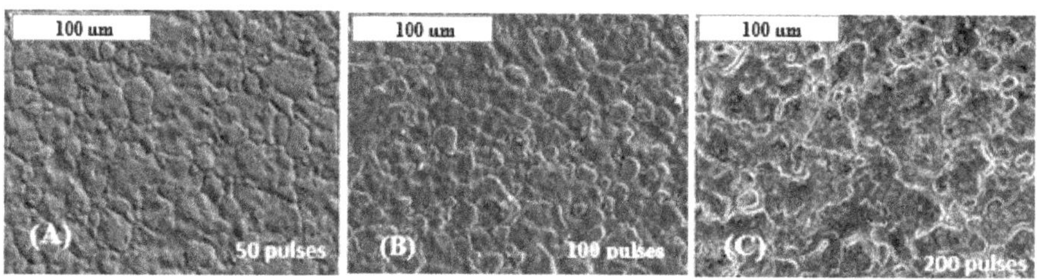

Figure 4. The microstructure of Nimonic 263 after LSP by laser pulse energy of 10 mJ (F = 0.2 J$\cdot$cm^{-2}), wavelength of 1064 nm, and number of pulses 50 (**A**); 100 (**B**) and 200 (**C**), respectively (power density 3.3 GW/cm^2).

At 1064 nm laser action, the pulse energy values were 2 mJ, 10 mJ and 15 mJ, and the number of pulses 50, 100 and 200. The finest microstructure was obtained at pulse energy of 10 mJ (presented in Figure 4), as pulse energy of 15 mJ induced not only mechanical but thermal changes as well. Increase in the pulse energy leads to decrease of the grains size.

LSP is not a thermal but a mechanical process for treating materials, as Berthe *et al.* [27] presented in their work. This process is accompanied by significant changes in microstructures and phases near the surface. The mechanism of microstructure changes is similar to those arisen during the laser shoot peening. According to [28,29], after the deformation the grain size decreases due to sliding mechanism and hardening deformation. When the peak pressure of the shock wave exceeds the dynamic yield strength of alloys or metallic materials, plastic deformation occurs, resulting in the improvement of the near-surface microstructure and mechanical properties, which is accompanied by significant changes in microstructures and phases. Shock waves can produce one or the combination of the following metallurgical effects: generation of point defects, dislocations and twins, phase transformations and precipitation [30,31]. The changes in microstructure, induced by laser beam action, are typical for this superalloy and can be summarized as formation of carbides, TiC, CrC and (Ti, Mo)C, that segregate both at the grains and at the grain boundaries [32].

Titanium monocarbide (TiC) is very hard, stable at both high and low temperatures. (Ti, Mo)C is characteristic structural component that strongly influences mechanical and physical properties of material, as discussed by Chena *et al.* [33] and Jang *et al.* [34]. SEM microphotographs shown in Figure 5 present phases occurring after 1064 nm picosecond LSP, 10 mJ laser beam energy and pulse number of 100 (Figure 5a,b) and 50 (Figure 5c,d).

Table 4 shows the results of EDS analysis of the microconstituents denoted in Figure 5. These results suggest formation of various phases, both desirable and undesirable, depending on their morphology, size and the place/area of appearance. According to EDS results in Spectrum 1 and 3, as well as morphology of these phases which is rather regular pyramid, it can be assumed these phases are Ti carbides.

According to Table 4 and EDS results, carbides in Spectrum 2 and 4 are TiMo carbides. These types of compounds are characterized by the irregular shape and micrometer scale dimensions [34]. In this case, the presence of TiMo carbides is undesirable because their presence can cause the crack formation in the material.

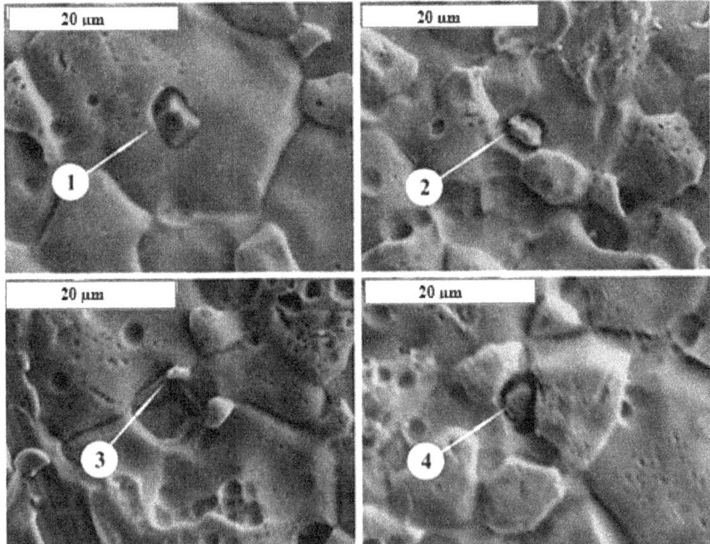

Figure 5. SEM of the microconstituents formed during LSP at 1064 nm, E = 10 mJ, and number of accumulated pulses: (**a**) 100; (**b**) 100; (**c**) 50 and (**d**) 50. Power density: 3.3 GW/cm^2.

Table 4. EDS analysis of the spots denoted in Figure 5.

Spectrum	C	Al	Si	Ti	Cr	Mn	Fe	Co	Ni	Mo
Spectrum 1	11.79	0.11	0.15	49.87	6.96	0.20	0.19	7.41	18.79	4.52
Spectrum 2	14.56	0.13	0.19	13.36	11.56	1.00	0.33	13.22	34.04	12.19
Spectrum 3	12.84	0.25	0.20	4.67	14.25	0.54	0.30	16.82	44.75	5.38
Spectrum 4	18.29	1.02	0.94	21.59	9.62	0.35	0.24	10.67	27.73	9.54

3.2. Laser Shock Peening of Nimonic 263 at 532 nm Wavelength

LSP is performed with the action of laser light at 532 nm wavelength, as well. Figure 6 shows the microstructure of Nimonic 263 obtained by the stated wavelength, 25 mJ energy and 50, 100 and 200 number of accumulated pulses, respectively. In contrast to the 1064 nm wavelength, the pulse energy of 2 mJ, 10 mJ and 15 mJ could not be applied with the wavelength of 532 nm, as it does not produce sufficient fluence required to make changes in a surface layer. When the pulse energy is increased, the pores start to form, while the increase in the number of applied pulses results in a "coral" structure, as shown in Figure 6. The grain boundaries are more pronounced with higher number of applied pulses. Unlike the treatment with the wavelength of 1064 nm, melting did not occur after 200 pulses (Figure 6c).

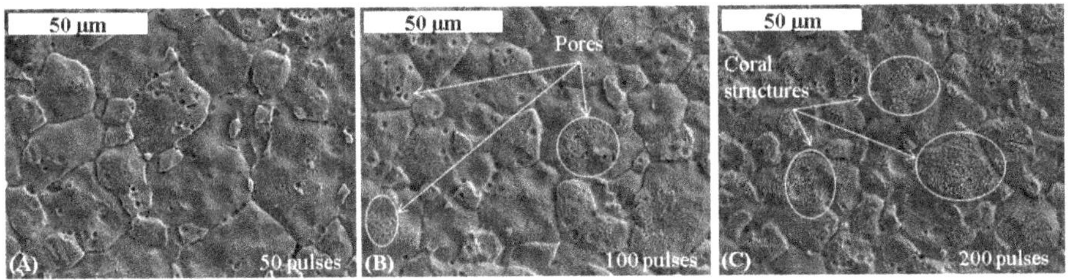

Figure 6. The microstructure of Nimonic 263 after LSP processing by laser energy of 25 mJ and wavelength of 532 nm (power density 2.6 GW/cm^2), and number of pulses 50 (**A**); 100 (**B**) and 200 (**C**).

As the higher number of pulses and pulse energy caused unfavorable "coral" structure formation, the best results, for the material processing at 532 nm, are considered to be for 50 number of pulses and 25 mJ pulse energy action. Characteristic microstructures and microconstituents arisen after picosecond LSP with the laser wavelength of 532 nm, laser beam energy of 25 mJ and 50 laser beam pulses are shown in Figure 7. Corresponding EDS analysis listed in Table 5 shows the presence of oxygen in Figure 7a and suggests that oxides are formed. The creation of different grains can be noticed in Figure 7b. The grain boundaries are clearly pronounced. Some of the grains consist of many pores, sized up to 1.2 µm. Figure 7c shows the twins arisen by deformation induced by mechanical treatment of the surface. Due to the high strain rates involved in the laser shock peening process, the shock waves induce plastic deformation, through the sliding and deformation hardening, which results in twins' formation.

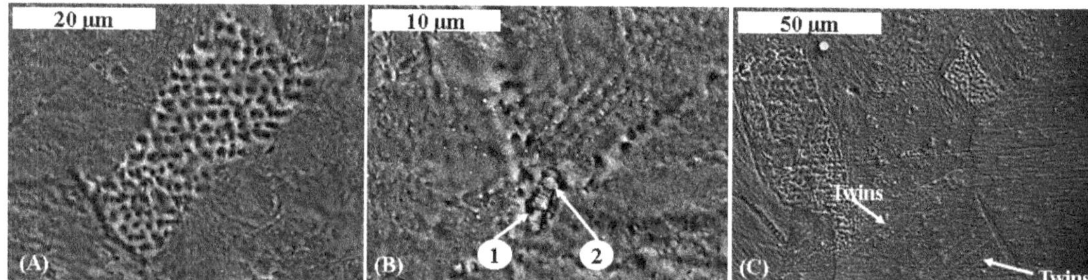

Figure 7. Nimonic 263 surface, processed by laser energy of 25 mJ at wavelength of 532 nm and 50 accumulated pulses (power density 2.4 GW/cm^2): (**A**) microstructure; (**B**) grains; and (**C**) twinned grains.

Table 5. EDS analysis of the area presented in Figure 7a and the spots shown in Figure 7b.

Spectrum	C	O	Al	Si	Ti	Cr	Mn	Fe	Co	Ni	Mo
Figure 7a	11.49	6.92	0.28	0.23	1.58	15.59	0.45	0.43	16.69	41.79	4.57
Spectrum 1	13.25	-	0.13	0.2	25.94	9.66	0.34	0.24	8.82	21.98	19.45
Spectrum 2	12.96	-	0.18	0.25	20.15	11.50	0.37	0.29	11.12	27.43	15.77

Table 5 presents EDS results of the whole area shown in Figure 7a, and the spectrums presented in Figure 7b. In Figure 7a, degasation is noticed in some of the grains, and EDS analysis shows the higher content of carbon and oxide due to mechanism of the process. The presence of the triple grain boundary is presented in Figure 7b. The EDS results presented in Table 5 show the high content of Ti and Mo and, along with the morphology, suggest that the (TiMo) carbides are formed. The size of these carbides is up to 2 µm. At the grain boundaries the carbides are in the favorable form of chain, with higher hardness [34].

Alloy materials usually have some special mechanical and physical characteristics under high strain rates; their structures may generate dislocation, grains and twins, as it was presented by Ren *et al.* [35]. In Figure 7c, the twinned grains are noticed, formed due to the mechanical treatment of a material. As was discussed by Matijasevic *et al.* [29], the twinned grains have been formed at the high stress concentration when the large number of barriers interferes with the dislocation movement. As the dislocations pile up at obstacles in local area, the lattice strain increases. Internal stresses are added on the outside ones and cause formation of the twinned grain. It is well known that all face-centered cubic (FCC) metals, under the action of impact load (high strain rate), can be deformed by twinning [29].

3.3. Comparison of Results Obtained by Laser Operating at 1064 nm and 532 nm

Figure 8a–c presents the microstructures of the base material and laser shock peened surface by 1064 nm and 532 nm light, respectively.

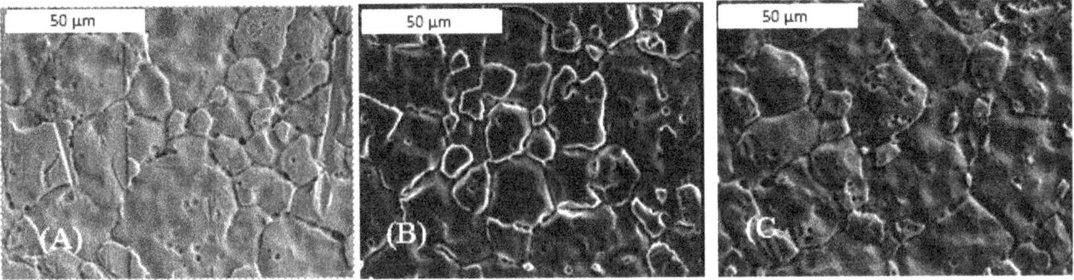

Figure 8. The microstructures of Nimonic 263, recorded by SEM: (**A**) base material; (**B**) laser treated material with 1064 nm, 50 pulses, 10 mJ (power density 3.3 GW/cm^2); and (**C**) laser treated material with 532 nm, 50 pulses, 25 mJ.

After laser shock peening by picosecond laser the microstructure is finer (the grains are smaller) and more uniform. The finest structure with the average grain size Fm = 163.24 μm^2 is obtained by the following process parameters: pulse energy of 10 mJ, 50 pulses and wavelength of 1064 nm (Figure 8b).

The above shown microstructural changes introduced by the optimized LSP may cause improvements of mechanical properties such as hardness, tensile strength, and fatigue strength, as it was previously discussed by Warren *et al.* [36] and Ren *et al.* [35].

In Table 6, the estimated values of peak pressure are presented for energy at 532 nm and at 1064 nm wavelength. The laser pulse diameter at the Nimonic surface is ~1.6 mm for 1064 nm wavelength and ~3 mm for 532 nm beam. Laser fluence is calculated by dividing laser energy (mJ) and modified surface area (cm^2). For the calculated fluences and pulse duration of 150 ps, the values presented in Table 6 are obtained using the formula: $P\,(\mathrm{GPa}) = \sqrt{I_0}(\mathrm{GW/cm^2})$ in water-confined ablation mode [16,37].

Table 6. Estimated values of peak pressure for different pulse energies at two different wavelength.

Wavelength (nm)	532			1064		
Pulse energy (mJ)	25	30	35	2	10	15
Power density(GW/cm^2)	0.6	3.3	5.0	1.9	2.4	2.8
Peak pressure (GPa)	2.2	2.4	2.7	1.3	2.9	3.6

Surface morphology/topography plays an important role in the performance of parts of various machines. Surface roughness implies that the surface is not perfectly flat, and consequently, small sized stress concentrations along the material surface occur. Under fatigue loading, cracks always nucleate from the free surface. Cracks nucleate at positions where the plastic strain concentrations are high. High surface roughness generates local stress concentration and accelerates crack initiation. For wear resistance applications, removal of the roughened surface is necessary, as previously shown by Grinspan and Gnanamoorthy [38]. This is why a significant part of this research was dedicated to the surface roughness analysis.

Two-dimensional profiles and 3D maps of the areas after LSP processing at both wavelengths and after 50 accumulated pulses are presented in Figure 9. The noncontact profilometry measurements are based on the interference between white light reflected from the sample surface and the reference surface.

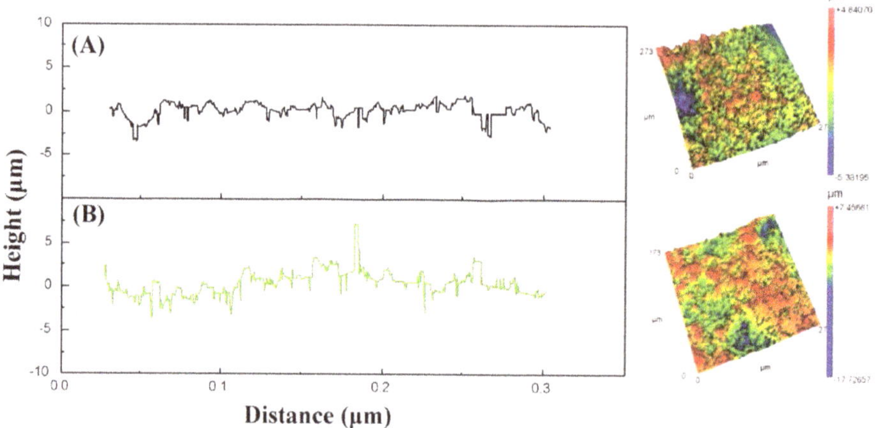

Figure 9. 3D maps and 2D profiles of the areas modified with 50 accumulated laser pulses: (**A**) at 1064 nm wavelength and output laser energy of 10 mJ (power density 3.3 GW/cm^2); and (**B**) at 532 nm wavelength and pulse laser energy of 25 mJ.

Two-dimensional surface profiles show that LSP processing of the alloy surface caused relatively homogeneous modification of the surface throughout the interaction area, as shown in Figure 9. No significant ablation or hydrodynamic effects were detected due to low value of fluence and presence of the protective black paint.

Generally, LSP modifies the surface and accordingly, surface parameters of the non-irradiated sample. For more detailed analysis, the effect of process conditions on the surface parameters is presented in Figure 10. Figure 10a,b shows the influence of pulse energy and wavelength on the average roughness (R_a) and peak to valley (PV) ratio, respectively.

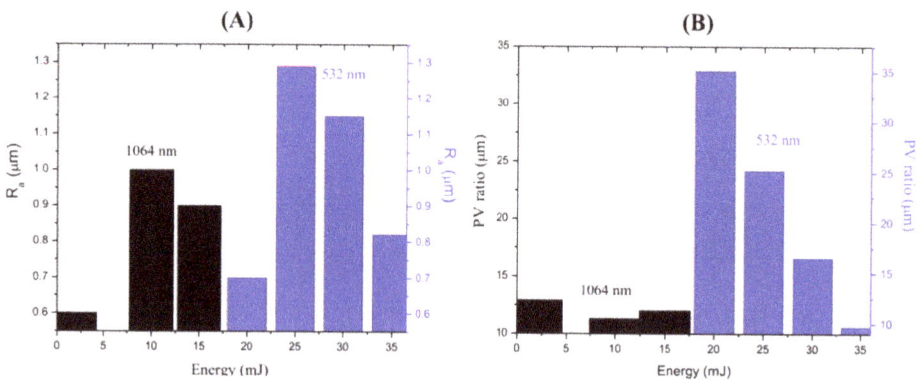

Figure 10. Diagrams of Nimonic 263 surface morphology parameters with respect to the laser energy, after 50 accumulated pulses: (**A**) average roughness (R_a) and (**B**) peak to valley (PV) ratio.

Diagram in Figure 10a shows that the main increase of the R_a value is after 25 mJ laser action at 532 nm. With additional increase of laser energy, the average surface roughness is decreasing. Analysis of the results of surface parameters with an increase in the number of pulses shows that there are no significant changes in the R_a value; therefore, they are not discussed further. Concerning PV ratio, the profilometry analysis shows slight fluctuations of PV ratio with an increase of the pulse energy at 1064 nm, as it could be seen from Figure 10b. At 532 nm laser action, the descending trend of PV ratio is noticed when higher energy is applied (Figure 10b).

Hardening effect is expected to be present after surface treatment by the laser light. The results of microhardness measurement (HV0.5) for the base material (BM) and the material treated by laser process parameters: wavelength at 1064 nm, $E1 = 2$ mJ, $E2 = 10$ mJ and $E3 = 15$ mJ, and the number of pulses 50, 100 and 200, are given in Figure 11.

As can be observed in Figure 11, at the experimental conditions of low fluence laser irradiation of the Nimonic 263 alloy, there are no significant variations in the microhardness values between different irradiated areas of the sample. But, comparing with the non-irradiated sample, there is an overall increase of microhardness due to LSP processing. There is a relationship between the hardness and the strength of austenitic materials, where normally the hardness increases while the strength increases, and the ductility decreases.

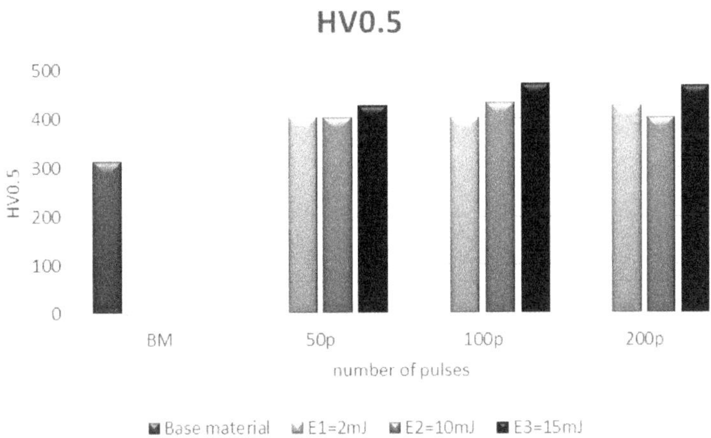

Figure 11. The microhardness measurements of LSP processed areas by laser energies: $E1 = 2$ mJ, $E2 = 10$ mJ and $E3 = 15$ mJ, at wavelength 1064 nm.

Figure 12 presents the microhardness values of the material treated by the following laser process parameters: wavelength at 532 nm; $E1 = 25$ mJ, $E2 = 30$ mJ and $E3 = 35$ mJ; and the number of pulses 50, 100 and 200. Comparing Figures 11 and 12 it can be concluded that laser beam wavelength does not significantly affect the microhardness values. Figure 13 shows the microhardness measurements of areas treated by laser energy of 25 mJ and number of pulses 50 and 100, at wavelength 532 nm. As can be seen in the graph, the microhardness value slightly decreases in depth as the influence of shock waves is fading away.

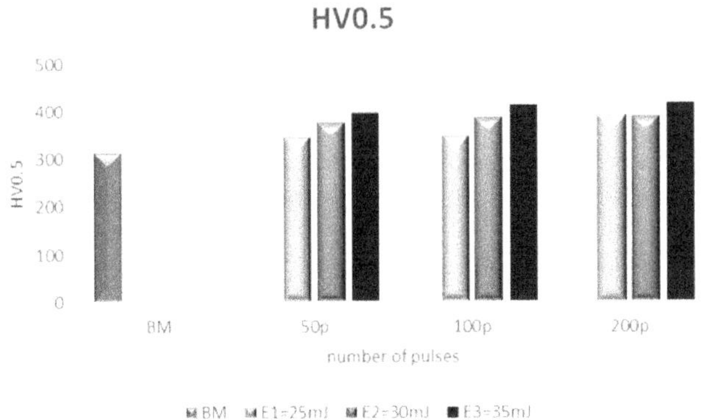

Figure 12. The microhardness measurements of LSP processed areas by laser energies: $E1 = 25$ mJ, $E2 = 30$ mJ and $E3 = 35$ mJ, at wavelength 532 nm.

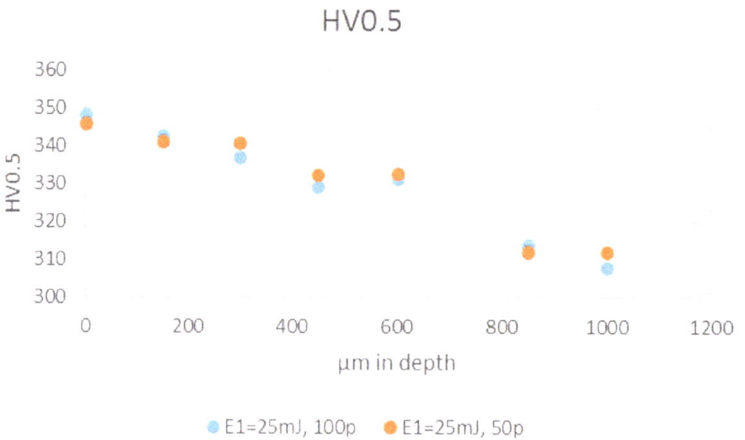

Figure 13. The microhardness measurments in depth of LSP areas treated by laser energy $E1 = 25$ mJ and number of pulses 50 and 100, at wavelength 532 nm.

4. Conclusions

In this paper, the surface microstructural changes of superalloy Nimonic 263 induced by picosecond laser mechanical treatment were discussed. Laser shock peening was performed in different regimes. Special attention was dedicated to the effects of 532 nm and 1064 nm wavelength laser treatments. The following conclusions can be drawn from this study.

After LSP by picosecond laser, the microstructure is finer (the grains are smaller) and more uniform, compared to the base material.

The laser action at two different wavelengths has different impact on material characteristics/morphology/surface. The laser at 1064 nm wavelength produces finer structure and lower roughness than the laser action at 532 nm wavelength. At both wavelengths, LSP processing induced favorable phases, but during 1064 nm treatment unwanted phases, Ti carbides, occurred that might cause initiation of crack formation.

At wavelength 532 nm, with an increase in pulse energy and number of accumulated pulses, the pores start to form and create a coral-like structure. The grain boundaries are more pronounced, they start to segregate, and carbides are formed at the grain boundaries.

At wavelength 1064 nm, the finest structure is obtained for pulse energy 10 mJ, while the further increase in energy causes the formation of pores. By increasing the pulse numbers, the grains become finer; however, at 200 pulses, melting occurs.

Overall, picosecond LSP treatment of Nimonic 263 improved microhardness of superalloy's surface. These results show that LSP by picosecond laser presents a promising technique for superalloy's surface processing, which aims to obtain beneficial mechanical characteristics of the observed material.

Further research might include residual stresses measurements and analysis, as well as the investigation of the same process using different lasers and processing different materials.

Acknowledgments: The work was supported by the Ministry of Education and Science of the Republic of Serbia, under grant numbers TR 35040 and ON 172019.

Author Contributions: S. Petronic and M. Burzic conceived and designed the experiments; S. Polic performed the experiments; T. Sibalija and K. Colic analyzed the data; D. Milovanovic contributed materials and analysis tools; and S. Petronic and T. Sibalija wrote the paper.

Conflicts of Interest: The authors declare no conflict of interest.

References

1. Pollock, M.T.; Tin, S. Nickel-Based Superalloys for Advanced Turbine Engines: Chemistry, Microstructure, and Properties. *J. Propuls. Power* **2006**, *22*, 361–374. [CrossRef]
2. Shahriari, D.; Sadeghi, M.H.; Akbarzadeh, A.; Cheraghzade, M. The influence of heat treatment and hot deformation conditions on γ' precipitate dissolution of Nimonic 115 superalloy. *J. Adv. Manuf. Technol.* **2009**, *45*, 841–850. [CrossRef]
3. Singh, P.N.; Singh, V. Influence of ageing treatment on work hardening behavior of Ni-base superalloy. *Scr. Mater.* **1996**, *34*, 1861–1865. [CrossRef]
4. Park, N.K.; Kim, I.S.; Na, Y.S.; Yeom, J.T. Hot forging of a nickel-base superalloy. *J Mater. Process. Technol.* **2001**, *111*, 98–102. [CrossRef]
5. Milovanovic, D.S.; Radak, B.B.; Gakovic, B.M.; Batani, D.; Momcilovic, M.D.; Trtica, M.S. Surface morphology modifications of titanium based implant induced by 40 picosecond laser pulses at 266 nm. *J. Alloy Compd.* **2010**, *501*, 89–92. [CrossRef]
6. Petronic, S.; Milovanovic, D.; Milosavljevic, A.; Momcilovic, M.; Petrusko, D. Influence of picosecond laser irradiation on nickel-based superalloy surface microstructure. *Phys Scr.* **2012**. [CrossRef]
7. Yilbas, B.S.; Shuja, S.Z.; Arif, A.; Gondal, M.A. Laser-shock processing of steel. *J. Mater. Proc. Technol.* **2003**, *135*, 6–17. [CrossRef]
8. Peyre, P.; Fabbro, R. Laser shock processing: A review of the physics and applications. *Opt. Quantum Electr.* **1995**, *27*, 1213–1229.
9. Devaux, D.; Fabbro, R.; Tollier, L.; Bartnicki, E. Generation of shock waves by laser-induced plasma in confined geometry. *J. Appl. Phys.* **1993**, *74*, 2268–2273. [CrossRef]
10. Peyre, P.; Fabbro, R.; Berthe, L.; Dubouchet, C. Laser shock processing of materials, physical processes involved and examples of application. *J. Laser Appl.* **1996**, *8*, 135–141. [CrossRef]
11. Zhong, H.; Yu, C. Laser shock processing of 2024-T62 aluminum alloy. *Mater. Sci. Eng. A* **1998**, *257*, 322–327.
12. Clauer, A.H.; Holbrook, J.H.; Fairand, B.P. Effects of laser induced shock waves. In *Shock Waves and High-Strain, Phenomena in Metals*; Meyers, M.A., Murr, L.E., Eds.; Plenum Press: New York, NY, USA, 1981; pp. 675–703.
13. Amarchinta, H.K.; Grandhi, R.V.; Clauer, A.H.; Langer, K.; Stargel, D.S. Simulation of residual stress induced by a laser peening process through inverse optimization of material models. *J. Mater. Process. Technol.* **2010**, *210*, 1997–2006. [CrossRef]
14. Peyre, P.; Chaieb, I.; Braham, C. FEM calculation of residual stresses induced by laser shock processing in stainless steels. *Model. Simul. Mater. Sci. Eng.* **2007**, *15*, 205–221. [CrossRef]
15. Sibalija, T.; Petronic, S.; Majstorovic, V.; Milosavljevic, A. Modelling and optimisation of laser shock peening using an integrated simulated annealing-based method. *Int. J. Adv. Manuf. Technol.* **2014**, *73*, 1141–1158. [CrossRef]
16. Ding, K.; Ye, L. *Laser Shock Peening, Performance and Process Simulation*; Woodhead Publishing Limited: Cambridge, UK, 2006; pp. 50–53.
17. Schijve, J. *Fatigue of Structures and Materials*; Kluwer Academic Publisher: Dordrecth, The Netherlands, 2001; pp. 71–78.
18. Lee, D. Feasibility Study on Laser Microwelding and Laser Shock Peening Using Femtosecond Laser Pulses. Ph.D. Thesis, University of Michigen, Ann Arbor, MI, USA, 2008.
19. Semaltianos, N.G.; Perrie, W.; French, P.; Sharp, M.; Dearden, G.; Logothetidis, S.; Watkins, K.G. Femtosecond laser ablation characteristics of nickel-based superalloy C263. *Appl. Phys. A* **2009**, *94*, 999–1009. [CrossRef]
20. Gomez-Rosas, G.; Rubio-Gonzalez, C.; Ocaña, J.L.; Molpeceres, C.; Porro, J.A.; Morales, M.; Casillas, F.J. Laser Shock Processing of 6061-T6 Al alloy with 1064 nm and 532 nm wavelengths. *Appl. Surf. Sci.* **2010**, *256*, 5828–5831. [CrossRef]
21. Schumann, H. *Metallographie*; Deutscher Verlag fuer Grundstoffindustrie: Leipzig, Germany, 1974; p. 15.
22. Petronic, S.; Burzic, M.; Milovanovic, D.; Colic, K.; Radovanovic, Z. Picosecond laser shock peening of base material Nimonic 263 material and laser welded Nimonic 263 alloy. *Weld. Weld. Struct.* **2015**, *60*, 149–155.
23. Petronic, S.; Kovacevic, A.G.; Milosavljevic, A.; Sedmak, A. Microstructural changes of Nimonic 263 superalloy caused by laser beam action. *Phys. Scr.* **2012**. [CrossRef]
24. Bauerle, D. *Laser Processing and Chemistry*; Springer Verlag: Berlin, Germany, 2003; pp. 13–256.

25. Peyre, P.; Berthe, L.; Scherpereel, X.; Fabbro, R.; Bartniki, E. Experimental Study of Laser-Driven Shock Waves in Stainless Steels. *J. Appl. Phys.* **1998**, *84*, 5985–5992. [CrossRef]
26. Niehofff, H.S.; Vollertsen, F. Laser induced shock waves in deformation processing. *J. Metall.* **2005**, *11*, 183–194.
27. Berthe, L.; Fabbro, R.; Peyre, P.; Tollier, L.; Bartnicki, E. Shock waves from a water-confined laser-generated plasma. *J. Appl. Phys.* **1997**, *82*, 2826–2832. [CrossRef]
28. Drobnjak. *Physical metallurgy*; Faculty of Technology and Metallurgy, University of Belgrade: Belgrade, Serbia, 1981; pp. 75–77.
29. Matijasevic, B.; Kinder, J.; Radovic, N.A.; Volkov-Husovic, T. Shot-peening induced twinning in ship-building steel. *Metalurgija* **2002**, *8*, 149–155.
30. Zhang, Y.K.; Lu, J.Z.; Ren, X.D.; Yao, H.B.; Yao, H.X. Effect of laser shock processing on the mechanical properties and fatigue lives of the turbojet engine blades manufactured by LY2 aluminum alloy. *Mater. Des.* **2009**, *30*, 1697–1703. [CrossRef]
31. Schneider, M.S.; Kad, B.; Kalantar, D.H.; Remington, B.A.; Kenik, E.; Jarmakani, H.; Meyers, M.A. Laser shock compression of copper and copper-aluminum alloys. *Int. J. Impact Eng.* **2005**, *32*, 473–507. [CrossRef]
32. Reed, R. *The Superalloys—Fundamentals and Applications*; Cambridge University Press: New York, NY, USA, 2006.
33. Chena, C.Y.; Yena, H.W.; Kaoa, F.H.; Li, W.C.; Huang, C.Y.; Yanga, J.R.; Wang, S.H. Precipitation hardening of high-strength low-alloy steels by nanometer-sized carbides. *Mater. Sci. Eng. A* **2009**, *499*, 162–166. [CrossRef]
34. Jang, J.H.; Lee, C.H.; Heo, Y.U.; Suh, D.H. Stability of (Ti, *M*)C (*M* = Nb, V, Mo and W) Carbide in Steels using First-Principles Calculations. *Acta Mater.* **2012**, *60*, 208–217. [CrossRef]
35. Ren, X.D.; Zhang, Y.K.; Zhang, T.; Jiang, D.W.; Yongzhuo, H.F.; Jiang, Y.F.; Chen, K.M. Comparison of the simulation and experimental fatigue crack behaviours in the nanoseconds laser shocked aluminum alloy. *Mater. Des.* **2011**, *32*, 1138–1143. [CrossRef]
36. Warren, A.W.; Guo, Y.B.; Chen, S.C. Massive parallel laser shock peening: Simulation, analysis, and validation. *Int. J. Fatigue* **2008**, *30*, 188–197. [CrossRef]
37. Fabbro, R.; Fournier, J.; Ballard, P.; Devaux, D.; Virmont, J. Physical study of laser-produced plasma in confined geometry. *J. Appl. Phys.* **1990**. [CrossRef]
38. Grinspan, A.S.; Gnanamoorthy, R. Surface modification by oil jet peening in Al alloys, AA6063-T6 and AA6061-T4 Part 2: Surface morphology, erosion, and mass loss. *Appl. Surf. Sci.* **2006**, *253*, 997–1005. [CrossRef]

16

Probing Interfaces in Metals Using Neutron Reflectometry

Michael J. Demkowicz [1,2,*] and Jaroslaw Majewski [3,4,*]

Academic Editor: Klaus-Dieter Liss

1 Department of Materials Science and Engineering, Massachusetts Institute of Technology, Cambridge, MA 02139, USA
2 Materials Science and Engineering, Texas A & M University, College Station, TX 77843, USA
3 MPA-CINT/Los Alamos Neutron Scattering Center, Los Alamos National Laboratory, Los Alamos, NM 87545, USA
4 Department of Chemical Engineering, University of California at Davis, Davis, CA 95616, USA
* Correspondence: demkowicz@tamu.edu (M.J.D.); jarek@lanl.gov (J.M.)

Abstract: Solid-state interfaces play a major role in a variety of material properties. They are especially important in determining the behavior of nano-structured materials, such as metallic multilayers. However, interface structure and properties remain poorly understood, in part because the experimental toolbox for characterizing them is limited. Neutron reflectometry (NR) offers unique opportunities for studying interfaces in metals due to the high penetration depth of neutrons and the non-monotonic dependence of their scattering cross-sections on atomic numbers. We review the basic physics of NR and outline the advantages that this method offers for investigating interface behavior in metals, especially under extreme environments. We then present several example NR studies to illustrate these advantages and discuss avenues for expanding the use of NR within the metals community.

Keywords: neutron reactor; spallation source; metals; extreme conditions

---

1. Interfaces in Metals

Metals form a wide variety of interfaces, including grain and phase boundaries [1], surface-liquid interfaces [2,3], solidification fronts [4], and mechanical contacts [5]. Although they typically occupy a small fraction of the total volume, interfaces play an outsized role in determining the properties of metals [6–9]. Understanding interfaces is therefore critical to predicting and controlling the behavior of metals.

Experimental investigation of interfaces presents significant challenges. Because they are often buried within the material, accessing them frequently requires destructive characterization or sample preparation methods, such as transmission electron microscopy (TEM) [10] or atom probe tomography (APT) [11]. Interfaces in metals typically have low thickness; indeed, some are atomically sharp [12]. Thus, characterizing them requires high—sometimes Å-level—spatial resolution. Moreover, certain interfaces only exist at high temperatures and pressures [13–15] or under contact with external media, such as gases or liquids [2,16]. Investigating such interfaces requires special *in situ* characterization methods.

An expanded experimental toolbox promises to accelerate progress in understanding metal interfaces, especially in extreme environments. This paper offers a primer on neutron reflectometry (NR): a characterization method with several advantages for studying metal interfaces [17–21]. NR is a mature experimental tool. The first NR experiments were conducted by Fermi and Zinn [22] and

Fermi and Marshall [23]. The technique has experienced continuous improvement since then [17,24–26]. Nevertheless, use of this method within the metals community has been relatively limited. We illustrate the potential benefits of NR to investigations of interfaces in metals by explaining the physics of the NR experiment and by presenting several example studies.

2. The Physics of Neutron Reflectometry (NR)

Figure 1 shows a schematic of a typical NR measurement. The sample is a thin, planar film on a substrate. The experiment is usually conducted in air or vacuum, but may also be carried it out in other media (e.g., see Section 6.4) [27]. The neutron source may be a fission nuclear reactor or a spallation source. In the nuclear reactor, the sustained nuclear fission of ^{235}U- or ^{239}Pu-rich fuels immersed in H_2O, D_2O, or solid graphite produces neutrons that may be used for scattering experiments. Spallation sources usually utilize pulsed high-energy (~GeV) protons to bombard targets made of heavy elements (such as W, Hg, U) to extract neutrons [28,29].

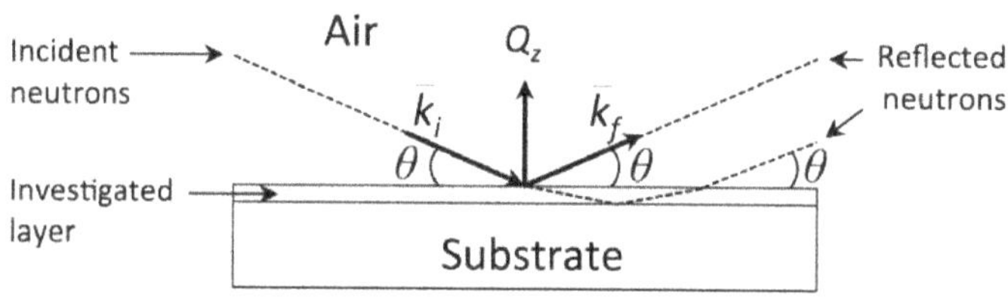

Figure 1. Schematic of a neutron reflectometry measurement.

In material property studies, the neutrons extracted from nuclear reactors or spallation sources are moderated to decrease their energies and therefore increase their wavelengths to Å ranges. Such moderation, depending on the final wavelength of neutrons required, is usually achieved by passing high-energy neutrons through H_2O, liquid H_2, or solid methane. Low energy neutrons are typically detected indirectly through absorption reactions with materials of high cross-sections for such reactions. Typically, ^{3}He, ^{6}Li, or ^{10}B is used to emit high-energy particles, whose ionization signatures may be detected by a number of means.

Upon interacting with the sample at a particular angle of incidence, θ (or a particular value of the neutron momentum transfer vector, Q_z), the incoming neutron beam can undergo absorption, reflection, transmission, or refraction. Consequently, there is a difference between the intensity of the outgoing, specularly reflected neutron beam and that of the incident beam. This difference—measured as a function of Q_z—encodes information about the distribution of the nuclear scattering length density along the direction normal to the sample surface. Moreover, the neutron is a ½-spin fermion and possesses a magnetic moment oppositely oriented to the spin. Therefore, its interaction with matter may depend on the sample's spin or magnetic field. Neutrons interact both with nuclear spins and the magnetic moments of unpaired electrons via dipole-dipole processes. Interactions with unpaired electrons may be of similar magnitude as nuclear scattering. However, they are not inherently isotropic. Rather, they depend on the orientation of the sample's magnetization *vis-a-vis* the direction of the neutron momentum wavevector transfer Q_z: only the component of sample's magnetization which is perpendicular to Q_z affects the neutron scattering. Therefore, the intensity of specularly reflected neutrons measured as a function of Q_z also encodes information about the distribution of the magnetization in the sample as a function of depth [30].

Depending on the specific NR technique, NR can take advantage of a range of different neutron-sample interactions [31]. However, this short review focuses on elastic specular NR, which is by far the most widely used NR technique. Elastic scattering conserves energy. Thus, we exclude any energy-dissipating neutron-matter interactions, except neutron absorption. In specular NR, the detector is positioned so as to measure outgoing neutrons at the same angle of incidence as the incoming neutrons, as illustrated in Figure 1. The experiment measures reflectivity, R, defined as the ratio of the number of reflected neutrons to the number of incoming neutrons.

The de Broglie expression, $\lambda = \frac{h}{m_n v}$, relates the neutron's wavelength λ with its momentum, $p = m_n v$ (where h is Planck's constant and $m_n = 1.6749 \times 10^{-27}$ kg is the neutron mass). Based on this formula, some simple relationships between the wavelength λ (Å), energy E (meV), and speed v (m/s) of the neutron can be developed: $E = 81.89/\lambda^2$ and $v = 3960/\lambda$. Thus, for example, a neutron with de Broglie wavelength of 1.5 Å has an energy $E = 36.4$ meV and velocity $v = 2640$ m/s. NR utilizes neutrons with wavelengths from sub-Å to tens of Å. By contrast, the size of an atomic nucleus is on the order of ~10 fm (1 fm = 10^{-15} m = 10^{-5} Å). Thus, to an excellent approximation, the incoming neutrons may be thought of as waves interacting with a uniform medium whose properties are determined by the density and type of atoms it contains. Their behavior may be described using Schrödinger's equation.

The assumption of elastic and specular conditions greatly simplifies the analysis the NR measurement. We further assume that the scattering properties of the sample vary only in one direction—namely, along the sample's normal—and therefore the components of the neutron wavevector parallel to the sample surface are not affected. Under these conditions, the component of the neutron wavevector parallel to the sample surface is conserved and the magnitude of the outgoing wavevector, $\vec{k}_f$, equals that of the incoming wavevector, $\vec{k}_i$: $\left|\vec{k}_i\right| = \left|\vec{k}_f\right| = \frac{2\pi}{\lambda}$. The difference between them, $\vec{Q}_z = \vec{k}_f - \vec{k}_i$, is known as the "momentum wavevector transfer" and lies perpendicular to the sample surface. From the geometry of the measurement (Figure 1), we calculate $\left|\vec{Q}_z\right| = Q_z = \frac{4\pi \sin(\theta)}{\lambda}$.

Quantum mechanics describes the incoming and outgoing neutron beams as a wavefunction, ψ, consisting of a superposition of plane waves:

$$\psi\left(\vec{x}\right) = e^{i\vec{k}_i\cdot\vec{x}} + re^{i\vec{k}_f\cdot\vec{x}} \tag{1}$$

Here, r is the amplitude of the outgoing (reflected) wave, normalized by the amplitude of the incoming wave (taken as unity). Knowing r, we may calculate reflectivity as $R = |r|^2$. To compute r, however, we must model the interaction of the incoming neutron beam with the sample and substrate.

Because the component of the wavevector parallel to the sample surface is conserved, we may rewrite ψ as solely a function of z—the distance perpendicular to the sample surface—and $k_i^{\perp}$—the component of $\vec{k}_i$ in the z-direction:

$$\psi(z) = e^{ik_i^{\perp}z} + re^{-ik_i^{\perp}z} \tag{2}$$

where we have used $k_f^{\perp} = -k_i^{\perp}$. Indeed, the entire NR measurement may be analyzed as a one-dimensional problem in the z-direction [32]. We write down wavefunctions of the type shown in Equation (2) for every distinct layer of material in the experiment, including air (or any other external medium), every layer of material in the sample, and the substrate (though in the substrate there is no reflected wave).

In free space, the neutron has kinetic energy $E_k = \frac{1}{2}m_n v^2$ and zero potential energy (if gravity is neglected). By contrast, within a material, it has a potential energy given by the Fermi pseudopotential [33]:

$$V_{\text{Fermi}} = \frac{h^2\beta}{2\pi m_n} \tag{3}$$

where β is the nuclear scattering length density (SLD). β describes the mean neutron scattering effectiveness of the material, which depends on the number of nuclei of type i per unit volume, N_i, and the coherent neutron scattering length, b_i:

$$\beta = \sum_i N_i b_i \tag{4}$$

b_i may be complex with its imaginary component describing absorption. Its real part may be either positive or negative, depending on the isotope [34,35]. Since neutrons have spin, their interactions with magnetic materials require an extended description that tracks changes in spin polarization [30]. The interaction of a neutron's spin with atomic magnetism (or other source of magnetic induction) can lead to magnetic scattering length density distributions that may be of the same order of magnitude as the nuclear scattering length densities. However, depending on the neutron's spin orientation (spin-up or spin-down) *vis-à-vis* the magnetic field of reference, the magnetic component is either added or subtracted from the nuclear one.

Often, samples investigated by NR may be described as a stack of discrete layers: each with its own composition, density, and thickness. In such cases, the NR experiment may be described with a 1D, time-independent Schrödinger equation on a piece-wise linear potential [19,26,32]. By matching wave functions and their derivatives at the interfaces between successive layers (as well as in the surrounding medium and in the substrate), one may solve for the amplitudes of all the waves in the setup. In particular, r—the amplitude of the outgoing wave measured at the sample surface—may be found. This amplitude comes about by coherent interference of partial waves on all interfaces in the film. From it, we determine the quantity measured by the neutron detector in Figure 1: reflectivity, R.

Thus, the SLD and thickness of the sample and substrate determine R. Since SLD in turn depends on composition and density, R is an indirect measure of these characteristics as well as the thickness of the individual material layers in the sample and substrate [17,26]. The goal of the reflectivity experiment is to measure $R(Q_z)$ and then infer $\beta(z)$ by fitting a model of the SLD distribution to the data. Q_z may be varied by changing the angle of incidence, θ (if the neutron beam is monochromatic, *i.e.*, λ = const.), or by changing the neutron wavelength, λ. The latter method of varying Q_z is typical of NR measurements at facilities where different neutron wavelengths, λ, are distinguished by the time of flight method. Figure 2 shows calculated reflectivity curves corresponding to a 500 Å Ni film on a quartz substrate in air. To illustrate the sensitivity of NR to isotopic composition, the calculation is carried out for two different isotopes: ^{58}Ni and ^{62}Ni.

Figure 2 illustrates some of the common features of reflectivity curves. Whenever the energy of the neutron is at or below the potential of the substrate (*i.e.*, whenever $k_i^2 \leqslant 4\pi\beta_{\text{substrate}}$), the neutrons are totally reflected from the surface. The onset of total reflection is called the critical edge and the value of Q_z at that point is referred to as Q_{critical}. The fringes in Figure 2 arise from interference between waves reflected from the top surface and the buried interface between the substrate and the layer. For this simple case, the spacing of the fringes may be calculated analytically: $d_{\text{fringe}} = \dfrac{2\pi}{t_{\text{layer}}}$, where t_{layer} is the thickness of the Ni layer. The amplitude of the fringes relates to the contrast between the layer and the substrate. The overall falloff of the curve obeys the Fresnel law: $R \sim Q_z^{-4}$. Most interfaces are not discontinuous, but rather graded due to chemical mixing or surface roughness. The surface roughness (which can be characterized by the root mean square displacement from the average interface, σ) may also be obtained from the reflectivity curve [36,37]. In general, the falloff of $R(Q_z)$ for rough or diffuse interfaces is even faster than that given by Fresnel's law.

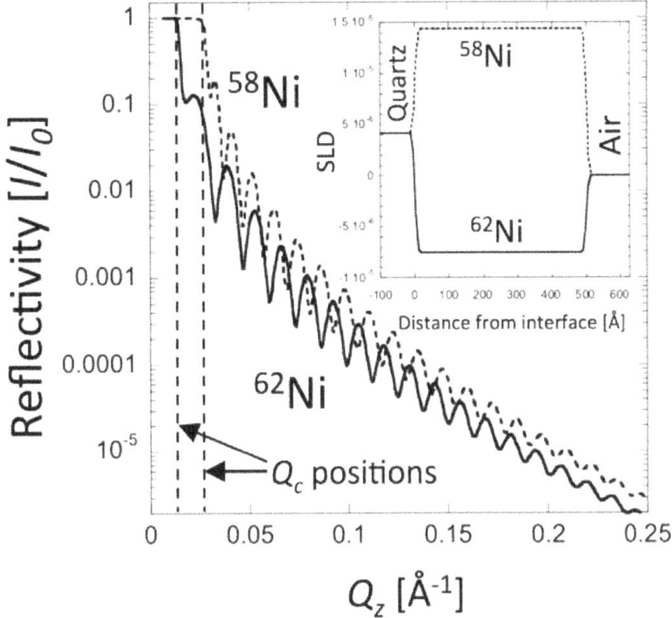

Figure 2. Calculated neutron reflectivity curves for 500 Å films of ^{58}Ni (dashed line) and ^{62}Ni (solid line) on a quartz substrate. The inset shows the SLD distributions for each isotope. The simulation was performed using MOTOFIT assuming RMS roughness parameters of 5 Å and the experimental resolution $\Delta Q_z/Q_z = 3\%$.

3. Interpreting Reflectivity Curves

$R(Q_z)$ contains information about the SLD distribution perpendicular to the sample surface, $\beta(z)$. Inferring $\beta(z)$ from $R(Q_z)$, however, is not trivial [38]. Because the NR measurement only collects the intensity of the reflected beam and not its phase, there is no unique mathematical transformation from $R(Q_z)$ to $\beta(z)$. Therefore, NR data is usually interpreted by iteratively adjusting a trial SLD distribution, $\overline{\beta}(z)$, until the reflectivity it predicts, $\overline{R}(Q_z)$, matches the measured reflectivity, $R(Q_z)$, to within a specified tolerance.

The continuous function $\beta(z)$ may often be approximated by a series of discrete layers—referred to as "boxes" or "slabs"—each with a constant SLD. Inter-layer roughness may be taken into account using an error function centered at each interface [36] or any other relevant functional form. A theoretical NR curve, $\overline{R}(Q_z)$, may be calculated from a trial SLD distribution, $\overline{\beta}(z)$, using the Parratt recursion formula [39,40], which relates the amplitudes of the reflected and transmitted waves at each interface. A number of approaches have been developed for adjusting $\overline{\beta}(z)$ to minimize the difference between $\overline{R}(Q_z)$ and $R(Q_z)$. One example is the Levenburg-Marquardt nonlinear least-squares method used in open-source reflectivity package, MOTOFIT, which runs in the IGOR Pro environment [41]. This method seeks the least-squares fit of reflectivities, corresponding to a minimum χ^2 value. SLD models with the least number of boxes are usually preferred as they involve the smallest number of fitting parameters.

Once a best-fit set of model parameters is achieved, the uncertainties of these parameters may also be quantified by measuring the increase in χ^2 that comes about from perturbing each individual fitting parameter. For example, Reference [42] defines $\tilde{\chi}^2$ as the deviation of the reflectivity calculated using the perturbed parameter values from the best-fit reflectivity:

$$\tilde{\chi}^2 = \sum_{i=1}^{N} \left(\frac{y_i^{bf} - y_i^{p}}{y_i^{bf}} \right) \tag{5}$$

Here, y_i^{bf} is the best-fit to the measured reflectivity, y_i^p is the reflectivity value obtained by perturbing one parameter of the structural model, and N is the number of data points. The uncertainties on the fitting parameters are then defined as bounds within which $\tilde{\chi}^2$ is 5% or less.

Equation (4) shows that, at any given z, $\beta(z)$ depends on the number and types of isotopes in the sample as well as their volume density. Thus, $\beta(z)$ provides information concerning the composition and density at a depth z. Inferring these quantities from $\beta(z)$ typically requires prior knowledge of some of the variables (e.g., nominal compositions or densities) or further input from other characterization methods. It should be re-emphasized that the SLD density profiles obtained from the fitting procedures described above are not unique. Due to the fact that only the intensities of scattered neutrons are measured in the NR experiment, but not their amplitudes and phases, there is no unique mathematical transformation leading from $R(Q_z)$ to SLD profile. Therefore, to resolve this problem, other data (e.g., from complementary characterization techniques) are often needed. In some cases, the phase of scattered neutrons may be resolved, as described by the work of Majkrzak and Berk [43–45] and others [46–49], enabling better inferences of SLD profiles.

4. Advantages of NR

Neutron reflectometry offers unique advantages for characterizing solid-state interfaces in metals, including in extreme environments. Some of these advantages are easily deduced by considering the dependence of coherent scattering length, b_i, on atomic number, Z. Figure 3 shows that b_i is rather weakly dependent on Z. Indeed, the scattering lengths of almost all elements (in their natural isotopic abundance) are of the same order of magnitude. Therefore, in general, no one element can dominate the scattering of a multi-component sample, drowning out the contributions of other elements. In particular, light elements—such as H/D or He—may be detected, even when embedded in a matrix of heavy elements, e.g., of actinides. Moreover, NR is often able to distinguish elements with small differences in atomic number.

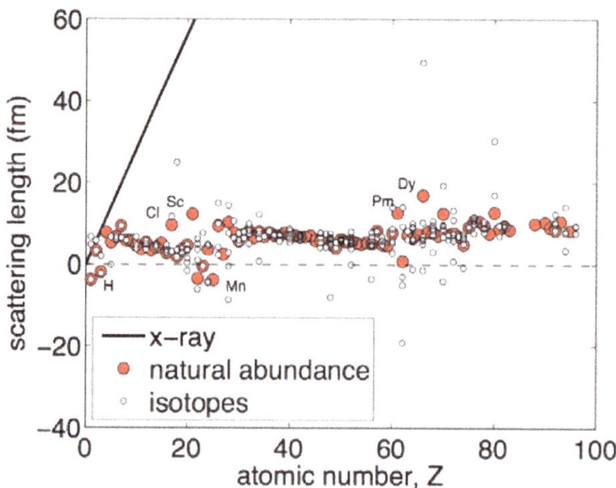

Figure 3. Neutron coherent scattering length, b_i, as a function of atomic number, Z. The X-ray scattering length is computed using Equation (6).

Figure 3 also shows that there are marked differences in b_i between different isotopes of certain elements. Thus, NR is well suited to investigations that require tracking of isotopes, such as tracer diffusion studies. Isotopic substitution may also aid in the interpretation of $\beta(z)$ profiles. For example, Ni has five different isotopes. The b_i for ^{58}Ni (abundance 68.3%) is 14.4×10^{-5} Å, whereas for ^{62}Ni (abundance 3.6%) b_i is negative and equals -8.7×10^{-5} Å. Substitution of ^{58}Ni (bulk SLD = 13.3×10^{-6} Å^{-2}) for ^{62}Ni (SLD = -7.5×10^{-6} Å^{-2}) significantly changes NR curves, as illustrated in Figure 2. Moreover, because neutron beams may be spin polarized, NR is especially well suited to investigations of magnetic properties of materials [50,51].

To appreciate the above-mentioned qualities, it is useful to compare NR with X-ray reflectometry (XRR) [52–54]. Although the mathematical description of XRR is similar to NR, the underlying physics of scattering is different: X-rays scatter from atoms' electrons while neutrons scatter off of atoms' nuclei. To interpret XRR experiments, we can replace the neutron coherent scattering lengths, b_i, in Equation (4) with the X-ray scattering lengths:

$$b_Z^{\text{X-ray}} = Z \cdot (2.81 fm) \tag{6}$$

where Z is the atomic number and 2.81 *fm* is the classical radius of the electron. Because X-rays scatter primarily from electrons, $b_Z^{\text{X-ray}}$ is directly proportional to the number of electrons per atom, *i.e.*, to the atomic number, Z, in charge-neutral materials. Consequently, XRR cannot detect differences between isotopes. Moreover, because it is a linearly increasing function of Z, $b_Z^{\text{X-ray}}$ provides very little contrast between elements with small Z differences. Finally, light elements—especially when embedded in high-Z matrices—are essentially undetectable with XRR, as the X-ray scattering length of the latter dominates reflectivity curves. However, it is often advantageous to use XRR and NR in tandem, as they may provide complementary information.

Equation (4) shows that scattering length density, β, is not only a function of b_i, but also of N_i: the number density of isotope *i*. Thus, depth profiles obtained by NR (and XRR) are sensitive not only to composition, but also to density. By contrast, depth profiles obtained through Rutherford backscattering are not sensitive to density [55]. NR is therefore capable of characterizing the evolution of porosity and of detecting displacive phase transformations that involve changes in density.

NR is also remarkable for its depth resolution, which is much greater than for XRR, especially for high-Z materials. Usually, NR techniques enable investigations of structures with total thickness up to ~3000 Å. For such thickness, the spacing between the scattering fringes (Figure 2) is very small, requiring very high $\Delta Q_z/Q_z$ resolution in the neutron detection. Typical $\Delta Q_z/Q_z$ values for existing neutron reflectometers are in the range from 2% to 5%. An instrumental resolution of $\Delta Q_z/Q_z = 2\%$ will result in the ability to distinguish between two thickness values which differ by 2%. Therefore differences of the film thickness on the order of Å can be readily detected. Smaller $\Delta Q_z/Q_z$ values may be achieved using detectors with higher spatial resolutions, better beam collimation, or better discrimination of the neutron wavelengths. However, increasing the resolution may result in smaller incident beam intensities, which can lead to longer measurements times and therefore higher scattering background. Thus, a proper balance between the two must be found. In general, it is advisable to adjust the resolution to match the expected thickness of the investigated films. Thick films, which give rise to dense oscillations of the interference fringes in $R(Q_z)$, require higher resolutions. By contrast, thin ones with broad interference oscillations in $R(Q_z)$ can be measured with lower resolution and therefore higher intensities, which can result in shorter measurement times and higher Q_z^{max} values.

Another advantage of NR is due to the ability of neutrons to penetrate deeply into solid matter. Several *mm* thick aluminum, quartz, silicon, or stainless steel windows absorb only a small fraction of incident neutrons with wavelengths in the Å range. Thus, NR measurements may be carried out to investigate the structure of buried interfaces as well as samples immersed in liquids or shielded from their environments by neutron-transparent containers. This quality is especially useful for investigating materials exposed to volatile media or under high pressure. Finally, NR frequently requires straightforward sample preparation and is not destructive. Therefore, samples investigated with NR may be subsequently further analyzed using other characterization methods. However, certain materials may be activated through interactions with the neutron beam, requiring some time for the radioactivity to decay before further characterization may be performed. It is also important to note that, since NR data are normalized to the incident neutron intensity, the measured SLD values are absolute.

5. Practical Considerations

Because neutron sources are inherently week (fluxes of ~10^{6-7} n/s/cm^2), the samples used for NR must be large. Samples with an area as small as 1 cm^2 may be measured, but at the expense of longer time of data acquisition and increased background noise. Neutron spallation sources usually provide some advantages enabling faster NR measurements. This is due to the polychromatic nature of the pulsed neutron beams they generate and the time of flight method used to discriminate between different neutron wavelengths, λ. NR spectra within a limited Q_z^{max} range (<0.1 Å^{-1}) can be obtained in 5–10 min. However, to obtain a full-spectrum NR data set (usually with $Q_z^{max} \approx 0.3$ Å^{-1} and $R \approx 10^{-6}$) requires from one to several hours of measurement time, regardless which neutron source is used.

Existing NR beamlines usually provide point-, line-, or 2-D neutron scattering detectors. Line- and 2-D detectors enable recording of scattering signals beyond the specular reflection: the so-called "off-specular" reflections. The "off-specular" data provides the neutron intensity distribution as function of the components of the neutron momentum transfer vector parallel to the sample's surface. This information can provide additional insight to extend the interpretation of the specular reflectivity measurements regarding in-plane correlations of the samples studied [56]. For example, these data allow correlations between the roughness of different interfaces or the growth of in-plane islands to be addressed. Reflectivities at high Q_z values are of great interest as they allow access to shorter length scales, which are important for characterizing the detailed structure of the investigated films. However, as already mentioned, the reflectivity R rapidly decays as Q_z^{-4}, making it difficult to acquire data at high Q_z. This challenge may be mitigated to some extent through the preparation of high quality samples: by minimizing the roughness of the sample surface as well as the roughness of its internal interfaces, high quality NR data at high Q_z and low R values may be collected. RMS roughness parameters up to 20 Å are usually tolerable, but detailed NR investigations typically require RMS roughness below 5 Å. Samples with such low roughness are most conveniently prepared using vapor deposition techniques. For such samples, $R \approx 10^{-6}$ and $Q_z^{max} \approx 0.2$–0.3 Å^{-1} can be routinely achieved for sample areas of several cm^2.

At the time of writing, there are several world-class NR instruments available worldwide, e.g., at the Spallation Neutron Source at Oak Ridge National Laboratory, the Lujan Center at Los Alamos National Laboratory, NCNR at NIST, the Institute Laue-Langevin in France, J-PARC in Japan, ANSTO in Australia, FRM-II in Germany, and several others. Several neutron sources are currently under construction or discussion. For example, the European Spallation Source in Sweden and the Second Target Station at SNS/ORNL will provide excellent capabilities for NR.

6. Example Applications of NR to Metals

This section provides examples of NR measurements conducted on metals. The examples are chosen to illustrate the unique advantages of the NR, namely its ability to detect density changes (Section 6.1), its sensitivity to magnetic moments and complementarity to X-ray reflectometry (XRR, Section 6.2), its sensitivity to light elements (Section 6.3), and its ability to penetrate through container walls (Section 6.4).

6.1. He in fcc/bcc Composites: Detecting Density Changes

Some nuclear transmutation reactions give rise to alpha particles, *i.e.*, nuclei of ^{4}He. When implanted into solids, these particles rapidly come to rest, pick up two electrons, and become regular He atoms. Since He is a noble gas and does not bond with surrounding atoms, it usually has negligible solubility within solids [57]. Thus, it precipitates out of solution into nanometer-scale bubbles [58]. These precipitates are usually deleterious to the properties of the solid, e.g., they lead to embrittlement in Ni-base alloys [59] and surface damage in plasma-facing materials [60]. Much effort has been invested into mitigating damage induced by implanted He, especially in materials for nuclear energy [61–64].

One way of controlling implanted He is to trap it at specially designed internal interfaces in composite materials [61]. However, investigations of this effect are limited by the difficulty of characterizing He precipitates at internal interfaces. NR (and XRR) provides a distinct advantage within this context: its sensitivity to local density changes enables detection of the onset of He precipitate formation [42,65,66].

Kashinath *et al.* investigated He precipitation at interfaces between copper (Cu) and one of three body-centered cubic (bcc) metals: niobium (Nb), vanadium (V), and molybdenum (Mo) [42]. They found that each of these interfaces has a distinct critical He dose at which precipitates begin to form. Figure 4 illustrates the findings of this study. Upper and lower bounds on best-fit SLD profiles were estimated by superimposing the upper and lower error bounds for each individual fitting parameter, as defined in Section 3. All SLD profiles with $\tilde{\chi}^2$ less than or equal to 5% are contained within these bounds, but the converse is not true: not all SLD profiles within these bounds have $\tilde{\chi}^2$ less than or equal to 5%. Therefore, these uncertainty estimates for best-fit SLD profiles are conservative.

The target is a Cu/Nb bilayer deposited on a Si substrate. Both the Cu and the Nb layer are approximately 20 nm thick. After implantation of 20 keV $^4He^+$ ions to a dose of $3 \times 10^{16}/cm^2$, the reflectivity of the sample is consistent with an unaltered Cu/Nb bilayer structure, as shown in Figure 4a. However, upon implantation to a slightly higher He dose of $4 \times 10^{16}/cm^2$, there is a clear change in the reflectivity, indicated by arrows in Figure 4a,b. This change may be explained by the formation of a layer of reduced density on the Cu side of the Cu–Nb interface, as illustrated in Figure 4b.

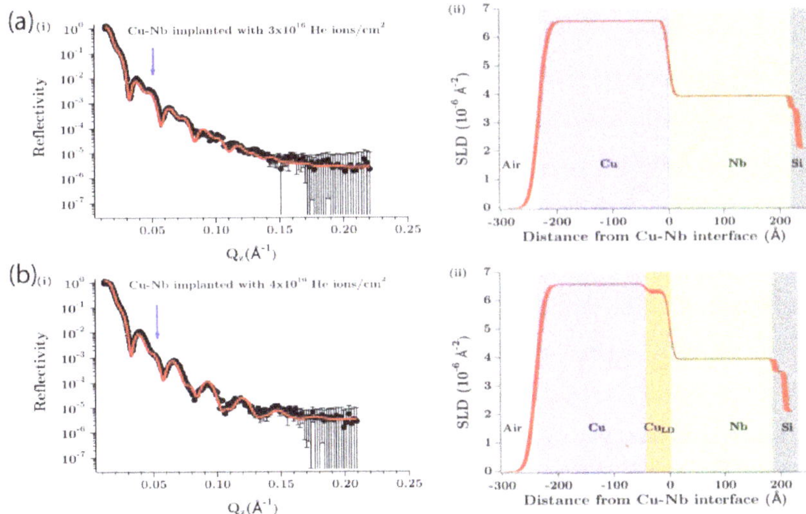

Figure 4. Reflectivity curves (**left** column) and SLD profiles (**right** column) for (**a**) $3 \times 10^{16}/cm^2$ and (**b**) $3 \times 10^{16}/cm^2$ He ions implanted in a Cu/Nb bilayer on a Si substrate. Reprinted with permission from Reference [42]. Copyright (2013), AIP publishing LLC.

At the He doses used in this study, nearly all the implanted He is believed to either escape through the Cu free surface or become trapped at the Cu–Nb or Nb–Si interfaces [42]. Precipitation within the Cu or Nb layers themselves is thought to be minimal. Thus, the low-density layer adjacent to the Cu–Nb interface in Figure 4b is thought to arise from the formation of He precipitates there. The critical He dose of $4 \times 10^{16}/cm^2$ is consistent with preceding transmission electron microscopy (TEM) studies [61,67] as well as atomistic simulations [68]. However, whereas those previous investigations merely inferred interfacial precipitation, NR is able to observe it directly.

6.2. Fe/Y_2O_3 Interface: Sensitivity to Magnetization and Complementary to XRR

The structure of interfaces between low solubility metals—such as those discussed in the previous section—is easy to describe, as these interfaces are usually atomically sharp [12]. By contrast, the structure of oxide/oxide or metal/oxide interfaces is much more difficult to assess. Such interfaces are often several nanometers wide [69], exhibit transitions in structure reminiscent of phase changes [8], and contain intrinsic defects with distinct local compositions [70]. NR provides several advantages for investigating such interfaces, including high depth resolution and sensitivity to composition.

Watkins *et al.* used NR to study the structure of an interface between α-Fe and Y_2O_3 [71]. They found that this interface is a ~64 Å-thick transitional zone containing mixtures or compounds of Fe, Y, and O. By comparing their NR data to XRR and X-ray diffraction (XRD) measurements, they further determined that the interface was likely compositionally sharp upon synthesis and only later broadened as the neighboring crystals reacted. Finally, since α-Fe is ferromagnetic while Y_2O_3 is not, Watkins *et al.* were able to track changes in magnetization across this interface. Figure 5 shows that to model the reflectivity of this interface, contributions of spin-up and spin-down states of the neutron beam must be averaged. By using comparing the SLD profiles of these two states, the exact depth at which the ferromagnetic ordering is lost may be found (marked with an "x" in the right panel in Figure 5).

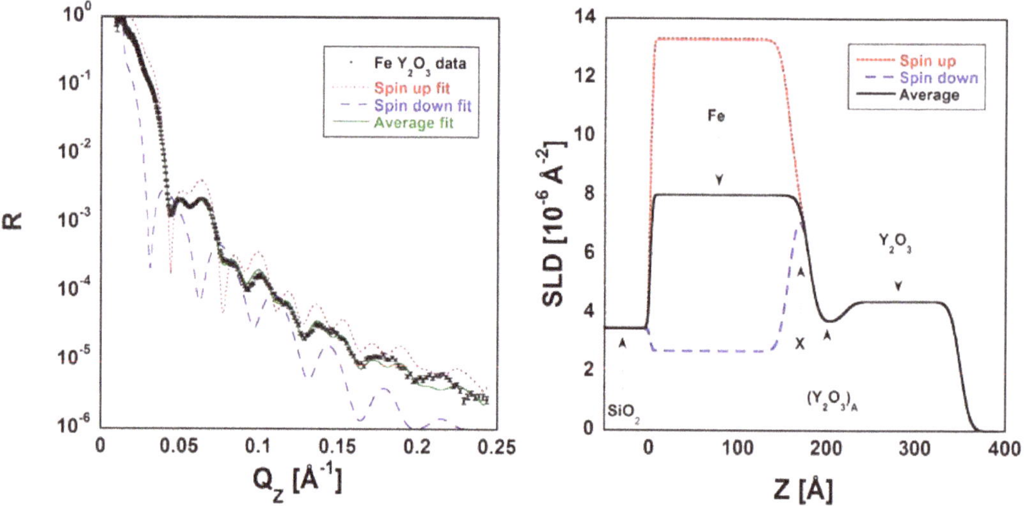

Figure 5. The neutron reflectivity (**left**) of the α-Fe and Y_2O_3 interface investigated by Watkins *et al.* is the average of SLD contributions from spin-up and spin-down states of the neutron beam (**right**). The depth at which ferromagnetic ordering is lost is marked with an "x" in the **right** panel. Also in the **right** panel, $(Y_2O_3)_A$ refers to a distinctive Y–O layer forming at this metal/oxide interface. Reprinted with permission from Reference [71]. Copyright (2014), AIP publishing LLC.

6.3. Actinides: Sensitivity to Light Elements

Actinides and their oxides exhibit some of the most intriguing and challenging chemistry known [72]. Frequently, the composition of these materials is not stoichiometrically precise. Moreover, their oxide structures can change dramatically under different environmental conditions. Neutrons provide a distinct advantage over X-rays in structural characterization of hydrides and oxides of heavy metals because they are better able to detect the lighter elements, such as H/D and O, within their actinide matrices. Figure 6 illustrates neutron scattering length densities for different uranium oxide phases, showing that NR is able to distinguish between them.

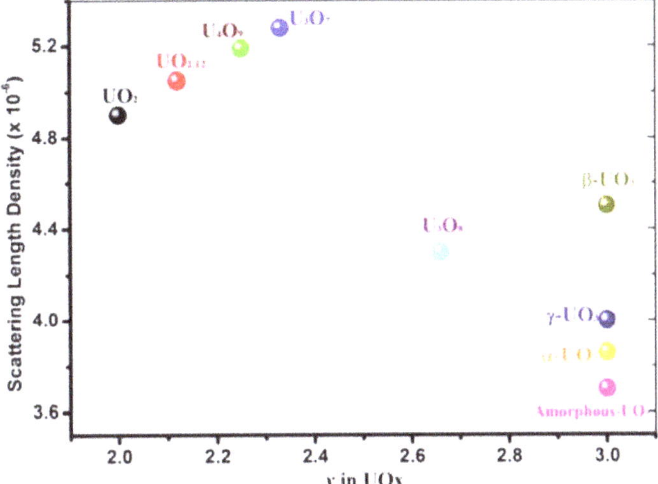

Figure 6. The calculated values of the nuclear scattering length density (SLD) for some common phases in uranium-oxide system. The calculations are based on specific densities published in literature [73,74].

The work of He *et al.* illustrates the utility of NR for investigating imperfect uranium oxide films [75]. They deposited uranium oxide on silicon substrates (with a thin native layer of oxide) using a combination of DC magnetron and reactive sputtering. In this technique, U atoms generated from a solid target by sputtering are readily oxidized by residual O_2 present in the Ar/O_2 mixture under moderate vacuum (approximately 1–3 $\times$ 10^{-4} torr) and then deposited on the substrate above the target. Several steps were taken to ensure high film quality: a multi-step sequential reactive deposition was used to minimize preferential film growth, the substrate was rotated to even out source distribution anomalies, and the partial pressures of Ar and O_2 were adjusted to control the composition of the uranium oxide. Nevertheless, the resulting film has a non-uniform, depth-dependent stoichiometry and structure.

The NR data for these films along with the best-fit curve according to the real-space SLD profile are shown in Figure 7a. According to these results, the total thickness of the UO_x film is about 630 Å. Figure 7b illustrates schematically the real-space structure represented by the best-fit SLD profile. The simplest model that fits the NR data has a three-layer structure. There is no heteroepitaxial growth of uranium oxide on the substrate/film interface due to the (~10 Å) native amorphous Si oxide layer on top of silicon wafer. The SLD of the layer at the film/air interface (~5.0 $\times$ 10^{-6} Å^{-2}) suggests the presence of hyper-stoichiometric phases. Meanwhile, the SLD of middle layer of the film (~3.8 $\times$ 10^{-6} Å^{-2}) together with the fact that no sharp X-ray diffraction peaks were observable (data not shown) indicates that this layer consists of amorphous α-UO_3. Overall, NR demonstrates a remarkably rich variation in structure and stoichiometry in this nominally uniform sample.

Another example of the utility of NR for studies of heavy metal (lanthanide) oxides arises from recent work on Dysprosium (Dy) oxidation [76]. They deposited Dy films on silicon substrates using the same DC magnetron sputtering technique as discussed above and characterized their structures using NR after exposure to air at two different temperatures: 25 °C (ambient temperature) and 150 °C. Figure 8 shows that, under both conditions, the film may be described three-slab model. Under ambient temperature, it consists of 20 Å silicon oxide on top of the Si substrate, 418 Å Dy, and 43 Å Dy_2O_3. After exposure to air in 150 °C for ~0.5 h, the thickness of the Dy_2O_3 increased to 114 Å while simultaneously the thickness of the Dy layer decreased to 363 Å. The total thickness of Dy and Dy_2O_3 layers increased from 461 Å to 477 Å, indicating an overall swelling of the sample. The roughness parameters of the air-Dy_2O_3 and Dy_2O_3-Dy interfaces decreased, making the top surface facing the air and the interface between the metallic Dy and its oxide smoother. For the two cases of uranium oxide and dysprosium studies described above, the approximate errors for the thickness, SLD, and roughness parameters vales were ±5 Å, 0.1 $Å^{-2}$, and 2 Å, respectively.

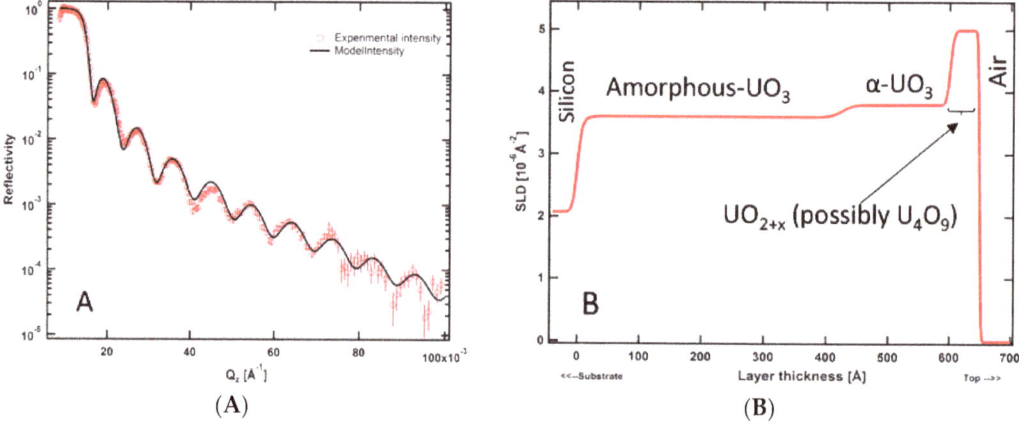

Figure 7. (**A**) NR data obtained from a UO_x deposited on a silicon substrate (open circles). Error bars indicate one standard deviation. The solid line through the data points corresponds to the best-fit SLD profile shown in (**B**).

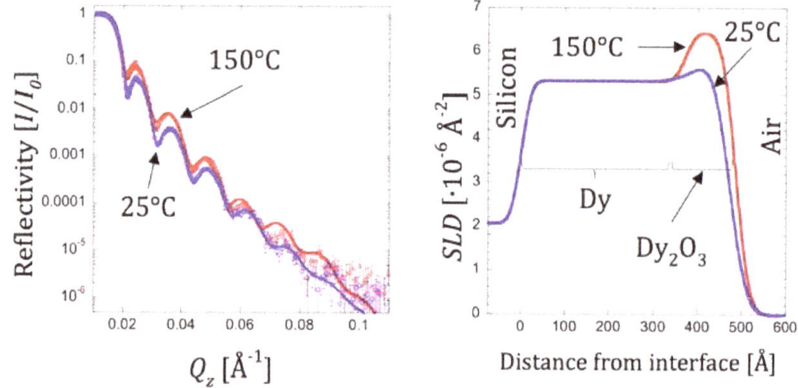

Figure 8. (**Left**) NR data (open circles) from a film of Dy covered with capping layer of Dy_2O_3 at ambient temperature (25 °C) and after expose to air at 150 °C for ~0.5 h. Error bars in the NR measurements correspond to one standard deviation. The solid line through the data points corresponds to the best-fit SLD profile shown on the (**right**).

6.4. Surfaces in Pressurized Liquids: Penetration of Neutrons through Containers

Neutrons penetrate through thick sections of solid matter with low attenuation. Thus, they are able to "see through" the walls of a high-pressure cell, enabling examination of metal surfaces in pressurized media [77]. Junghans *et al.* used this capability to study the corrosion of oxidized aluminum (Al) surfaces in pressurized seawater [27]. The corrosion of structural materials in the deep sea depends on numerous chemical and physical factors, including pH, dissolved oxygen, chloride ion activity, salinity, ocean currents, temperature, and hydrostatic pressure [78–82]. Al and its alloys find widespread use in marine environments, including in civil and defense vessels, offshore rigs, drill pipes for deep wells, and diving suits. There are several reports on corrosion of Al and its alloys at shallow depths, but investigations at high pressures are limited [78–80,82–84].

When in contact with air, Al develops a thin passivation layer of Al_2O_3. The surface oxide has a higher SLD than pure Al, providing neutron scattering contrast between the two materials (Al:

2.08×10^{-6} Å^{-2}; crystalline Al_2O_3: 5.74×10^{-6} Å^{-2}). Junghans *et al.* deposited uniform, ~900 Å-thick Al films on monocrystalline quartz wafers using DC magnetron sputtering and collected a total of 14 NR spectra over the course of 50 h (~3 h per spectrum). The film was in contact with 3.5 wt. % NaCl solution at pressures ranging from 1 to 600 atm in a specially developed solid/liquid, high pressure/temperature cell [77]. This cell provides the capabilities of solid/fluid interface investigations up to 2000 atm (~30,000 psi) and 200 °C. The cell's simple aluminum construction makes it easy to operate at high pressures and elevated temperatures, while the 13 mm thick neutron windows allow up to 74% neutron transmission. Figure 9 shows five representative NR measurements from this study [27].

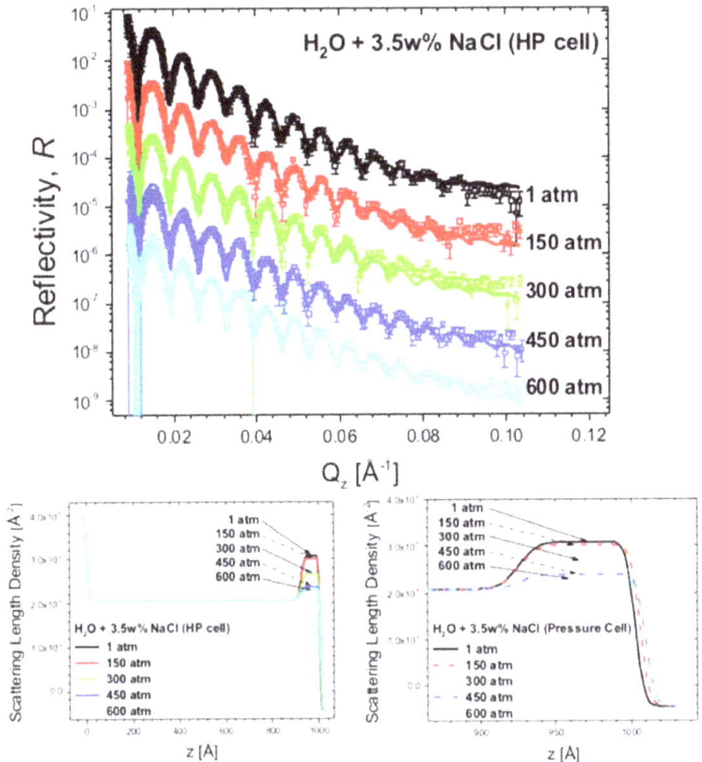

Figure 9. (**Top** panel) Five NR measurements of Al film deposited on quartz substrate and investigated at 25 °C in contact with H_2O + 3.5 wt. % of NaCl at different pressures. Solid curves are fits corresponding to the SLD profiles shown in the middle and right panels. Both the NR data and fits are offset by a decade along *y*-axis for clarity; (**Bottom left** panel) SLD distribution of the Al/Al_2O_3/liquid system and (**Bottom right** panel) magnified SLD distribution in the contact region. In both of the bottom panels, $z = 0$ at the quartz substrate/Al interface.

The NR results show virtually no corrosion of the Al layers. The observed decrease in the SLD of the starting Al_2O_3 passivation layer cannot be explained by the formation of new chemical compounds by the highly scattering Na, Cl, and O ions. However, this decrease is consistent with formation of stable Al–Cl–H_2O (or Al–O–Cl–H_2O) complexes or hydration of Al_2O_3 to $Al(OH)_3$. These results suggest that for the time scale of 50 h the influence of hydrostatic pressure only slightly influences interactions of the Al oxide film with Cl^- ions and H_2O. The corrosion rate is lower than reported by Beccaria *et al.* [78–80], suggesting slower kinetics for the reactions involved.

7. Conclusions

Neutron reflectometry (NR) is a mature experimental technique that has been used extensively in condensed-matter physics. However, its potential for investigating interfaces in metals has not yet been utilized widely. The present overview is intended to raise awareness of NR in the metals community with the hope of motivating wider use of this technique. NR provides several unique advantages for investigating interfaces in metals:

- **It is non-destructive**. Thus, NR results may be combined with other, follow-on investigations, e.g., using XRR, TEM, or APT.
- **Å-level depth resolution** enables detailed investigation of interface structure: thickness, SLD, and roughness of the layers.
- **Sensitivity to composition, isotopic distribution, density, and magnetic moment** allow multiple physical characteristics to be measured simultaneously.
- **Ability to detect low-Z elements**, such as H/D, He, and other light isotopes.
- **Suitability for *in situ* studies** due to the high penetrability of neutrons through container walls and surrounding media. This capability enables investigations of a variety of buried interfaces, including solid-liquid ones, which are otherwise very difficult to access with X-rays.
- **The measured SLD values are absolute** due to the fact that the reflected beam is normalized by the incident intensity of the neutron beam.
- **Ready access** thanks to the availability of several suitable neutron sources (reactors and spallation facilities) worldwide.

Several example applications of NR to metal surfaces and buried interfaces have been discussed above. As the metals community continues to explore the structure and properties of interfaces in ever-greater detail [7–9], NR stands poised to contribute valuable new additions to this ever-growing list of examples.

Acknowledgments: J.M. thanks Heming He, Kirk Rector, Peng Wang, Ann Junghans, Erik Watkins (LANL), and David Allred (BYU). M.J.D. acknowledges Abishek Kashinath and support from the Center for Materials in Irradiation and Mechanical Extremes (CMIME), an Energy Frontier Research Center funded by the U.S. Department of Energy, Office of Science, Office of Basic Energy Sciences under Award No. 2008LANL1026. This work benefited from the use of the Lujan Neutron Scattering Center at LANSCE funded by the DOE Office of Basic Energy Sciences and Los Alamos National Laboratory under DOE Contract DE-AC52-06NA25396.

Author Contributions: Both authors contributed equally to the composition of this article.

Conflicts of Interest: The authors declare no conflict of interest.

References

1. Sutton, A.P.; Balluffi, R.W. *Interfaces in Crystalline Materials*; Oxford University Press: Oxford, UK, 1995.
2. Furukawa, S.; de Feyter, S. Two-dimensional crystal engineering at the liquid-solid interface. In *Templates in Chemistry III*; Broekmann, P., Dotz, K.H., Schalley, C.A., Eds.; Springer-Verlag Berlin: Berlin, Germany, 2009; Volume 287, pp. 87–133.
3. Kaplan, W.D.; Kauffmann, Y. Structural order in liquids induced by interfaces with crystals. *Ann. Rev. Mater. Res.* **2006**, *36*, 1–48. [CrossRef]
4. Boettinger, W.J.; Warren, J.A.; Beckermann, C.; Karma, A. Phase-field simulation of solidification. *Ann. Rev. Mater. Res.* **2002**, *32*, 163–194. [CrossRef]
5. Stachowiak, G.W.; Batchelor, A.W. *Engineering Tribology*; Butterworth-Heinemann: Boston, MA, USA, 2000.
6. Hirth, J.P.; Pond, R.C.; Hoagland, R.G.; Liu, X.Y.; Wang, J. Interface defects, reference spaces and the frank-bilby equation. *Prog. Mater. Sci.* **2013**, *58*, 749–823. [CrossRef]
7. Beyerlein, I.J.; Demkowicz, M.J.; Misra, A.; Uberuaga, B.P. Defect-interface interactions. *Prog. Mater. Sci.* **2015**, *74*, 125–210. [CrossRef]
8. Cantwell, P.R.; Tang, M.; Dillon, S.J.; Luo, J.; Rohrer, G.S.; Harmer, M.P. Grain boundary complexions. *Acta Mater.* **2014**, *62*, 1–48. [CrossRef]

9. Mishin, Y.; Asta, M.; Li, J. Atomistic modeling of interfaces and their impact on microstructure and properties. *Acta Mater.* **2010**, *58*, 1117–1151. [CrossRef]
10. Colliex, C.; Bocher, L.; de la Pena, F.; Gloter, A.; March, K.; Walls, M. Atomic-scale stem-eels mapping across functional interfaces. *JOM* **2010**, *62*, 53–57. [CrossRef]
11. Schmitz, G.; Ene, C.; Galinski, H.; Schlesiger, R.; Stender, P. Nanoanalysis of interfacial chemistry. *JOM* **2010**, *62*, 58–63. [CrossRef]
12. Mitchell, T.E.; Lu, Y.C.; Griffin, A.J.; Nastasi, M.; Kung, H. Structure and mechanical properties of copper/niobium multilayers. *J. Am. Ceram. Soc.* **1997**, *80*, 1673–1676. [CrossRef]
13. Shaha, K.P.; Pei, Y.T.; Chen, C.Q.; Turkin, A.A.; Vainshtein, D.I.; de Hosson, J.T.M. On the dynamic roughening transition in nanocomposite film growth. *Appl. Phys. Lett.* **2009**. [CrossRef]
14. Olmsted, D.L.; Foiles, S.M.; Holm, E.A. Grain boundary interface roughening transition and its effect on grain boundary mobility for non-faceting boundaries. *Scr. Mater.* **2007**, *57*, 1161–1164. [CrossRef]
15. Bellon, P. Nonequilibrium roughening and faceting of interfaces in driven alloys. *Phys. Rev. Lett.* **1998**, *81*, 4176–4179. [CrossRef]
16. Wiesler, D.G.; Majkrzak, C.F. Neutron reflectometry studies of surface oxidation. *Phys. B* **1994**, *198*, 181–186. [CrossRef]
17. Russell, T.P. X-ray and neutron reflectivity for the investigation of polymers. *Mater. Sci. Rep.* **1990**, *5*, 171–271. [CrossRef]
18. Dietrich, S.; Haase, A. Scattering of X-rays and neutrons at interfaces. *Phys. Rep. Rev. Sec. Phys. Lett.* **1995**, *260*, 1–138. [CrossRef]
19. Zhou, X.L.; Chen, S.H. Theoretical foundation of X-ray and neutron reflectometry. *Phys. Rep. Rev. Sec. Phys. Lett.* **1995**, *257*, 223–348. [CrossRef]
20. Penfold, J.; Thomas, R.K. The application of the specular reflection of neutrons to the study of surfaces and interfaces. *J. Phys. Condes. Matter* **1990**, *2*, 1369–1412. [CrossRef]
21. Van der Lee, A. Grazing incidence specular reflectivity: Theory, experiment, and applications. *Solid State Sci.* **2000**, *2*, 257–278. [CrossRef]
22. Fermi, E.; Zinn, W.H. Reflection of neutrons on mirrors. *Phys. Rev.* **1946**, *70*, 103.
23. Fermi, E.; Marshall, L. Interference phenomena of slow neutrons. *Phys. Rev.* **1947**, *71*, 666–677. [CrossRef]
24. Lekner, J. Theory of reflection of electromagnetic and particle waves. In *Developments in Electromagnetic Theory and Applications 3*; Springer Netherlands: Dordrecht, The Netherlands, 1987.
25. Russell, T.P. The characterization of polymer interfaces. *Annu. Rev. Mater. Sci.* **1991**, *21*, 249–268. [CrossRef]
26. Smith, G.S.; Majkrzak, C.F. Neutron reflectometry. In *International Tables for Crystallography*, 1st ed.; International Union of Crystallography, Ed.; Springer: Chester, UK; New York, NY, USA, 2006.
27. Junghans, A.; Chellappa, R.; Wang, P.; Majewski, J.; Luciano, G.; Marcelli, R.; Proietti, E. Neutron reflectometry studies of aluminum-saline water interface under hydrostatic pressure. *Corros. Sci.* **2015**, *90*, 101–106. [CrossRef]
28. Carpenter, J.M. Pulsed spallation neutron sources for slow-neutron scattering. *Nucl. Instrum. Methods* **1977**, *145*, 91–113. [CrossRef]
29. Bauer, G.S. Physics and technology of spallation neutron sources. *Nucl. Instrum. Methods Phys. Res. Sect. A Accel. Spectrom. Dect. Assoc. Equip.* **2001**, *463*, 505–543. [CrossRef]
30. Hughes, D.J.; Burgy, M.T. Reflection of neutrons from magnetized mirrors. *Phys. Rev.* **1951**, *81*, 498–506. [CrossRef]
31. Lauter, V.; Ambaye, H.; Goyette, R.; Lee, W.T.H.; Parizzi, A. Highlights from the magnetism reflectometer at the SNS. *Phys. B* **2009**, *404*, 2543–2546. [CrossRef]
32. Merzbacher, E. *Quantum Mechanics*; Wiley: New York, NY, USA, 1961; p. 544.
33. Sears, V.F. Neutron scattering lengths and cross sections. *Neutron News* **1992**, *3*, 29–37. [CrossRef]
34. Lovesey, S.W. *Theory of Neutron Scattering from Condensed Matter*; Clarendon Press: Oxfordshire, UK, 1984.
35. Neutron SLDs. Available online: http://www.ncnr.nist.gov/resources/n-lengths/ (accessed on 23 December 2015).
36. Nevot, L.; Croce, P. Characterization of surfaces by grazing X-ray reflection—Application to study of polishing of some silicate-glasses. *Rev. Phys. Appl.* **1980**, *15*, 761–779.
37. Als-Nielsen, J.; McMorrow, D. *Elements of Modern X-ray Physics*, 2nd ed.; Hoboken, N.J., Ed.; Wiley: Hoboken, NJ, USA, 2011.

38. Lovell, M.R.; Richardson, R.M. Analysis methods in neutron and X-ray reflectometry. *Curr. Opin. Colloid Interface Sci.* **1999**, *4*, 197–204. [CrossRef]
39. Parratt, L.G. Surface studies of solids by total reflection of X-rays. *Phys. Rev.* **1954**, *95*, 359–369. [CrossRef]
40. Yasaka, M. X-ray thin film measurement techniques. *Rigaku J.* **2010**, *26*, 1–9.
41. Nelson, A. Co-refinement of multiple-contrast neutron/X-ray reflectivity data using motofit. *J. Appl. Crystallogr.* **2006**, *39*, 273–276. [CrossRef]
42. Kashinath, A.; Wang, P.; Majewski, J.; Baldwin, J.K.; Wang, Y.Q.; Demkowicz, M.J. Detection of helium bubble formation at fcc-bcc interfaces using neutron reflectometry. *J. Appl. Phys.* **2013**. [CrossRef]
43. Majkrzak, C.F.; Berk, N.F. Exact determination of the phase in neutron reflectometry. *Phys. Rev. B* **1995**, *52*, 10827–10830. [CrossRef]
44. Berk, N.F.; Majkrzak, C.F. Inverting specular neutron reflectivity from symmetric, compactly supported potentials. In Proceedings of the International Symposium on Advance in Neutron Optics and Related Research Facilities, Kumatori, Osaka, Japan, 19–21 March 1996; p. 107.
45. Majkrzak, C.F.; Berk, N.F. Exact determination of the phase in neutron reflectometry by variation of the surrounding media. *Phys. Rev. B* **1998**, *58*, 15416–15418. [CrossRef]
46. Dehaan, V.O.; Vanwell, A.A.; Adenwalla, S.; Felcher, G.P. Retrieval of phase information in neutron reflectometry. *Phys. Rev. B* **1995**, *52*, 10831–10833. [CrossRef]
47. Kasper, J.; Leeb, H.; Lipperheide, R. Phase determination in spin-polarized neutron specular reflection. *Phys. Rev. Lett.* **1998**, *80*, 2614–2617. [CrossRef]
48. Lipperheide, R.; Kasper, J.; Leeb, H. Surface profiles from polarization measurements in neutron reflectometry. *Phys. B* **1998**, *248*, 366–371. [CrossRef]
49. Leeb, H.; Grotz, H.; Kasper, J.; Lipperheide, R. Complete determination of the reflection coefficient in neutron specular reflection by absorptive nonmagnetic media. *Phys. Rev. B* **2001**. [CrossRef]
50. Majkrzak, C.F. Neutron scattering studies of magnetic thin films and multilayers. *Phys. B* **1996**, *221*, 342–356. [CrossRef]
51. Williams, W.G. *Polarized Neutrons*; Clarendon Press: Oxford, UK; New York, NY, USA; Oxford University Press: Oxford, UK; New York, NY, USA, 1988.
52. Renaud, G.; Lazzari, R.; Leroy, F. Probing surface and interface morphology with grazing incidence small angle X-ray scattering. *Surf. Sci. Rep.* **2009**, *64*, 255–380. [CrossRef]
53. Sinha, S.K. Reflectivity using neutrons or X-rays—A critical comparison. *Phys. B* **1991**, *173*, 25–34. [CrossRef]
54. Stoev, K.N.; Sakurai, K. Review on grazing incidence X-ray spectrometry and reflectometry. *Spectroc. Acta B Atom. Spectr.* **1999**, *54*, 41–82. [CrossRef]
55. Feldman, L.C.; Mayer, J.W. *Fundamentals of Surface and Thin Film Analysis*; North-Holland: New York, NY, USA, 1986.
56. Jablin, M.S.; Zhernenkov, M.; Toperverg, B.P.; Dubey, M.; Smith, H.L.; Vidyasagar, A.; Toomey, R.; Hurd, A.J.; Majewski, J. In-plane correlations in a polymer-supported lipid membrane measured by off-specular neutron scattering. *Phys. Rev. Lett.* **2011**. [CrossRef] [PubMed]
57. Laakmann, J.; Jung, P.; Uelhoff, W. Solubility of helium in gold. *Acta Metall.* **1987**, *35*, 2063–2069. [CrossRef]
58. Trinkaus, H.; Singh, B.N. Helium accumulation in metals during irradiation—Where do we stand? *J. Nucl. Mater.* **2003**, *323*, 229–242. [CrossRef]
59. Judge, C.D.; Gauquelin, N.; Walters, L.; Wright, M.; Cole, J.I.; Madden, J.; Botton, G.A.; Griffiths, M. Intergranular fracture in irradiated inconel X-750 containing very high concentrations of helium and hydrogen. *J. Nucl. Mater.* **2015**, *457*, 165–172. [CrossRef]
60. Baldwin, M.J.; Doerner, R.P. Helium induced nanoscopic morphology on tungsten under fusion relevant plasma conditions. *Nucl. Fusion* **2008**. [CrossRef]
61. Demkowicz, M.J.; Misra, A.; Caro, A. The role of interface structure in controlling high helium concentrations. *Curr. Opin. Solid State Mat. Sci.* **2012**, *16*, 101–108. [CrossRef]
62. Misra, A.; Demkowicz, M.J.; Zhang, X.; Hoagland, R.G. The radiation damage tolerance of ultra-high strength nanolayered composites. *JOM* **2007**, *59*, 62–65. [CrossRef]
63. Odette, G.R.; Miao, P.; Edwards, D.J.; Yamamoto, T.; Kurtz, R.J.; Tanigawa, H. Helium transport, fate and management in nanostructured ferritic alloys: *In situ* helium implanter studies. *J. Nucl. Mater.* **2011**, *417*, 1001–1004. [CrossRef]

64. Odette, G.R.; Hoelzer, D.T. Irradiation-tolerant nanostructured ferritic alloys: Transforming helium from a liability to an asset. *JOM* **2010**, *62*, 84–92. [CrossRef]
65. Zhernenkov, M.; Gill, S.; Stanic, V.; DiMasi, E.; Kisslinger, K.; Baldwin, J.K.; Misra, A.; Demkowicz, M.J.; Ecker, L. Design of radiation resistant metallic multilayers for advanced nuclear systems. *Appl. Phys. Lett.* **2014**. [CrossRef]
66. Zhernenkov, M.; Jablin, M.S.; Misra, A.; Nastasi, M.; Wang, Y.-Q.; Demkowicz, M.J.; Baldwin, J.K.; Majewski, J. Trapping of implanted he at Cu/Nb interfaces measured by neutron reflectometry. *Appl. Phys. Lett.* **2011**. [CrossRef]
67. Demkowicz, M.J.; Bhattacharyya, D.; Usov, I.; Wang, Y.Q.; Nastasi, M.; Misra, A. The effect of excess atomic volume on he bubble formation at fcc-bcc interfaces. *Appl. Phys. Lett.* **2010**. [CrossRef]
68. Kashinath, A.; Misra, A.; Demkowicz, M.J. Stable storage of helium in nanoscale platelets at semicoherent interfaces. *Phys. Rev. Lett.* **2013**. [CrossRef] [PubMed]
69. Kaplan, W.D.; Chatain, D.; Wynblatt, P.; Carter, W.C. A review of wetting *versus* adsorption, complexions, and related phenomena: The rosetta stone of wetting. *J. Mater. Sci.* **2013**, *48*, 5681–5717. [CrossRef]
70. Dholabhai, P.P.; Pilania, G.; Aguiar, J.A.; Misra, A.; Uberuaga, B.P. Termination chemistry-driven dislocation structure at $SrTiO_3$/MgO heterointerfaces. *Nat. Commun.* **2014**. [CrossRef] [PubMed]
71. Watkins, E.B.; Kashinath, A.; Wang, P.; Baldwin, J.K.; Majewski, J.; Demkowicz, M.J. Characterization of a Fe/Y_2O_3 metal/oxide interface using neutron and X-ray scattering. *Appl. Phys. Lett.* **2014**. [CrossRef]
72. Morss, L.R.; Edelstein, N.M.; Fuger, J.; Katz, J.J. *The Chemistry of the Actinide and Transactinide Elements*, 3rd ed.; Springer: Dordrecht, The Netherlands, 2006.
73. Hoekstra, H.R.; Siegel, S. The uranium-oxygen system—U_3O_8–UO_3. *J. Inorg. Nucl. Chem.* **1961**, *18*, 154–165. [CrossRef]
74. Loopstra, B.O.; Cordfunk, E. On structure of α-UO_3. *Recl. Trav. Chim. PaysBas* **1966**, *85*, 135–142. [CrossRef]
75. He, H.M.; Wang, P.; Allred, D.D.; Majewski, J.; Wilkerson, M.P.; Rector, K.D. Characterization of chemical speciation in ultrathin uranium oxide layered films. *Anal. Chem.* **2012**, *84*, 10380–10387. [CrossRef] [PubMed]
76. Watkins, E.B.; Scott, B.; Allred, D.D.; Majewski, J. Unpublished work, 2015.
77. Wang, P.; Lerner, A.H.; Taylor, M.; Baldwin, J.K.; Grubbs, R.K.; Majewski, J.; Hickmott, D.D. High-pressure and high-temperature neutron reflectometry cell for solid-fluid interface studies. *Eur. Phys. J. Plus* **2012**. [CrossRef]
78. Beccaria, A.M.; Poggi, G. Influence of hydrostatic-pressure on pitting of aluminum in sea-water. *Br. Corros. J.* **1985**, *20*, 183–186. [CrossRef]
79. Beccaria, A.M.; Poggi, G. Effect of some surface treatments on kinetics of aluminum corrosion in NaCl solutions at various hydrostatic pressures. *Br. Corros. J.* **1986**, *21*, 19–22. [CrossRef]
80. Beccaria, A.M.; Fiordiponti, P.; Mattogno, G. The effect of hydrostatic-pressure on the corrosion of nickel in slightly alkaline-solutions containing $C1^-$ ions. *Corros. Sci.* **1989**, *29*, 403–413. [CrossRef]
81. Heusler, K.E. Untersuchungen der korrosion von aluminium in wasser bei hohen temperaturen und drucken. *Mater. Corros.* **1967**, *18*, 11–15. [CrossRef]
82. Dexter, S.C. Effect of variations in sea-water upon the corrosion of aluminum. *Corrosion* **1980**, *36*, 423–432. [CrossRef]
83. Venkatesan, R.; Venkatasamy, M.A.; Bhaskaran, T.A.; Dwarakadasa, E.S.; Ravindran, M. Corrosion of ferrous alloys in deep sea environments. *Br. Corros. J.* **2002**, *37*, 257–266. [CrossRef]
84. Sawant, S.S.; Wagh, A.B. Corrosion behaviour of metals and alloys the waters of the arabian sea. *Corros. Prev. Control* **1990**, *37*, 154–157.

17

Onset Frequency of Fatigue Effects in Pure Aluminum and 7075 (AlZnMg) and 2024 (AlCuMg) Alloys

Jose I. Rojas [1,*] **and Daniel Crespo** [2]

1 Department of Physics-Division of Aerospace Engineering, Universitat Politècnica de Catalunya, c/ Esteve Terradas 7, 08860 Castelldefels, Spain

2 Department of Physics, Universitat Politècnica de Catalunya, c/ Esteve Terradas 7, 08860 Castelldefels, Spain; daniel.crespo@upc.edu

* Correspondence: josep.ignasi.rojas@upc.edu

Academic Editor: Nong Gao

Abstract: The viscoelastic response of pure Al and 7075 (AlZnMg) and 2024 (AlCuMg) alloys, obtained with a dynamic-mechanical analyzer (DMA), is studied. The purpose is to identify relationships between the viscoelasticity and fatigue response of these materials, of great interest for structural applications, in view of their mutual dependence on intrinsic microstructural effects associated with internal friction. The objective is to investigate the influence of dynamic loading frequency and temperature on fatigue, based on their effect on the viscoelastic behavior. This research suggests that the decrease of yield and fatigue behavior reported for Al alloys as temperature increases may be associated with the increase of internal friction. Furthermore, materials subjected to dynamic loading below a given threshold frequency exhibit a static-like response, such that creep mechanisms dominate and fatigue effects are negligible. In this work, an alternative procedure to the time-consuming fatigue tests is proposed to estimate this threshold frequency, based on the frequency dependence of the initial decrease of the storage modulus with temperature, obtained from the relatively short DMA tests. This allows for a fast estimation of the threshold frequency. The frequencies obtained for pure Al and 2024 and 7075 alloys are 0.001–0.005, 0.006 and 0.075–0.350 Hz, respectively.

Keywords: aluminum alloys; AlZnMg; AlCuMg; viscoelasticity; dynamic-mechanical analysis; internal friction; loading frequency; fatigue

---

1. Introduction

Fatigue is a form of failure that may occur in structures subjected to dynamic loading, even at stress levels significantly lower than the ultimate tensile strength under static loading [1]. Failure results from a gradual process of damage accumulation and local strength reduction, which is manifested by crack initiation and propagation, after relatively long periods of dynamic loading. It is particularly dangerous in structural applications, because of its brittle, catastrophic nature and because it occurs suddenly and without warning, since very little plastic deformation is observed in the material prior to failure [1,2]. The fatigue fracture behavior of materials is dominated by the microstructure [3]. When a material is subjected to dynamic loading, energy is dissipated due to internal friction phenomena. Most of this energy manifests as heat and causes temperature increases in the material, a process termed hysteresis heating. It has been suggested that all metals, when subjected to hysteresis heating, are prone to fatigue [4].

In previous investigations, the viscoelastic response (including the internal friction behavior) of Al alloys (AA) 7075 and 2024 was measured with a dynamic-mechanical analyzer (DMA) [5,6]. In this work, experimental results on the viscoelastic response of pure Al are presented, first. Second, these results for pure Al and the aforementioned alloys are analyzed with the purpose of identifying relationships between the viscoelastic response and the fatigue behavior of these materials, in view of their mutual dependence on intrinsic microstructural effects associated with internal friction [7]. Particularly, the objective is to investigate the influence of the dynamic loading frequency and temperature on fatigue, based on the effect of these variables on the viscoelastic behavior. The results seem to support the work by Amiri and Khonsari [4], as per the correlation between the fatigue life and the initial hysteresis heating during dynamic loading. Namely, it is likely that the decrease of yield and fatigue response observed in some metals as temperature increases is associated with the increase of internal friction with temperature. Moreover, following previous investigations by other researchers, suggesting the existence of a threshold frequency marking the transition from a static-like response of the material to the advent of fatigue effects, in this work, an alternative procedure is proposed to estimate this threshold frequency based on experimental data obtained with the relatively short DMA tests. These findings are of remarkable importance, especially for the alloys, in view of their widespread use in structural applications under dynamic loading. Particularly, AA 7075 and 2024 are key representatives of the AlZnMg and AlCuMg alloy families (or 7xxx and 2xxx series, respectively), belonging to the group of age-hardenable alloys. These alloys feature excellent mechanical properties and are highly suitable for a number of industrial applications, especially in the aerospace sector and transport industry [8].

1.1. Influence of the Loading Frequency on the Fatigue Response of Metals

Fatigue may be sensitive, for instance, to the strength, the manufacturing conditions and the surface treatment of the material, but also to the loading frequency and loading environment, the displacement rates and the stress amplitude [1,9–11]. In this work, we address the effects of the dynamic loading frequency and temperature, in conjunction with the microstructure.

Much research has been devoted to ascertain whether accelerated laboratory tests (*i.e.*, with loading frequencies higher than those in service conditions) affect the fatigue response and how, but this is yet a controversial issue. This is particularly true for the study of high cycle fatigue (HCF) and very high cycle fatigue (VHCF) behavior by means of very high frequency tests. Tests in VHCF and very low crack growth rates are time consuming with conventional fatigue testing techniques, like rotating bending, with a maximum frequency of 100 Hz. Hence, accelerated laboratory tests are very interesting because a significant reduction of testing time is possible using high speed servo-hydraulic machines [12], which may work at frequencies of 600 Hz, or especially using ultrasonic equipment, which may reach frequencies of 20 kHz [13].

Zhu *et al.* [14] state that environmental effects need to be considered, and Mayer *et al.* [15] explain that this is so because the time-dependent interaction with the environment may cause an extrinsic frequency influence on fatigue properties, on top of the intrinsic strain rate effects. Furuya *et al.* [12] state that frequency generally affects high frequency fatigue tests because: (1) fatigue limits and lives decrease due to the temperature increase caused by plastic deformation [16]; (2) dislocations may not match the applied frequency because dislocation movement is slow compared to sonic velocity [17]; and (3) provided that embrittlement by hydrogen diffusion had an effect [18], fatigue lives would depend on both the number of loading cycles and time. However, Mayer [19] reported also that the HCF behavior of metallic alloys is relatively insensitive to the test frequency, provided that the ultrasonic testing procedure is appropriate (e.g., adequate cooling) and that fatigue-creep interaction and the time-dependent interaction with the environment are negligible. The reasons suggested are, on the one hand, that cyclic plastic straining is limited near the fatigue limit or the threshold of fatigue crack growth (FCG), and thus, plastic strain rates are low, even at high frequencies; and on the other hand, the fact that shear stress has little sensitivity to strain rate [20]. Mayer *et al.* [15] also commented that the influence of frequency becomes significantly smaller if the dynamic stress amplitude is lower, maybe because cyclic loading is almost perfectly elastic.

For body-centered cubic (bcc) metals and metallic alloys, the HCF behavior is reported to be more sensitive to frequency than for face-centered cubic (fcc) metals [13]. However, Furuya *et al.* [12] observed that the fatigue behavior of high strength steels is independent of frequency. The argued cause was their extremely high strength and, thus, reduced plasticity and dislocation mobility. The hysteresis energy is low in low plasticity materials, and thus, the frequency effects on fatigue associated with the temperature increase are minimized. Likewise, Yan *et al.* [21] observed very little variation of the fatigue strength of high strength steel when testing at a conventional frequency (52.5 Hz) and at an ultrasonic frequency (20 kHz).

For an Al alloy similar to AA 7075 tested in the HCF regime at room temperature (RT), samples tested at 100 Hz were reported to fail earlier than those tested at 20 kHz. However, the effect of frequency on fatigue behavior was not statistically significant [15]. On the contrary, for E319 cast Al alloy at 293, 423 and 523 K, fatigue life at 20 kHz was 5–10-times longer than that at 75 Hz [14], but this author states that fatigue crack initiation is not influenced either by temperature or frequency. Rather, the observed difference in fatigue life is attributed to environmental effects on FCG rate. The fact that the moisture of ambient air deteriorates the fatigue life of high strength Al alloys by increasing the FCG rate has also been suggested by other authors [15,22]. Namely, Menan and Henaff [23] suggest for AA 2024-T351 that fatigue and corrosion may interact, such that FCG rates are enhanced. These synergistic effects are more notorious at low frequencies, for a given number of cycles at RT. Finally, Benson and Hancock [24] observed strain rate effects on cyclic plastic deformation of AA 7075-T6, provided that cyclic stresses were close to the yield stress.

As per low frequency loading, on the one hand, Nikbin and Radon [25] proposed a method to predict the frequency region of interaction between creep and FCG using static data (obtained at 423 K for Al alloy RR58) and RT high frequency fatigue data and assuming a linear cumulative damage law. The results showed that the interaction region is 0.1–1 Hz for the Al alloy (see Figure 4 in [25]). In the intermediate (steady state) stage of cracking for static and low frequency tests, crack growth is sensitive to frequency, and the fracture mode is time dependent inter-granular in nature, suggesting that creep dominates. Conversely, for high frequency tests, crack growth is insensitive to frequency, and the fracture mode is trans-granular, suggesting that pure fatigue mechanisms dominate. The results indicated also little interaction between these processes.

On the other hand, Henaff *et al.* [26] analyzed creep crack growth (CCG) rates, FCG rates and creep-fatigue crack growth (CFCG) rates of AA 2650-T6 at 293, 403 and 448 K, for frequencies of 0.05 and 20 Hz. The objective was to enable the prediction of crack growth resistance of that alloy under very low frequency loading at elevated temperatures. It was concluded that, in the studied frequency range, frequency has only a slight effect on FCG rates at 448 K. In particular, under low frequency loading, a high increase was observed in the fracture surface fraction of the inter-granular type, similar to that corresponding to CCG. This shows that creep damage might occur during loading at low frequency, in accordance with the findings in [25]. Henaff *et al.* [26] reported also that, for a given temperature, CFCG is unaffected by frequency above a critical value of the loading frequency (see Figure 12b in [26]). Below, CFCG is inversely proportional to excitation frequency, *i.e.*, a time-dependent crack growth processes take place. This researcher suggested the existence of a creep-fatigue-environment interaction, as CFCG is affected by the environment at low frequency loading, and proposed an alternative method to predict CFCG rates at very low frequencies, using a superposition model and results obtained at higher frequencies.

1.2. *Influence of Temperature on the Fatigue Response of Al Alloys*

For E319 cast Al alloy tested at 293, 423 and 523 K, Zhu *et al.* [14] observed that the fatigue strength decreases with temperature and that the temperature dependence of the fatigue resistance at 108 cycles follows the temperature dependence of the yield and tensile strength for this alloy closely. Furthermore, by integration of a universal version of a modified superposition model, the effects of temperature, frequency and the environment on the S-N curve of this alloy can be predicted, and it is possible also to extrapolate ultrasonic data to conventional fatigue behavior [14]. Henaff *et al.* [26] concluded that the temperature has almost no influence on FCG rates for AA 2650-T6, after conducting tests at 293, 403 and 448 K and frequencies of 0.05 and 20 Hz. Amiri and Khonsari [4] state that the initial slope of the temperature rise due to hysteresis heating observed at the beginning of fatigue tests is a characteristic of metals. Capitalizing on this, they developed an empirical model that predicts fatigue life based on that slope, thus preserving testing time. Indeed, the correlation of the temperature evolution with fatigue has been used successfully in many ways, aside from for predicting fatigue life, as in the previous example. Namely, it has been used for providing information on FCG [27] and the endurance limit of materials [28] or for quantification of the cumulative damage in fatigue [16]. Furthermore, the heat dissipated during ultrasonic cycling can be used to calculate the cyclic plastic strain amplitude [15].

1.3. Influence of the Microstructure on the Fatigue Response of Al Alloys

There is abundant research in the literature on the effect of the microstructure on the fatigue response of materials. For example, it is proposed that the mechanisms responsible for the fatigue fracture behavior are associated with the competing and synergistic influences of intrinsic microstructural effects and interactions between dislocations and the microstructure [3]. Indeed, researchers claim that the prediction of fatigue life should be possible based on the knowledge of the microstructure prior to the beginning of service, without the need for expensive, time-consuming fatigue experiments [29]. This would enable optimization of the material properties by controlling the microstructure. Accordingly, a model based on dislocation stress was proposed to predict S-N curves using microstructure/material-sensitive parameters instead of constitutive equation parameters [29]. The model is successful for low cycle fatigue life prediction.

2. Materials and Methods

The tested specimens were machine cut from a sheet of as-received pure Al (99.5 wt. % purity according to the supplier, Alu-Stock, S.A., Vitoria-Gasteiz, Spain) in the H24 temper. The H24 temper consists of cold-working (*i.e.*, strain hardening) beyond the desired hardness, followed by a softening treatment consisting of annealing up to halfway of the peak hardness. The specimens were rectangular plates 60 mm long, 8–12 mm wide and 2 mm thick. Half of these plates were annealed at 750 K for 30 min and immediately quenched in water to RT, to remove the strain hardening. A TA Instruments Q800 DMA (TA Instruments, New Castle, DE, USA) was used to measure the viscoelastic response of the samples in N_2 atmosphere. Namely, the DMA measured the storage modulus E' (*i.e.*, the elastic-real-component of the dynamic tensile modulus, accounting for the deformation energy stored by the material), the loss modulus E'' (*i.e.*, the viscous-imaginary component of the dynamic tensile modulus, accounting for the energy dissipation due to internal friction during relaxation processes) and the loss tangent (also termed mechanical damping or tanδ) [7]. The 3-point bending clamp was used, and the DMA was set to sequentially apply dynamic loading with frequencies ranging from 1–100 Hz, at temperatures from RT to 723 K in step increments of 5 K. More details on the procedure, as well as the viscoelastic data of AA 7075-T6 and 2024-T3 used in this work, can be found in [5,6].

3. Results and Discussion

3.1. Storage Modulus

Figure 1a shows E' for pure Al in the H24 temper, from RT to 648 K, while Figure 1b shows E' for pure Al, from RT to 723 K, in both cases as obtained from DMA tests at frequencies ranging from 1–100 Hz. The behavior of E' for pure Al is similar in some aspects to that observed for AA 7075-T6 and 2024-T3 [5,6] and AA 6082 [30] (an Excel file including the values of E', E'' and loss tangent measured for pure Al in the H24 temper, pure Al, AA 7075-T6 and AA 2024-T3 is provided as Supplementary Material). For example, E' also decreases initially. The slope at low temperatures (below the beginning of the dissolution of Guinier-Preston (GP) and Guinier-Preston-Bagariastkij (GPB) zones, for the alloys) is what is most interesting in this study, as explained in Section 3.5. Furthermore, a significant decrease in E' is observed, with the beginning of this drop shifted to higher temperatures (from around 423–523 K) as the loading frequency increases. Thus, at a given temperature, the alloys show a stiffer response (*i.e.*, E' is larger) at higher frequencies, as expected. Furthermore, E' depends more significantly on frequency at high temperatures (above 423 K). The fact that the viscoelastic behavior becomes more prominent with temperature has already been observed in amorphous alloys [31], aside from AA 7075-T6 and 2024-T3 [5,6] and AA 6082 [30].

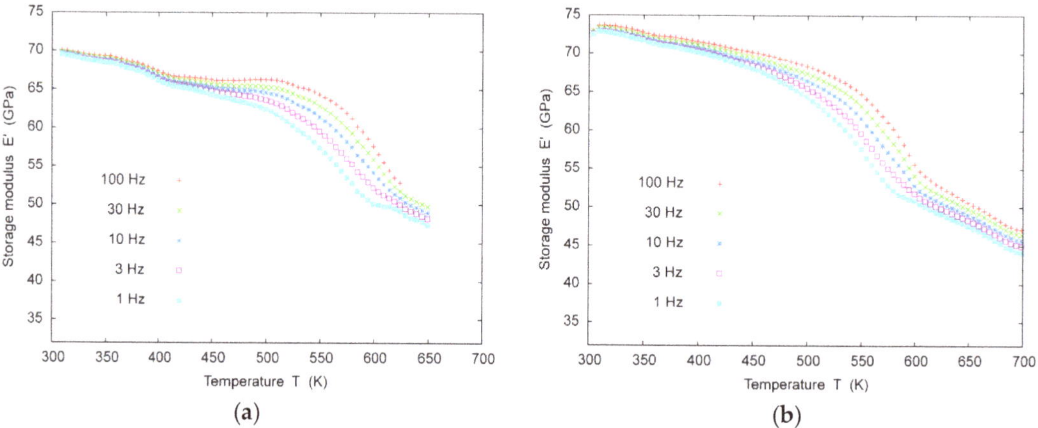

Figure 1. Storage modulus E' *vs.* temperature T from dynamic-mechanical analyzer tests at frequencies ranging from 1–100 Hz: (**a**) for pure Al in the H24 temper, from room temperature (RT) to 648 K; (**b**) for pure Al, from RT to 723 K.

3.2. Loss Modulus

Figure 2a shows E'' for pure Al in the H24 temper, from RT to 648 K, while Figure 2b shows E'' for pure Al, from RT to 723 K, in both cases as obtained from DMA tests at frequencies ranging from 1–100 Hz. In this case, the behavior of E'' for pure Al shows noticeable differences to that observed for AA 7075-T6 and 2024-T3 [5,6] and AA 6082 [30]. At low temperatures, the slopes of E'' are similar for all of the studied frequencies and not very steep (all show almost a plateau). At 393–533 K, the slopes increase sharply. This variation in the slope is shifted towards higher temperatures with increasing loading frequency. The observed behavior may be due to the viscous loss at higher frequencies competing with shorter relaxation times. Since the relaxation time decreases with temperature due to the Arrhenius-type behavior of the relaxation rate [7], this means that the temperature above which the viscous effect exceeds the relaxation is higher for higher frequency. In other words, higher frequency viscous loss curves rise at a higher temperature than lower frequency curves.

For AA 7075-T6, 2024-T3 and 6082, the sharp growth in E'' with temperature reaches very high values without showing a peak, which is usually explained by the presence of coupled relaxations [7]. On the contrary, for pure Al, E'' clearly exhibits a peak, which is achieved virtually at the same temperature for all of the frequencies (around 573 K). The peak is larger (both in width and height) as the loading frequency decreases. Previous works suggest that AlZnMg alloys, AlCuMg alloys and pure Al exhibit mechanical relaxation peaks associated with dislocations and grain boundaries [32,33]. For example, dislocation motions explain some internal friction peaks associated with semi-coherent precipitates for the alloys [34] and also the Bordoni peak, which has been extensively studied in cold-worked pure Al [7]. In this case, the observed peak corresponds to a typical internal friction peak in polycrystalline Al, related to grain boundaries [7]. In particular, the mechanism governing this relaxation is based on sliding at boundaries between adjacent grains. Upon application of stress, this process starts with the sliding of a grain over the adjacent one, caused by the shear stress acting initially across their mutual boundary. As a consequence, the shear stress is reduced gradually, and opposing stresses build up at the end of the boundary and into other adjacent grains. The process terminates when the shear stress has vanished across most of the boundary, and most of the total shearing force is sustained by the grain corners.

In addition, in Figure 2, a transition is observed around 473–513 K between the low temperature region where E'' is smaller for lower loading frequencies and the high temperature region where E'' is smaller for higher frequencies (a stiffer response in this case is expected). Finally, after the aforementioned peak, E'' seems to increase again in Figure 2b, particularly for the lower frequencies.

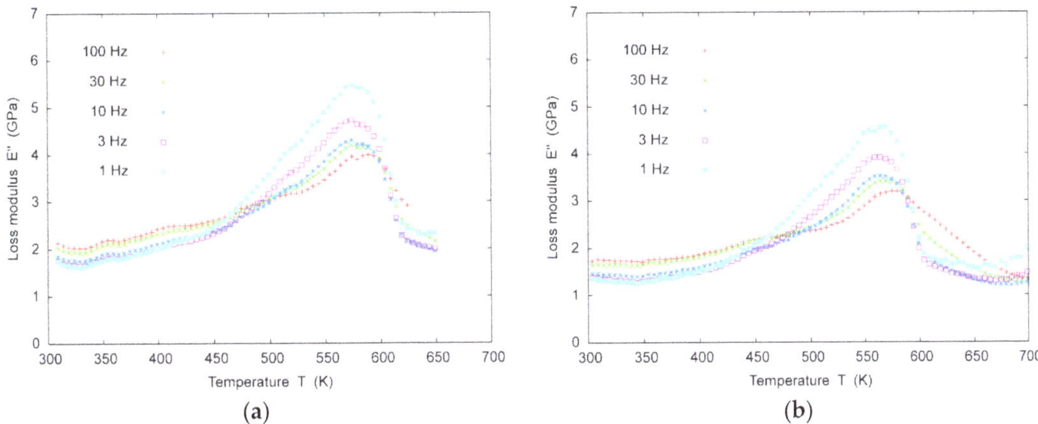

Figure 2. Loss modulus E'' *vs.* temperature T from dynamic-mechanical analyzer tests at frequencies ranging from 1–100 Hz: (**a**) for pure Al in the H24 temper, from room temperature (RT) to 648 K; (**b**) for pure Al, from RT to 723 K.

As usual, the loss tangent obtained from the DMA tests exhibits qualitatively the same behavior as E'' (the plots of loss tangent *vs.* temperature are included as Supplementary Material). The only remarkable comment is that there are no appreciable differences between the mechanical damping behavior for pure Al in the H24 temper and for pure Al, like for the E'' behavior (differences in the measured absolute values fall virtually within the instrument accuracy).

3.3. Temperature Dependence of the Storage Modulus

In absence of microstructural transformations, e.g., for pure Al and the alloys in the range RT to 373 K, E' decreases linearly with temperature. The reasons supporting this assumption are explained herein. There is abundant literature reporting a decrease with temperature of the elastic stiffness constants of metals, e.g., the static elastic modulus of pure Al and Al alloys [35–38]. This decrease can be assumed linear in a wide temperature range. Significant deviations from linearity are only observed close to 0 K and well above the temperature range of most interest for this research, *i.e.*, well above RT to 373 K, as explained in Section 3.5. Wolfenden and Wolla [39] observed also a highly quasi-linear decrease of the dynamic elastic modulus with temperature, measured at high frequency (80 kHz) from RT to 748 K, for pure Al and for AA 6061 reinforced with alumina.

Figure 3 in [6] shows a comparison of static and high frequency dynamic elastic modulus data available in the literature with data obtained with the DMA for AA 2024-T3 and for pure Al at 1 Hz. As expected, our DMA results for pure Al in the range of RT to 373 K fall between the static and high frequency dynamic values of elastic modulus in the literature, but they show a pronounced decrease of dynamic elastic modulus starting around 423–523 K. This is coincident with the observed E'' peak, as shown in Figure 2. Wolfenden and Wolla [39] assumed a linear behavior from RT to 748 K and explained this in terms of the Granato-Lücke theory for dislocation damping [40], but their data are too scattered to be explained with a single mechanism in the considered temperature range, as shown by our DMA results. Moreover, the low frequency dynamic elastic modulus of pure Al should also be sensitive to micrometric mechanisms, such as boundary migration during recrystallization and grain growth, as shown by Zhang *et al.* [41]. This mechanism is likely to be much more relevant in fatigue processes. However, the microstructural evolution and mechanical properties of AA 7075-T6 and 2024-T3 are controlled by the successive redissolution and precipitation of minority phases, and that is why it is likely that boundary migration is not the single most significant mechanism governing the behavior of their dynamic elastic moduli.

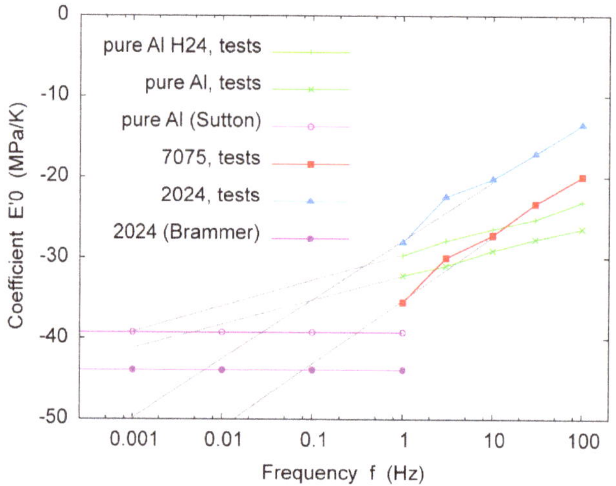

Figure 3. Temperature softening coefficient E_0' *vs.* frequency f for pure Al in the H24 temper, pure Al, Al alloy (AA) 7075-T6 and AA 2024-T3, as obtained from linear regression of test data from the dynamic-mechanical analyzer. Logarithmic tendency lines fitted to the data to extrapolate them to lower frequencies are also shown, as well as the rates of loss of static elastic modulus with temperature, as obtained by linear regression of data in the literature for pure Al [36] and AA 2024 [35].

Consequently, it can be assumed that, in absence of microstructural transformations, E' decreases linearly with temperature. Accordingly, Table 1 shows the slope of the decrease of E' with temperature for pure Al in the H24 temper and for pure Al, as obtained from linear regression of the DMA experimental results. In the following, to refer properly to this parameter on the basis of its physical meaning, it is termed "temperature softening coefficient", E_0'. Finally, it is important to note that the larger decrease of E' for low frequencies is typically explained by the Arrhenius-type behavior of the relaxation rate. That is, the mechanical relaxation time decreases with temperature [7], so that, at low frequencies, the shorter relaxation times lead to responses with larger values of E'' and smaller values of E'.

Table 1. Temperature softening coefficient E_0', for pure Al in the H24 temper and for pure Al, as obtained from linear regression of dynamic-mechanical analyzer test data.

Loading Frequency (Hz)	E_0' (Pure Al in H24 Temper) (MPa·K^{-1})	E_0' (Pure Al) (MPa·K^{-1})
100	−23.1	−26.4
30	−25.2	−27.7
10	−26.4	−29.1
3	−27.9	−31.0
1	−29.7	−32.2

3.4. Effect of Internal Friction on Fatigue Strength

Considering that viscoelasticity is linked to the fatigue and yield stress behavior [7], it is likely that the decrease of fatigue and yield stress behavior observed in metals as temperature increases is associated with the increase of internal friction. The reasons supporting this assumption are explained herein. Namely, Zhu *et al.* [14] observed that the fatigue strength of an Al alloy decreases with temperature, with a behavior that follows closely the temperature dependence of the yield and tensile strength. Furthermore, Liaw *et al.* [16] suggested that the fatigue limits and fatigue lives of steels decrease due to the temperature increase caused by plastic deformation. Finally, from the work by Amiri and Khonsari [4], it appears that the more intense the hysteresis heating of a metal during dynamic loading, the lower its fatigue resistance. Extrapolating the assumption stated above to AA 7075-T6, 2024-T3 and AA 6082, their yield and fatigue strength would decrease with temperature according to the observed increase in internal friction in the tested temperature range, as shown in [5,6] and [30], respectively. For pure Al, the yield and fatigue strength would decrease up to the E'' peak.

Furthermore, if considering the influence of loading frequency on internal friction, our results suggest that below 423 K, the yield and fatigue strength would decrease with increasing frequency, while at high temperatures, these properties would increase. These presumptions agree with some of the research reported in the literature, but still, there is controversy about the effect of frequency on the fatigue response, as shown in Section 1.1. For example, our findings agree with results from low frequency investigations for Al alloy RR58 at 423 K [25] and for AA 2650-T6 at 175 °C [26], pointing out that fatigue life is sensitive to frequency and that it increases with loading frequency. There is agreement also with the results for E319 cast Al alloy at 423 and 523 K, pointing out that fatigue life at 20 kHz was longer than at 75 Hz [14]. However, this behavior is attributed to environmental effects on the FCG rate, rather than intrinsic temperature or frequency effects.

3.5. Onset Frequency of Fatigue Effects

It is reasonable to accept that, at sufficiently low loading frequency, fatigue effects become negligible. Indeed, in a series of dynamic loading tests with loading frequency decreasing to very low values, the test time scale will eventually become much larger than the largest mechanical relaxation time for any of the possible relaxation processes for the tested material. That is, the reduction in the loading frequency is equivalent to an increase in the reaction time, and therefore, there is a threshold frequency below which the reaction time is longer than the relaxation time. Consequently, the viscous effect as the loss mechanism is negligible as compared to the relaxation. This means that eventually, E'' will decrease to a level that there will be no appreciable frictional energy loss due to relaxation effects. Recalling the statement by Amiri and Khonsari [4] that some degree of hysteresis heating (which is caused by the energy dissipated due to internal friction) is necessary for metals to experience fatigue, the conclusion is that, eventually, fatigue effects vanish. This hypothesis is in line with research by Henaff *et al.* [26] and Nikbin and Radon [25], suggesting the existence of a critical value of loading frequency below which, for a given temperature, the material exhibits a static-like response (relaxed case), crack growth is sensitive to frequency and static creep dominates instead of pure fatigue mechanisms.

Next, a procedure to estimate this threshold frequency is proposed based on the frequency dependence of the temperature softening coefficient E_0'. The procedure is illustrated in Figure 3, which shows E_0' as a function of the loading frequency for pure Al in the H24 temper and pure Al, as obtained from linear regression of DMA data, and for AA 7075-T6 and 2024-T3, as obtained in [5,6]. These values of E_0' are compared to average values of the rate of loss of static elastic modulus with temperature, as obtained by linear regression of data in the literature for pure Al [36] and AA 2024 [35]. To calculate these averages, data from Kamm and Alers [37] and from Varshni [38] were disregarded, as they correspond to temperatures below RT. Data from Wolfenden and Wolla [39] were also disregarded, since these data are too scattered in a broad temperature range, and thus, the computable slope is probably not representative.

Assuming that, at the threshold frequency, E_0' should become equal to the rate of loss of the elastic modulus with temperature under static loading conditions, this threshold frequency may be estimated by the intersection of the latter rate with a tendency line extrapolating the behavior of E_0' measured experimentally to lower frequencies. In the example shown in Figure 3, the intersection of a logarithmic tendency line with the rate of loss of the static elastic modulus for pure Al in the H24 temper and pure Al gives a threshold frequency of around 0.001 and 0.005 Hz, respectively. For AA 2024-T3, the threshold frequency obtained with the same procedure would be about 0.006 Hz. For AA 7075-T6, no data on the variation of the elastic stiffness constants with temperature were found in the literature, but using the data available for AA 2024 and pure Al, a threshold frequency of 0.075 and 0.350 Hz would be obtained, respectively. These results are similar to those in the literature: Henaff *et al.* [26] reported a critical frequency of 0.020 Hz for AA 2650-T6, and Nikbin and Radon [25] reported that the transition region is 0.100–1 Hz for cast Al alloy RR58 at 423 K. However, to better assess the performance of the proposed procedure, further comparison with experimental data on fatigue response at very low frequencies is necessary. Unfortunately, there is a lack of this type of test data [26], due to the extremely long testing time. It is interesting to note that, on one side, according to our results, fatigue effects would seem to appear at lower loading frequencies for pure Al, compared to the alloys. Thus, apparently, the precipitation structure in the alloys would cause not only hardening, but would also enable the alloys to be loaded at a wider range of low frequencies without experiencing fatigue. On the other side, fatigue would appear already at lower frequencies for pure Al in the H24 temper compared to pure Al. In this case, the reason may be the lower ductility (and, thus, lower resistance to FCG) associated with the H24 temper.

The proposed procedure for the determination of the threshold frequency is a major result of this work. Low frequency fatigue experiments are, by definition, very long. Furthermore, there is still controversy about the effects of the exposure to the environment on the experimental results, as it is not feasible to perform longstanding experiments in constant environment conditions. Besides, the environment during service life is likely to be different from that during the experiments, and thus, the estimations of the threshold frequency based on conventional, time-consuming fatigue experiments are likely to be inaccurate. The determination of the threshold frequency from the data obtained with the relatively short DMA tests (less than 3 h) reflects the intrinsic properties of the material only. This offers also a standardized method, which allows precise comparison between different alloys. Furthermore, it is quite insensitive to the specific instrumental range available for the tests, provided the range is large enough to allow a consistent regression fit.

4. Conclusions

It was suggested that some degree of hysteresis heating is necessary for metals to experience fatigue when subjected to dynamic loading. Hysteresis heating is caused by energy dissipated due to internal friction, and thus, an increase in the latter should have an effect on the fatigue response. In particular, the results of this research suggest that the decrease of yield and fatigue behavior reported for Al alloys as temperature increases may be associated with the increase of internal friction with temperature. Due to the Arrhenius-type behavior of relaxation processes, the relaxation time decreases with temperature. The reduction in the loading frequency is equivalent to an increase in the reaction time, and therefore, there is a threshold frequency below which the reaction time is longer than the relaxation time. Consequently, the viscous effect as the loss mechanism is negligible as compared to the relaxation. In other words, with the dynamic loading frequency decreasing to very low values, eventually there will be no appreciable frictional energy dissipation and, thus, no hysteresis heating, due to mechanical relaxation phenomena. Thus, below the threshold frequency, the material will exhibit a static-like response (relaxed case), such that creep mechanisms dominate and fatigue effects are negligible. In this work, an alternative procedure to the time-consuming conventional fatigue tests is proposed to estimate this threshold frequency, based on the frequency dependence of the slope of the initial decrease of E' with temperature, which in this work is termed the temperature softening coefficient. The interesting point of our approach comes from the fact that this coefficient is easily obtained from the relatively short DMA tests, hence allowing for a fast estimation of the threshold frequency. For pure Al, AA 2024-T3 and AA 7075-T6, the threshold frequencies obtained with this procedure are 0.001–0.005, 0.006 and 0.075–0.350 Hz, respectively. This suggests that fatigue effects start to appear at lower loading frequencies for pure Al, while the alloys may be loaded at a wider range of low frequencies without experiencing fatigue, probably due to effects related to the presence of precipitates. However, to better assess the performance of the proposed procedure, further comparison with experimental data on fatigue response at very low frequencies is necessary.

Acknowledgments: Work supported by the MINECO Grant FIS2014-54734-P and the Generalitat de Catalunya/Agència de Gestió d'Ajuts Universitaris i de Recerca (AGAUR) Grant 2014 SGR 00581. As part of these grants, we received funds for covering the costs to publish in open access.

Author Contributions: Jose I. Rojas conceived of, designed and performed the experiments. Jose I. Rojas and Daniel Crespo analyzed the data and wrote the paper.

Conflicts of Interest: The authors declare no conflict of interest.

Abbreviations

The following abbreviations are used in this manuscript:

AA	Aluminum alloy(s)
bcc	Body-centered cubic
CFCG	Creep-fatigue crack growth
CCG	Creep crack growth
DMA	Dynamic-mechanical analyzer
fcc	Face-centered cubic
FCG	Fatigue crack growth
GPBZ	Guinier–Preston–Bagariastkij zones
GPZ	Guinier–Preston zones
HCF	High cycle fatigue
RT	Room temperature
VHCF	Very high cycle fatigue

References

1. Callister, W.D.; Rethwisch, D.G. *Fundamentals of Materials Science and Engineering*, 4th ed.; John Wiley & Sons, Ltd.: Singapore, Singapore, 2013; pp. 1–910.
2. Van Kranenburg, C. *Fatigue Crack Growth in Aluminium Alloys*; Technische Universiteit Delft: Delft, The Netherlands, 2010; Volume 1, pp. 1–194.
3. Srivatsan, T.S.; Kolar, D.; Magnusen, P. Influence of temperature on cyclic stress response, strain resistance, and fracture behavior of aluminum alloy 2524. *Mater. Sci. Eng. A Struct. Mater. Prop. Microstruct. Process.* **2001**, *314*, 118–130. [CrossRef]
4. Amiri, M.; Khonsari, M.M. Life prediction of metals undergoing fatigue load based on temperature evolution. *Mater. Sci. Eng. A Struct. Mater. Prop. Microstruct. Process.* **2010**, *527*, 1555–1559. [CrossRef]
5. Rojas, J.I.; Aguiar, A.; Crespo, D. Effect of temperature and frequency of dynamic loading in the viscoelastic properties of aluminium alloy 7075-T6. *Phys. Status Solidi C* **2011**, *8*, 3111–3114. [CrossRef]
6. Rojas, J.I.; Crespo, D. Modeling of the effect of temperature, frequency and phase transformations on the viscoelastic properties of AA 7075-T6 and AA 2024-T3 aluminum alloys. *Metall. Mater. Trans. A* **2012**, *43*, 4633–4646. [CrossRef]

7. Nowick, A.S.; Berry, B.S. *Anelastic Relaxation in Crystalline Solids*, 1st ed.; Academic Press: New York, NY, USA, 1972; pp. 1–677.
8. Starke, E.A.; Staley, J.T. Application of modern aluminum alloys to aircraft. *Prog. Aerosp. Sci.* **1996**, *32*, 131–172. [CrossRef]
9. Hong, Y.; Zhao, A.; Qian, G. Essential characteristics and influential factors for very-high-cycle fatigue behavior of metallic materials. *Acta Metall. Sin.* **2009**, *45*, 769–780.
10. Braun, R. Transgranular environment-induced cracking of 7050 aluminium alloy under cyclic loading conditions at low frequencies. *Int. J. Fatigue* **2008**, *30*, 1827–1837. [CrossRef]
11. Nikitin, I.; Besel, M. Effect of low-frequency on fatigue behaviour of austenitic steel AISI 304 at room temperature and 25 °C. *Int. J. Fatigue* **2008**, *30*, 2044–2049. [CrossRef]
12. Furuya, Y.; Matsuoka, S.; Abe, T.; Yamaguchi, K. Gigacycle fatigue properties for high-strength low-alloy steel at 100 Hz, 600 Hz, and 20 kHz. *Scr. Mater.* **2002**, *46*, 157–162. [CrossRef]
13. Papakyriacou, M.; Mayer, H.; Pypen, C.; Plenk, H.; Stanzl-Tschegg, S. Influence of loading frequency on high cycle fatigue properties of b.c.c. and h.c.p. metals. *Mater. Sci. Eng. A Struct. Mater. Prop. Microstruct. Process.* **2001**, *308*, 143–152. [CrossRef]
14. Zhu, X.; Jones, J.W.; Allison, J.E. Effect of frequency, environment, and temperature on fatigue behavior of E319 cast aluminum alloy: Stress-controlled fatigue life response. *Metall. Mater. Trans. A Phys. Metall. Mater. Sci.* **2008**, *39A*, 2681–2688. [CrossRef]
15. Mayer, H.; Papakyriacou, M.; Pippan, R.; Stanzl-Tschegg, S. Influence of loading frequency on the high cycle fatigue properties of AlZnMgCu1.5 aluminium alloy. *Mater. Sci. Eng. A Struct. Mater. Prop. Microstruct. Process.* **2001**, *314*, 48–54. [CrossRef]
16. Liaw, P.K.; Wang, H.; Jiang, L.; Yang, B.; Huang, J.Y.; Kuo, R.C.; Huang, J.G. Thermographic detection of fatigue damage of pressure vessel steels at 1000 Hz and 20 Hz. *Scr. Mater.* **2000**, *42*, 389–395. [CrossRef]
17. Urabe, N.; Weertman, J. Dislocation mobility in potassium and iron single-crystals. *Mater. Sci. Eng.* **1975**, *18*, 41–49. [CrossRef]
18. Murakami, Y.; Nomoto, T.; Ueda, T. Factors influencing the mechanism of superlong fatigue failure in steels. *Fatigue Fract. Eng. Mater. Struct.* **1999**, *22*, 581–590. [CrossRef]
19. Mayer, H. Fatigue crack growth and threshold measurements at very high frequencies. *Int. Mater. Rev.* **1999**, *44*, 1–34. [CrossRef]
20. Laird, C.; Charsley, P. *Ultrasonic Fatigue*, 1st ed.; The Metallurgical Society of AIME: Philadelphia, PA, USA, 1982.
21. Yan, N.; Wang, Q.Y.; Chen, Q.; Sun, J.J. Influence of loading frequency on fatigue behavior of high strength steel. *Prog. Fract. Strength Mater. Struct.* **2007**, *353–358*, 227–230. [CrossRef]
22. Verkin, B.I.; Grinberg, N.M. Effect of vacuum on the fatigue behavior of metals and alloys. *Mater. Sci. Eng.* **1979**, *41*, 149–181. [CrossRef]
23. Menan, F.; Henaff, G. Influence of frequency and waveform on corrosion fatigue crack propagation in the 2024-T351 aluminium alloy in the S-L orientation. *Mater. Sci. Eng. A Struct. Mater. Prop. Microstruct. Process.* **2009**, *519*, 70–76. [CrossRef]
24. Benson, D.K.; Hancock, J.R. Effect of strain rate on cyclic response of metals. *Metall. Trans.* **1974**, *5*, 1711–1715. [CrossRef]
25. Nikbin, K.; Radon, J. Prediction of fatigue interaction from static creep and high frequency fatigue crack growth data. In Advances in Fracture Research, Proceedings of the 9th International Conference in Fracture (ICF9), Sydney, Australia, 1–5 April 1997; Karihaloo, B.L., Mai, Y.W., Ripley, M.I., Ritchie, R.O., Eds.; Pergamon Press, Ltd.: Kidlington, UK, 1997; Volume 1–6, p. 429.
26. Henaff, G.; Odemer, G.; Benoit, G.; Koffi, E.; Journet, B. Prediction of creep-fatigue crack growth rates in inert and active environments in an aluminium alloy. *Int. J. Fatigue* **2009**, *31*, 1943–1951. [CrossRef]
27. Botny, R.; Kaleta, K.; Grzebien, W.; Adamczewski, W. A method for determining the heat energy of the fatigue process in metals under uniaxial stress: Part 2. Measurement of the temperature of a fatigue specimen by means of thermovision camera-computer system. *Int. J. Fatigue* **1986**, *8*, 35–38. [CrossRef]
28. Luong, M.P. Fatigue limit evaluation of metals using an infrared thermographic technique. *Mech. Mater.* **1998**, *28*, 155–163. [CrossRef]
29. Chung, T.E.; Faulkner, R.G. Parametric representation of fatigue in alloys and its relation to microstructure. *Mater. Sci. Technol.* **1990**, *6*, 1187–1192. [CrossRef]

30. Rojas, J.I.; Lopez-Ponte, X.; Crespo, D. Effect of temperature, frequency and phase transformations on the viscoelastic behavior of commercial 6082 (Al-Mg-Si) alloy. *J. Alloys Compd.* in preparation.
31. Jeong, H.T.; Kim, J.H.; Kim, W.T.; Kim, D.H. The mechanical relaxations of a $Mm_{55}Al_{25}Ni_{10}Cu_{10}$ amorphous alloy studied by dynamic mechanical analysis. *Mater. Sci. Eng. A Struct. Mater. Prop. Microstruct. Process.* **2004**, *385*, 182–186. [CrossRef]
32. Belhas, S.; Riviere, A.; Woirgard, J.; Vergnol, J.; Defouquet, J. High-temperature relaxation mechanisms in Cu-Al solid-solutions. *J. Phys.* **1985**, *46*, 367–370. [CrossRef]
33. Riviere, A.; Gerland, M.; Pelosin, V. Influence of dislocation networks on the relaxation peaks at intermediate temperature in pure metals and metallic alloys. *Mater. Sci. Eng. A Struct. Mater. Prop. Microstruct. Process.* **2009**, *521–522*, 94–97. [CrossRef]
34. Mondino, M.; Schoeck, G. Coherency loss and internal friction. *Phys. Status Solidi A Appl. Res.* **1971**, *6*, 665–670. [CrossRef]
35. Brammer, J.A.; Percival, C.M. Elevated-temperature elastic moduli of 2024-aluminum obtained by a laser-pulse technique. *Exp. Mech.* **1970**, *10*, 245–250. [CrossRef]
36. Sutton, P.M. The variation of the elastic constants of crystalline aluminum with temperature between 63 K and 773 K. *Phys. Rev.* **1953**, *91*, 816–821. [CrossRef]
37. Kamm, G.N.; Alers, G.A. Low temperature elastic moduli of aluminum. *J. Appl. Phys.* **1964**, *35*, 327–330. [CrossRef]
38. Varshni, Y.P. Temperature dependence of the elastic constants. *Phys. Rev. B* **1970**, *2*, 3952–3958. [CrossRef]
39. Wolfenden, A.; Wolla, J.M. Mechanical damping and dynamic modulus measurements in alumina and tungsten fiber-reinforced aluminum composites. *J. Mater. Sci.* **1989**, *24*, 3205–3212. [CrossRef]
40. Granato, A.; Lucke, K. Theory of mechanical damping due to dislocations. *J. Appl. Phys.* **1956**, *27*, 583–593. [CrossRef]
41. Zhang, Y.; Godfrey, A.; Jensen, D.J. Local boundary migration during recrystallization in pure aluminium. *Scr. Mater.* **2011**, *64*, 331–334. [CrossRef]

Permissions

All chapters in this book were first published in Metals, by MDPI; hereby published with permission under the Creative Commons Attribution License or equivalent. Every chapter published in this book has been scrutinized by our experts. Their significance has been extensively debated. The topics covered herein carry significant findings which will fuel the growth of the discipline. They may even be implemented as practical applications or may be referred to as a beginning point for another development.

The contributors of this book come from diverse backgrounds, making this book a truly international effort. This book will bring forth new frontiers with its revolutionizing research information and detailed analysis of the nascent developments around the world.

We would like to thank all the contributing authors for lending their expertise to make the book truly unique. They have played a crucial role in the development of this book. Without their invaluable contributions this book wouldn't have been possible. They have made vital efforts to compile up to date information on the varied aspects of this subject to make this book a valuable addition to the collection of many professionals and students.

This book was conceptualized with the vision of imparting up-to-date information and advanced data in this field. To ensure the same, a matchless editorial board was set up. Every individual on the board went through rigorous rounds of assessment to prove their worth. After which they invested a large part of their time researching and compiling the most relevant data for our readers.

The editorial board has been involved in producing this book since its inception. They have spent rigorous hours researching and exploring the diverse topics which have resulted in the successful publishing of this book. They have passed on their knowledge of decades through this book. To expedite this challenging task, the publisher supported the team at every step. A small team of assistant editors was also appointed to further simplify the editing procedure and attain best results for the readers.

Apart from the editorial board, the designing team has also invested a significant amount of their time in understanding the subject and creating the most relevant covers. They scrutinized every image to scout for the most suitable representation of the subject and create an appropriate cover for the book.

The publishing team has been an ardent support to the editorial, designing and production team. Their endless efforts to recruit the best for this project, has resulted in the accomplishment of this book. They are a veteran in the field of academics and their pool of knowledge is as vast as their experience in printing. Their expertise and guidance has proved useful at every step. Their uncompromising quality standards have made this book an exceptional effort. Their encouragement from time to time has been an inspiration for everyone.

The publisher and the editorial board hope that this book will prove to be a valuable piece of knowledge for researchers, students, practitioners and scholars across the globe.

List of Contributors

Yuan Jin, Marc Bernacki, Andrea Agnoli and Nathalie Bozzolo
MINES ParisTech, PSL—Research University, CEMEF—Centre de mise en forme des matériaux, CNRS UMR 7635, CS 10207 rue Claude Daunesse, Sophia Antipolis Cedex 06904, France

Brian Lin, Gregory S. Rohrer and Anthony D. Rollett
Department of Materials Science and Engineering, Carnegie Mellon University, 5000 Forbes Avenue, Pittsburgh, PA 15213, USA

Teng-Shih Shih
Department of Mechanical Engineering, National Central University, Jhongli District, Taoyuan City 32001, Taiwan

Hwa-Sheng Yong and Wen-Nong Hsu
Department of Mechanical Engineering, National Central University, Jhongli District, Taoyuan City 32001, Taiwan
Graduate Student, National Central University, Jhongli District, Taoyuan City 32001, Taiwan

Uta Kühn and Hansjörg Klauß
Leibniz Institut für Festkörper- und Werkstoffforschung Dresden (IFW Dresden), Helmholtzstr. 20, Dresden D-01069, Germany

Jan Romberg and Ludwig Schultz
Leibniz Institut für Festkörper- und Werkstoffforschung Dresden (IFW Dresden), Helmholtzstr. 20, Dresden D-01069, Germany
Institut für Werkstoffwissenschaft, Technische Universität Dresden, Dresden D-01062, Germany

Jens Freudenberger
Leibniz Institut für Festkörper- und Werkstoffforschung Dresden (IFW Dresden), Helmholtzstr. 20, Dresden D-01069, Germany
Institut für Werkstoff wissenschaft, Technische Universität Dresden, Dresden D-01062, Germany
Institut für Werkstoff wissenschaft, Technische Universität Bergakademie Freiberg, Gustav-Zeuner-Str. 5, Freiberg D-09599, Germany

Hansjörg Bauder, Georg Plattner and Hans Krug
Carl Wezel KG, Industriestr. 95, Mühlacker D-75417, Germany

Frank Holländer
Lehrstuhl Metallkunde und Werkstofftechnik, Brandenburgische Technische Universität Cottbus, Konrad-Wachsmann-Allee 17, Cottbus D-03046, Germany

Juliane Scharnweber, Andy Eschke, Carl-Georg Oertel and Werner Skrotzki
Institut für Strukturphysik, Technische Universität Dresden, Dresden D-01062, Germany

Jürgen Eckert
Erich Schmid Institut für Materialwissenschaft, Österreichische Akademie der Wissenschaft, Jahnstraße 12, Leoben A-8700, Austria
Department Materialphysik, Montanuniversität Leoben, Jahnstraße 12, Leoben A-8700, Austria

Ludwig Schultz and Jan Romberg
Leibniz Institute for Solid State and Materials Research Dresden (IFW Dresden), Helmholtzstr. 20, 01069 Dresden, Germany
Institut für Werkstoffwissenschaft, Technische Universität Dresden, 01062 Dresden, Germany

Hiroyuki Watanabe
Osaka Municipal Technical Research Institute, 1-6-50 Morinomiya, Joto-ku, Osaka 5368553, Japan

Carl-Georg Oertel, Werner Skrotzki, Juliane Scharnweber and Andy Eschke
Institut für Strukturphysik, Technische Universität Dresden, 01062 Dresden, Germany

Goran Vukelic
Department of Marine Engineering and Ship Power Systems, Faculty of Maritime Studies Rijeka, University of Rijeka, Rijeka 51000, Croatia

Josip Brnic
Department of Engineering Mechanics, Faculty of Engineering, University of Rijeka, Rijeka 51000, Croatia

Jesús Toribio, Diego Vergara and Miguel Lorenzo
Fracture & Structural Integrity Research Group, University of Salamanca, 37008 Salamanca, Spain

Ning Wang
Gränges Technology AB, 612 33 Finspång, Sweden
Department of Materials Science and Engineering, NTNU, Trondheim NO-7491, Norway

Yanjun Li and Knut Marthinsen
Department of Materials Science and Engineering, NTNU, Trondheim NO-7491, Norway

Ke Huang
Department of Materials Science and Engineering, NTNU, Trondheim NO-7491, Norway
Thermomechanical Metallurgy Laboratory—PX Group Chair, Ecole Polytechnique Fédérale de Lausanne (EPFL), CH-2002 Neuchâtel, Switzerland

Janette Brezinová, Dagmar Draganovská, Anna Guzanová, Peter Balog and Ján Viňáš
Department of Mechanical Technology and Materials, Technical University of Košice, Mäsiarska 74, 040 01 Košice, Slovakia

Hyung Duk Yun, Duck Min Seo, Soon Yong Kwon and Lee Soon Park
School of Material Science and Engineering, Ulsan National Institute of Science and Technology (UNIST), Ulsan 44919, Korea

Min Yoeb Lee
Department of Polymer Science and Engineering, Kyungpook National University, Daegu 41566, Korea

Ashfaq Mohammad and Muneer Khan Mohammed
Princess Fatima Alnijiris's Research Chair for Advanced Manufacturing Technology (FARCAMT), Advanced Manufacturing Institute, King Saud University, Riyadh 11421, Saudi Arabia

Abdulrahman M. Alahmari
Princess Fatima Alnijiris's Research Chair for Advanced Manufacturing Technology (FARCAMT), Advanced Manufacturing Institute, King Saud University, Riyadh 11421, Saudi Arabia
Industrial Engineering Department, College of Engineering, King Saud University, Riyadh 11421, Saudi Arabia

Abdulrahman Alomar and Khaja Moiduddin
Industrial Engineering Department, College of Engineering, King Saud University, Riyadh 11421, Saudi Arabia

Ravi Kottan Renganayagalu
Structural Nanomaterials Laboratory, PSG Institute of Advanced Studies, Coimbatore 641004, India

Wenquan Zhang, Jinzhong Lu and Kaiyu Luo
School of Mechanical Engineering, Jiangsu University, Zhenjiang 212013, China

Iban Vicario and Patricia Caballero
Department of Foundry and Steel making, Tecnalia Research & Innovation, c/Geldo, Edif 700, E-48160 Derio, Spain

Ignacio Crespo and Luis Maria Plaza
Department of Aerospace, Tecnalia Research & Innovation, c/Mikeletegi 2, E-20009 Donostia, Spain

Ion Kepa Idoiaga
Industrias Lebario, c/Arbizolea 4, E-48213 Izurza, Spain

Jailson A. Da Nóbrega, Antonio A. Silva and Theophilo M. Maciel
Programa de Pós-Graduação em Engenharia Mecânica, Universidade Federal de Campina Grande (UFCG), Campina Grande-PB 58429-140, Brazil

Diego D. S. Diniz
Universidade Federal Rural do Semi-Árido (UFERSA), Campus Caraúbas, Caraúbas-RN 59700-000, Brazil

Victor Hugo C. de Albuquerque
Programa de Pós-Graduação em Informática Aplicada, Universidade de Fortaleza, Fortaleza-CE 60811-905, Brazil

João Manuel R. S. Tavares
Instituto de Ciência e Inovação em Engenharia Mecânica e Industrial, Departamento de Engenharia Mecânica, Universidade do Porto, Porto 4200-465, Portugal

Jing-Wen Feng, Li-Hua Zhan and Ying-Ge Yang
School of Mechanical and Electrical Engineering, Central South University, Changsha 410083, Hunan, China
State Key Laboratory of High Performance Complex Manufacturing, Central South University, Changsha 410083, Hunan, China
2011 Collaborative Innovation Center, Central South University, Changsha 410083, Hunan, China

Sanja Petronic, Meri Burzic and Katarina Colic
Innovation Center, Faculty of Mechanical Engineering, University of Belgrade, Kraljice Marije 16, 11000 Belgrade, Serbia

Tatjana Sibalija
Faculty of Information Technology, Faculty of Management, Metropolitan University, Tadeusa Koscuska 63, 11000 Belgrade, Serbia

Suzana Polic
Central Institute for Conservation in Belgrade, Terazije 26, 11000 Belgrade, Serbia

Dubravka Milovanovic
Vinca Institute of Nuclear Sciences, University of Belgrade, PO Box 522, 11001 Belgrade, Serbia

Michael J. Demkowicz
Department of Materials Science and Engineering, Massachusetts Institute of Technology, Cambridge, MA 02139, USA
Materials Science and Engineering, Texas A & M University, College Station, TX 77843, USA

Jaroslaw Majewski
MPA-CINT/Los Alamos Neutron Scattering Center, Los Alamos National Laboratory, Los Alamos, NM 87545, USA
Department of Chemical Engineering, University of California at Davis, Davis, CA 95616, USA

Jose I. Rojas
Department of Physics-Division of Aerospace Engineering, Universitat Politècnica de Catalunya, c/ Esteve Terradas 7, 08860 Castelldefels, Spain

Daniel Crespo
Department of Physics, Universitat Politècnica de Catalunya, c/ Esteve Terradas 7, 08860 Castelldefels, Spain

Index

www.ingramcontent.com/pod-product-compliance
Lightning Source LLC
Chambersburg PA
CBHW061946190326
41458CB00009B/2798

* 9 7 8 1 6 3 2 3 8 5 5 5 0 *